We work with leading authors to develop the
strongest educational materials in biology,
bringing cutting-edge thinking and best
learning practice to a global market.

Under a range of well-known imprints, including
Benjamin Cummings, we craft high-quality print and
electronic publications which help readers to understand
and apply their content, whether studying or at work.

To find out more about the complete range of our
publishing, please visit us on the World Wide Web at:
www.pearsoned.co.uk

CELL AND MOLECULAR BIOLOGY IN ACTION SERIES

Human Molecular Genetics

Third Edition

Peter Sudbery and
Ian Sudbery

PEARSON

Benjamin
Cummings

Harlow, England • London • New York • Boston • San Francisco • Toronto
Sydney • Tokyo • Singapore • Hong Kong • Seoul • Taipei • New Delhi
Cape Town • Madrid • Mexico City • Amsterdam • Munich • Paris • Milan

Pearson Education Limited
Edinburgh Gate
Harlow
Essex CM20 2JE
England

and Associated Companies throughout the world

Visit us on the World Wide Web at:
www.pearsoned.co.uk

First published under the Addison Wesley Longman imprint 1998
Second edition published 2002
Third edition published 2009

ISBN: 978-0-132-05157-6

British Library Cataloguing-in-Publication Data
A catalogue record for this book is available from the British Library

Library of Congress Cataloging-in-Publication Data
Sudbery, Peter.
   Human molecular genetics / Peter Sudbery and Ian Sudbery. – 3rd ed.
      p. ; cm.
   Includes bibliographical references and index.
   ISBN 978-0-13-205157-6 (pbk.)
   1. Human genetics.   2. Molecular genetics.   I. Sudbery, Ian.   II. Title.
[DNLM: 1. Genetics, Medical–methods.   2. Genetic Diseases, Inborn.
3. Genome, Human.   4. Molecular Biology–methods. QZ 50 S943h 2009]
QH431.S932 2009
611′01816–dc22

                                                                    2009009528

10 9 8 7 6 5 4 3 2 1
13 12 11 10 09

Typeset in Concorde BE by 35
Printed in Great Britain by Henry Ling Ltd., at the Dorset Press, Dorchester, Dorset

The publisher's policy is to use paper manufactured from sustainable forests.

Contents

Supporting resources

Visit **www.pearsoned.co.uk/sudbery** to find valuable online resources

For instructors
- PowerPoint slides of diagrams from the book

For more information please contact your local Pearson Education sales representative or visit **www.pearsoned.co.uk/sudbery**

Preface

This book is intended to provide an introduction to human molecular genetics that is more focused and up to date than that found in large, general textbooks on genetics, and yet remains accessible to second and third year science undergraduates, medical students and other healthcare specialists. It assumes the background knowledge of genetics and molecular biology that would normally be taught in the first year of a university course. Not only is this a necessity to keep both the size and cost down, but it also seems futile to duplicate the coverage in the generous selection of excellent general textbooks to which most readers will have access. Each chapter ends with a list of further reading for those who would like to investigate a topic in more detail. For the most part the references are also the sources for most of the material presented in the chapter.

An introductory text to such a large and complex field must restrict itself to illustrating general principles with selected examples. Inevitably, work on many important genes must be omitted, and we can only apologise to those who feel their favourite gene has not received the attention it deserves. Equally, a certain amount of simplification is essential to make the content accessible to an undergraduate audience. This is particularly true of those areas such as population genetics and epidemiological genetics that are mathematically based, which experience shows can be a cause of difficulty to biology students. We can only hope that we have succeeded in spreading enlightenment without oversimplifying.

Since the second edition of this book was completed in 2001, the sequence of the human genome has been finished to a high standard and the difficult task of annotation is now essentially complete. Chapter 4 has been re-written to give an account of the finished genome sequence and the exciting new horizons and technologies that have opened up as a result. The latest innovation in sequencing has been the development of a new generation of sequencing technologies that promises to allow the rapid and inexpensive sequencing of whole genomes and an account of these new methods is provided. In 2001, it was just becoming evident that the haplotype diversity of the human genome was less than expected. Subsequently, this observation has been exploited, along with the powerful new microarray technologies to allow the whole genomes to be scanned in populations of individuals to identify alleles that are associated with complex disease. These so-called 'genome wide association studies' have identified large numbers of risk alleles, and Chapter 6 has been completely rewritten to give an account of these developments. For the first time in this problematic field the positive findings have been consistently reproduced in independent studies. However, the combined risk of the alleles identified accounts for only a small part of the overall genetic variability observed in populations. The next development in this field is likely to be sequencing of whole genomes on a population-wide scale to allow the association of rare alleles to be tested. Such alleles are not captured by the microarrays currently used, which are limited by their design to genotype only those alleles that are relatively common. Since 2001, the use of siRNA as an experimental tool to silence human gene expression has developed from nowhere to become a powerful and much-used tool in human genetics. A new chapter has been added to give an account of siRNA and its applications.

The science of human molecular genetics develops at an ever-increasing pace, powered by the elucidation of the human genome sequence and the powerful new technologies that have developed to exploit this knowledge. The pace of these developments has made the objective of maintaining an up-to-date text increasingly difficult – major developments were transforming many fields even as we were writing. The 'data freeze' was applied at the end of August 2008.

We are grateful to the tolerance and support of our family, friends and colleagues in preparing this third edition. We very much appreciate the rapid and detailed reading of draft versions of Chapters 6 and 7 by Dr Mark E.S. Bailey, Professor Francis Choy, Dr Majid Hafezparast, Dr Louise Jones, Dr Mary Kimble, Professor Stephen M. Mount, Dr Harry Mountain, Professor Connie J. Mulligan, Dr David J. Price, Dr Mick Rae and Dr Arthur C. Robinson, respectively; of course, all deficiencies that remain are entirely our own responsibility. Anne Dalton, Steve Evans and Richard Kirk (North Trent Molecular Genetics Laboratory) generously provided examples of their work.

Acknowledgements

We are grateful to the following for permission to reproduce copyright material:

Figures
Figure 2.8 from Maltby, E., Langhill Centre for Human Genetics, Sheffield, UK; Figure 3.1 from Figure 1, p. 541: A gene map of the human genome, *Science*, 274, 540–1 (Schuler, G.D. et al. 1996); Figure 3.12 from Continuum of overlapping clones spanning the entire human chromosome *Nature*, 357, 540–1 (Chumakov et al. 1992), reprinted with permission from Nature, copyright 1992 with permission from Macmillan Magazines Limited; Figure 3.17 from http://www-genome.wi.mit.edu/, Dr T. Hudson; Figures 4.4, 4.10 from Initial sequencing and analysis of the human genome, *Nature*, 409, pp. 860–921 (The International Human Genome Sequencing Consortium 2001), Reprinted with permission from Nature, copyright 2001 Macmillan Magazines Limited; Figure 4.5 from Integration of cytogenetic landmarks into the draft sequence of the human genome *Nature*, 409, pp. 953–958 (The BAC Resource Consortium 2001), reprinted with permission from Nature, copyright 2001Macmillan Magazines Limited; Figures 4.9 and 4.11 from The sequence of the human genome, *Science*, 291, pp. 1304–1351 (Venter, J.C., Adams, M.D. and Myers, E.W. 2001); Figure 4.12 adapted from The sequence of the human genome, *Science*, 291, pp. 1304–1351 (Venter, J.C., Adams, M.D. and Myers, E.W. 2001); Figure 4.15 after A gene atlas of the mouse and human protein-encoding transcriptomes, *Proceedings of the National Academy of Sciences (USA)*, 101, 6062–6067 (Su, A.I. et al. 2004), Copyright 2004 National Academy of Sciences, USA.; Figure 4.20 from What is a gene, post-ENCODE. History and updated definition, *Genome Research*, 17, pp. 669–681 (Gerstein, M.B. et al. 2007); Figure on page 154 from Gain-of-function mutations in TRPV4 cause autosomal dominant brachyolmia, *Nature Genetics*, 40, pp. 999–1003 (Rock, M.J. et al. 2008); Figure 5.3 from Cystic Fibrosis: Molecular biology and therapeutic implications, *Science*, 256, 774–9 (Collins, F.S. 1992); Figure 5.6 from A novel gene containing a trinucleotide repeat that is expanded and unstable on Huntington's disease chromosomes, *Cell*, 72, 971–83 (The Huntington's disease collaborative research group 1993), with permission from Cell Press; Figure 5.12 from Pharmacogenetics and the practice of medicine, *Nature*, 405, pp. 857–865 (Roses, A.D. 2000), Reprinted with permission from Nature, copyright 2000 Macmillan Magazines Limited; Figure 6.7 from Complement Factor H Polymorphism and Age-Related Macular Degeneration, *Science*, 308, pp. 421–424 (Edwards, A.O. et al. 2005); Figures 6.9a, 6.9b from Genome-wide association studies for complex traits: consensus, uncertainty and challenges, *Nature Reviews Genetics*, 9, pp. 356–369 (McCarthy, M.I. et al. 2008); Figure 6.9c from A Genome-Wide Association Study Identifies IL23R as an Inflammatory Bowel Disease Gene, *Science*, 314, pp. 1461–1463 (Duerr, R.H. et al. 2006); Figure 7.12 after Interfering with disease: a progress report on siRNA-based therapeutics, *Nature Reviews Drug Discovery*, 6, pp. 443–453 (de Fougerolles, A., Vornlocher, H., Maraganore, J. and Lieberman, J. 2007); Figure 8.6 from Relative quantification of 40 nucleic acid sequences by multiplex ligation-dependent probe amplification, *Nucleic Acids Research*, 30, e.57 (Schouten, J.P. et al. 2002); Figure 10.1 from The structure and evolution of the human beta globin family, *Cell*, 21, 653–68 (Efstratiadis, A. et al. 1980), with permission from Cell Press; Figure 11.1 from Genetic fingerprinting, *Nature Medicine*, 11, pp. 1035–1039 (Jeffreys, A. 2005).

Tables
Table 1.1 from *The New Genetics and Clinical Practice*, 3rd ed., Oxford University Press (Weatherall, D.J. 1990); Table 6.10 from The Future of Genetic Studies of Complex Human Diseases *Science*, 273, pp. 1516–1517 (Risch, N. and Merikangas, K. 1996); Table 6.12 from Replication of Genome-Wide Association Signals in UK Samples Reveals Risk Loci for Type 2 Diabetes *Science*, 316, pp. 1336–1341 (Zeggini, E. et al. 2007).

In some instances we have been unable to trace the owners of copyright material, and we would appreciate any information that would enable us to do so.

Abbreviations

AAV	adeno-associated virus
ABC	ATP-binding cassette
ACGT	Advisory Committee for Gene Testing
AD	Alzheimer's disease
ADA	adenine deaminase
AGE	agarose gel electrophoresis
APM	affected pedigree member
APP	amyloid precursor protein
ARMS	amplification refractory mutation system
ASO	allele-specific oligonucleotide
ASP	affected sib pair
AT	ataxia telangiectasia
BAC	bacterial artificial chromosome
BMD	Becker muscular dystrophy
bp	base pairs
CBAVD	congenital bilateral absence of the vas deferens
CDK	cyclin-dependent kinase
CEPH	Centre d'Étude Polymorphism Humain
CF	cystic fibrosis
CFTR	cystic fibrosis transmembrane conductance regulator
cM	centimorgans
CMC	chemical mismatch cleavage
CMD	congenital muscular dystrophy
CNV	copy number variant
CPEO	chronic progressive external ophthalmoplegia
CRE	cyclic AMP response element
DASC	dystrophin-associated sarcoglycan complex
DGGE	denaturing gradient gel electrophoresis
DHHS	Department of Health and Human Services (USA)
DMAHP	dystrophia myotonica associated homeobox protein
DMD	Duchenne muscular dystrophy
DMPK	dystrophia myotonica protein kinase
DOE	Department of the Environment (USA)
EBI	European Bioinformatics Centre
ELSI	ethical, legal and social issues
EMBL	European Molecular Biology Laboratory
EMC	enzyme mismatch cleavage
EPC	European Patent Convention

ePCR	electronic PCR	LHON	Leber's hereditary optic neuropathy
EPO	European Patent Office		
ERV	endogenous retroviruses	LINE	long interspersed element
ESP	end sequence pair	LOAD	late-onset Alzheimer's disease
EST	expressed sequence tag		
ETS	external transcribed spacer unit	LOD	log ratio of odds
		LOH	loss of heterozygosity
FAP	familial adenomatous polyposis	LTR	long terminal repeat
		MAF	murior allele frequency
FBC	familial breast cancer	Mb	megabase pairs
FISH	fluorescence *in situ* hybridisation	MD	myotonic dystrophy
		MDR	multiple drug resistance
GAIC	Genetics and Insurance Committee	MELAS	mitochondrial encephalomyopathy, lactic acidosis and stroke-like episodes
GAP	GTP-activating protein		
GRR	genotype relative risk		
GTAC	Gene Therapy Advisory Committee	MHC	major histocompatibility complex
HBC	hereditary breast cancer	MIR	mammalian-wide interspersed elements
HD	Huntington's disease		
HGAC	Human Genetics Advisory Committee	MLS	maximum likelihood score
		MODY	maturity onset diabetes of the young
HGC	Human Genetics Commission		
		MRCA	most recent common ancestor
HLA	human leucocyte antigens		
HNPCC	hereditary non-polyposis colorectal cancer	mtDNA	mitochondrial DNA
		MVR-	multivariant repeat-
HnRNA	heterogeneous nuclear RNA	PCR	polymerase chain reaction
HPFH	hereditary persistence of fetal haemoglobin	mya	million years ago
		NARP	neurogenic muscle weakness, ataxia and retinitis pigmentosa
HRE	heat-shock response element		
HUGO	Human Genome Organisation	NBF	nucleotide binding fold
		NCBI	National Centre for Biotechnology Information
IBD	identical by descent		
IBS	identical by state	NCHGR	National Council for Human Genetic Research
IDDM	insulin-dependent diabetes mellitus		
		NF	neurofibromatosis
IDE	insulin degrading enzyme	NHEJ	non-homologous end joining
IGI	Initial Gene Index		
IHGSC	International Human Genome Sequencing Consortium	NIDDM	non-insulin-dependent diabetes mellitus
		NIH	National Institutes of Health (USA)
ITS	internal transcribed spacer		
kb	kilobase	NMR	nuclear magnetic resonance
kya	thousand years ago	NOD	non-obese diabetic (mouse)
LCN	low copy number	NRY	non-recombining region of the Y chromosome
LCR	locus control region		
LD	linkage disequilibrium	ORF	open reading frame

PAC	P1-derived artificial chromosome	snRNP	small nuclear ribonucleoprotein
PAR	pseudoautosomal region (of the Y chromosome)	SRE	serum response element
PC	principal component	SSCP	single-stranded conformational polymorphism
PCR	polymerase chain reaction		
PFGE	pulse field gel electrophoresis	SSTR	simple sequence tandem repeat
PIC	polymorphism information content	STC	sequence tagged connector
		STR	short tandem repeat
rDNA	DNA coding for ribosomal RNA molecules	STS	sequence tagged site
		TBF	TATA-box binding factor
RFLP	restriction fragment length polymorphism	TBP	TATA-binding protein
		TDT	transmission disequilibrium test
RH	radiation hybrid		
RT-PCR	reverse transcribed PCR	TNF	tumour necrosis factor
SACGT	Secretary's Advisory Committee on Genetic Testing (USA)	TRE	trinucleotide repeat expansion
		TSC	the SNP consortium
SCID	severe combined immune deficiency syndrome	USPTO	United States Patent Office
SD	standard deviation	UTR	untranslated region
SINE	short interspersed element	VNTR	variable number tandem repeat
SNP	single nucleotide polymorphism	YAC	yeast artificial chromosome

Human genetic disease

Key topics

- Frequency and types of genetic disease
- Single-gene disorders
  - Complexity in single-gene disorders
  - Autosomal recessive
  - Autosomal dominant
  - Sex-linked
- Multifactorial or complex disorders
  - Evidence for genetic factors in common diseases
  - Genetic influences on personality disorders and phenotypic traits
- Chromosomal imbalances
- Mitochondrial disorders
- The Human Genome Project

1.1 Introduction

About 5% of liveborn babies suffer from a significant medical disorder that may be life-threatening or, at the very least, require hospital treatment. A disorder present at birth is said to be **congenital**. Most congenital disorders will have a genetic component in their **aetiology**. This may take the form of a **single-gene** or **monogenic defect**, a mitochondrial disorder, a chromosomal imbalance or a multifactorial condition, which is partly genetic and partly environmental. Table 1.1 shows estimates of the genetic burden affecting the health of the newborn imposed by these various genetic factors.

A survey of admissions to a paediatric hospital in Montreal showed that one-third of all admissions were for diseases with a genetic component. Moreover, 70% of patients admitted more than once had a disorder with a genetic component. Such disorders are responsible for an immense amount of suffering, reduced quality of life, shortened life expectancy and distress to

Table 1.1 The genetic load in the newborn. It is difficult to give precise figures because of variation between different populations, so the figures are in the form of ranges. The genetic load of congenital defects is estimated as half the overall incidence of such disorders. No overall figures are available for mitochondrial disorders. (From *The New Genetics and Clinical Practice*, 3rd ed., Oxford University Press (Weatherall, D.J. 1990).)

Category	Frequency (per 1000 live births)
Single gene	
Autosomal dominant	1.8–9.5
Autosomal recessive	2.2–2.5
X-linked	0.5–2.0
Chromosomal	6.8
Congenital malformations	19–22

family members and carers. Moreover, it is becoming increasingly clear that many of the common diseases of later life, perhaps up to two-thirds, have a genetic component in their aetiology. Clearly our genetic constitution plays a major part in determining our lifetime health. This chapter considers the ways in which this comes about and introduces some of the major genetic disorders that are considered in more detail in later chapters.

1.2 Single-gene defects

Single-gene or monogenic disorders, as their name implies, are traceable to a defect in a single gene. They follow simple patterns of inheritance, predictable from the Mendelian laws of genetics. These patterns are classified according to whether the affected gene is located on the X chromosome or one of the 22 autosomes and whether the trait is recessive or dominant. That is, they are said to be **X-linked**, **autosomal recessive** or **autosomal dominant**, respectively. The Online Mendelian Inheritance in Man database (see Further reading at the end of the chapter) currently lists over 5000 disorders that have been definitely traced to single-gene defects. This represents a significant fraction of the total number of human genes, estimated to be between 30 000 and 35 000 (see Chapter 2). Single-gene disorders that are both severe and relatively common are known as **major disorders**.

The frequency of mutant alleles responsible for a monogenic disorder in a population is specified by the **Hardy–Weinberg distribution**, which states that:

$$p^2 + 2pq + q^2 = 1$$

where p is the allele frequency of the more common allele and q the frequency of the less common allele. This equation holds only if certain conditions are met, such as that mating is random and there is no migration into or from the population. This is discussed in more detail in Chapter 10. From this equation the carrier (heterozygote) frequency for recessive disorders can be calculated ($2pq$) from the observed frequency with which the disease occurs in a population. This can give surprising results. For example, the frequency of cystic fibrosis in the UK is 1 in 2000, from which the carrier frequency can be calculated as 1 in 22.

1.2.1 Autosomal recessive disorders

Autosomal recessive disorders require the inheritance of two defective alleles. This means that there are no functioning copies of the gene and implies that the disorder results from a loss of function. An example of an autosomal recessive pattern of inheritance is shown in Figure 1.1 and the frequencies of some major autosomal recessive disorders are shown in Table 1.2.

Typically more than one sibling of unaffected parents may be affected; children of cousin marriages are also at risk (Figure 1.1). Both parents must be carriers or **heterozygous** at the locus concerned (Figure 1.2); as a result, for each child there is a 25% chance that it will be affected, a 50% chance that it will be a carrier and a 25% chance that it will be completely normal. If the normal allele is dominant, the ratio of unaffected to affected children will be 3:1. To put it another way, for each child born there is a 25% chance that it will be affected.

As shown in Figure 1.1, the children of a cousin marriage may be **homozygous** for a deleterious allele that is heterozygous in a common grandparent. In fact, the chance that this will happen is 1 in 64 for each such allele. Because everyone is likely to be heterozygous for about five different deleterious alleles, cousin marriages are more likely to result in children affected by an autosomal recessive disorder than marriages

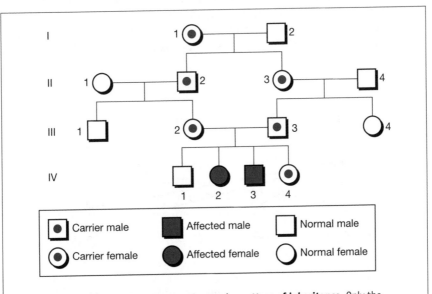

Figure 1.1 An example of an autosomal recessive pattern of inheritance. Only the homozygous recessive individuals (IV-2 and IV-3) are affected. Note how the heterozygous state in I-1 becomes homozygous in IV-2 and IV-3 as a result of the **consanguineous** marriage of III-2 and III-3.

Table 1.2 Some major monogenic recessive disorders.

Disease	Frequency	Symptoms
Cystic fibrosis	1 in 2000 (North Europeans)	Recurrent lung infections, pancreatic exocrine deficiency, male sterility
α_1-Antitrypsin deficiency	1 in 10 000 to 1 in 5000 (North Europeans)	Liver failure, emphysema
Phenylketonuria	1 in 5000 to 1 in 2000 (Europeans)	Mental retardation
Tay–Sachs disease	1 in 3000 (Ashkenazi Jews)	Neurological degeneration, blindness and paralysis
Sickle cell anaemia	1–2 in 100 (Africa where malaria is endemic)	Anaemia
Haemochromatosis	1 in 500	Excessive iron accumulation in adults, resulting in diabetes, liver cirrhosis and heart failure
Thalassaemias	1–2 in 100 (Mediterranean and Asia where malaria is endemic)	Anaemia

between unrelated parents. Indeed, many rare monogenic disorders are observed only in the children of cousin marriages.

Some diseases caused by single-gene defects are present in certain populations at a much higher frequency than would be expected by mutation alone. Important examples of this are cystic fibrosis in European populations, sickle cell anaemia and thalassaemias in Asian and African populations, and Tay–Sachs disease in **Ashkenazi Jews**. One reason for an elevated frequency

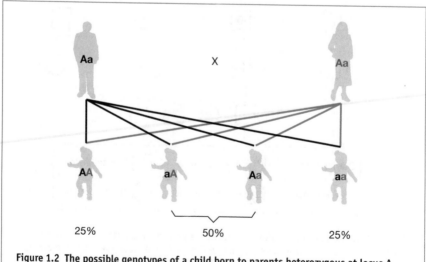

Figure 1.2 The possible genotypes of a child born to parents heterozygous at locus A.

is that the heterozygote may enjoy some advantage, resulting in the allele frequency being elevated by selection. In the case of sickle cell anaemia or thalassaemias, it has been demonstrated that heterozygotes are more resistant to malaria, which is endemic in those countries where such disorders are prevalent. This variation means that the ethnic origin of a patient may be relevant in arriving at a correct diagnosis. Other possible reasons for elevated disease frequencies are discussed in Chapter 10.

1.2.1.1 Cystic fibrosis

Cystic fibrosis (CF) is a common single-gene disorder among people of European extraction. It affects approximately 1 in 2000 babies, and about 1 in 22 are carriers. The affected gene encodes a protein known as the **cystic fibrosis transmembrane conductance regulator** (CFTR; see Chapter 5). The CFTR protein is responsible for the export of chloride ions across the plasma membrane of epithelial cells that line the lung airways. CF alleles impair or obliterate this function and, as water molecules follow the ions because of osmosis, insufficient water is secreted onto the cell surface. This results in a thick and sticky mucus that causes congestion in the lung airways. In a healthy person, the cilia of the epithelial cells continually move the mucus to the top of the airways and into the digestive system. In this way foreign bodies, including bacteria, are removed from the lungs. In a person with CF, the mucus is too thick to be moved by the cilia. Consequently this cleansing mechanism fails, resulting in repeated bacterial infections. The cumulative effect of these causes long-term damage to the lungs, and even with the best treatment the maximum life expectancy of someone with CF is 30 years. CF also affects other bodily systems. The pancreatic duct may become blocked, resulting in digestive difficulties. This is known as pancreatic exocrine deficiency. In some male patients, the vas deferens does not form properly, resulting in sterility. This is known as **congenital bilateral absence of the vas deferens** (CBAVD); sometimes this can occur without any other obvious symptoms of CF. Finally, CF results in an excess of chloride ion secretion in sweat glands, resulting in abnormally salty sweat. This can be simply recognised by measuring sweat electrolyte levels.

1.2.1.2 The haemoglobinopathies

Sickle cell anaemia, α and β **thalassaemias** and glucose-6-phosphate dehydrogenase deficiency (in some environments) affect the formation of haemoglobin, causing a class of disease known as the **haemoglobinopathies**. Haemoglobin is a tetramer of two β-globin molecules and two α-globin molecules, each complexed with a molecule of haem. Sickle cell anaemia results from a single-base mutation affecting the gene encoding β-globin. Study of this disease played an important part in the development of modern molecular genetics, because in 1957 Ingram demonstrated that the β-globin of sickle cell patients differed from the normal protein by one amino acid, providing experimental evidence for the first time that the sequence of amino acids in proteins was determined by genes. It is also important because it was the first, and is still the clearest, example in humans of heterozygous advantage. Carriers of the sickle cell gene are more resistant to malaria. This results in selection for the mutation and consequently an increased occurrence of the recessive homozygotes who suffer from the disease. Such a situation is known as a **balanced polymorphism**. The α and β thalassaemias result in a decrease or absence of α-globin and β-globin, respectively. As in sickle cell anaemia, heterozygotes are more resistant to malaria, so a balanced polymorphism results in thalassaemias being more common where malaria is, or was, endemic.

1.2.1.3 Tay–Sachs disease

Tay–Sachs disease is a progressive neurological degeneration that starts in the first year of life, characterised by developmental and mental retardation, progressive muscle weakness and paralysis, and blindness. Death usually results by the age of 5 years. It is caused by a mutation in the *HEXA* gene, which encodes the α-subunit of hexosaminidase A, a lysosomal enzyme required for the breakdown of a complex **glycolipid** called **ganglioside** GM_2 to a simpler molecule called ganglioside GM_3. The build-up of ganglioside GM_2 impairs neurone function and results in the disease. It is unusually common in the Ashkenazi Jew population, where it affects 1 in 3600 of the population.

1.2.1.4 Phenylketonuria

Phenylketonuria is characterised by mental subnormality. It results from a lack of phenylalanine hydroxylase, which converts phenylalanine to tyrosine, the first step in the phenylalanine degradation pathway. Phenylalanine consequently accumulates in the bloodstream. The damage to the developing nervous system is actually caused by phenylpyruvic acid, to which phenylalanine spontaneously converts. Most babies in the UK and the USA are now screened at birth for this deficiency because it can be almost completely controlled by a low-phenylalanine diet. It is thus an interesting example of a disorder that may be either completely genetically determined or completely environmentally determined.

1.2.1.5 α_1-Antitrypsin deficiency

Antitrypsin is an inhibitor of elastase. Deficiency results in unregulated breakdown of connective tissue that particularly affects the elasticity of the lungs, resulting in emphysema. It is another disorder that is particularly common among people of European descent. The protein can be readily made by recombinant methods; for example, transgenic sheep have been produced that secrete large amounts of the protein in their milk. This provides the prospect of a cheap and effective treatment of the disease.

1.2.1.6 Haemochromatosis

This is probably the most common recessive disorder in the UK. It is caused by excessive iron accumulation and can be treated by blood letting. The symptoms are very variable and it is often not correctly diagnosed. It commonly affects women after the menopause, because the blood loss during menstruation is protective.

1.2.2 Autosomal dominant disorders

Autosomal dominant disorders result from the inheritance of only one mutant allele. This usually results in a protein that has gained a novel function or is expressing its normal function in an unregulated fashion. If affected individuals are able to reproduce, the disorder is likely to be manifested in every generation of a pedigree in which it is segregating. Thus 50% of the children of an affected person are likely to suffer from the disease (Figure 1.3).

If the disease is sufficiently serious to prevent reproduction, it follows that most cases will arise from *de novo* mutation during **gametogenesis** and the occurrence of the disease will be apparently sporadic. Some autosomal dominant disorders are only manifested later in life, after an individual is likely to have finished reproducing. Such mutations will escape the effect of selection. Examples of this are Huntington's disease, familial breast cancer (see Chapter 5) and hereditary Alzheimer's disease (see Chapter 6). Some examples of major autosomal dominant disorders are shown in Table 1.3.

1.2.2.1 Huntington's disease

Huntington's disease (HD) is a progressive neurological degeneration that affects patients in middle and later life, from their fifth decade onwards. Typical symptoms include dementia, severe depression and a characteristic involuntary, dance-like movement known as chorea. It is a particularly distressing condition, because by the time the symptoms are evident an affected person is likely to have already had children, who each have a 50% chance of suffering from the disease. The HD gene proved difficult to clone and this was only achieved in 1993. The mutation that causes the

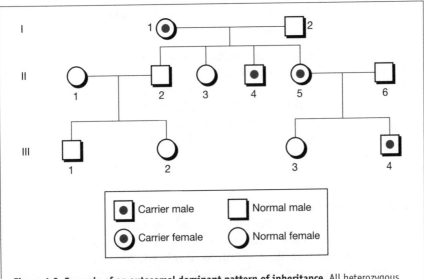

Figure 1.3 Example of an autosomal dominant pattern of inheritance. All heterozygous individuals are affected (I-1, II-4, II-5 and III-4). Because 50% of children will be affected, the disease is likely to be manifested in each generation.

Table 1.3 Examples of some major autosomal dominant disorders.

Disorder	Frequency	Symptoms
Familial hypercholesterolaemia	1 in 500	Premature heart disease
Familial breast cancer		
BRCA1	1 in 800 (USA)	High lifetime risk of breast and
BRCA2	1 in 100 (Ashkenazi Jews[1])	ovarian cancer. Earlier onset than sporadic cases
Familial Alzheimer's disease	10% lifetime risk at age 80[2]	Dementia. Earlier onset than sporadic cases
Hereditary non-polyposis colorectal cancer	1 in 400[3]	Colon cancer not associated with polyps
Familial adenomatous polyposis	1 in 8000	Bowel polyps that may become malignant
Neurofibromatosis type 1	1 in 4000	Tumours of peripheral nerves. *Café-au-lait* pigment spots on skin
Huntington's disease	1–2 in 10 000	Involuntary choreiform movements. Dementia. Late onset
Myotonic dystrophy	1 in 8500	Myotonia, heart defects and cataracts
Familial retinoblastoma	1 in 14 000	Multiple unilateral and bilateral tumours of the retina

[1] *BRCA1* and *BRCA2* each have a frequency of 1 in 100 in Ashkenazi Jews.

[2] Alzheimer's disease is normally a complex disease with genetic and environmental components. A subset of cases show early onset and simple autosomal dominant inheritance. What proportion of total cases is represented in this category is not known.

[3] Wide range of estimates from 1 in 10 000 to 1 in 200 (see Section 5.8.4).

disease is known as a trinucleotide repeat expansion (see Chapter 5). In normal individuals a sequence is found in which the triplet CAG is repeated about 15 times (the actual number can vary slightly from this value). In patients with HD the number of repeats increases to 36 or more. This type of mutation has been found in a number of other genes that have been cloned recently. The existence of this type of mutation was entirely unexpected and could only have been discovered by the cloning and characterisation of the genes involved.

1.2.2.2 Familial breast cancer

Genetic influences are not thought to be an important factor in the occurrence of most breast cancer cases (see below). However, a subset, perhaps about 5% of all cases, follow a pattern of inheritance characteristic of an autosomal dominant mutation. Two genes, *BRCA1* and *BRCA2*, have been identified and cloned (see Chapter 5). They were recognised by studying cases of breast cancer that occurred before the patient was 40 years old, an unusually early age for the onset of the disease. Inheritance of *BRCA1* results in an 80% lifetime risk of breast cancer. As well as leading to a predisposition to breast cancer, both *BRCA1* and *BRCA2* (to a lesser extent) also cause a predisposition to other cancers, particularly ovarian cancer. Inheritance of the *BRCA2* allele in males carries a 1 in 100 risk of male breast cancer. Although responsible for only a small fraction of breast cancer cases, *BRCA1* is a relatively common disease allele, estimated in US populations to occur at a frequency of 1 in 800. In Ashkenazi Jewish populations the frequency for both *BRCA1* and *BRCA2* is much higher (1 in 100).

1.2.2.3 Myotonic dystrophy

Myotonic dystrophy (MD) is characterised by myotonia (delayed muscle relaxation) and degeneration of various organs, including heart and eyes. It is variable in its expression, ranging from minimally affected late onset through classic adult onset to congenitally affected children of affected mothers. Although characterised as a Mendelian disorder of the autosomal dominant type, its pattern of inheritance shows some deviation from the classic pattern. In particular, it may show **anticipation**, where a mother may be only slightly affected but have a severely affected child. Often the mother's symptoms are so mild that the disease is diagnosed only when her child is found to be affected. MD is another example of a trinucleotide repeat expansion. It is now known that anticipation is a result of the instability of the expansion (see Chapter 5). MD is a relatively common disorder, the allele frequency being 1 in 8500.

1.2.3 X-linked disorders

X-linked disorders affect genes located on the X chromosome. Because males only have one copy of this chromosome and therefore one copy of each of the genes located on it, they are much more likely to suffer from such disorders. Thus X-linked disorders affect male children of unaffected parents. Most X-linked disorders are recessive, so for the most part females are unaffected but act as carriers. Thus, in a pedigree in which a sex-linked mutation is segregating, the mutation is inherited through the female line and the resultant disorder can apparently skip generations (Figure 1.4). Another characteristic of X-linked disorders is that uncles (on the maternal side) and nephews are often affected. One rare documented example of an X-linked dominant condition is hypophosphataemic (vitamin D-independent) rickets. In this case an affected father always passes the disorder to his daughters, but never to his sons.

Of course, it is possible for a female to be homozygous for an X-linked mutation and therefore suffer from the disease. Sometimes, a heterozygous female may show some symptoms of the disease. This may happen because of **X-chromosome inactivation** or **Lyonisation**. Named after its discoverer, Mary Lyon, this is a process that occurs early in development to randomly inactivate one or other of the X chromosomes. Cells descended from this progenitor cell abide by the decision made. Thus the body of a heterozygous female is a mosaic, consisting of patches of different clones, some of which will lack the product of the defective gene. If the function is cell autonomous, i.e. the function cannot be supplied by another cell expressing

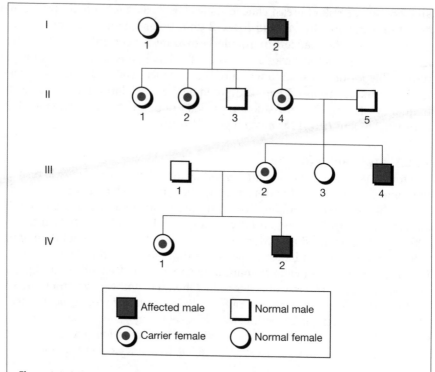

Figure 1.4 An example of an X-linked pattern of inheritance. Males are **hemizygous** for the X chromosome and have only one copy of genes on the X chromosome; males that inherit the affected gene are shown with a filled square. Note that the great-grandfather of IV-2 was affected, but not his mother or grandmother. The disease therefore skipped generations II and III with respect to IV-2 and the disorder was passed through the female line. However, IV-2's uncle was also affected, a typical pattern in X-linked disorders. Finally, note that daughters of affected males are obligate carriers.

the gene, it is possible that the lack of function will affect the overall phenotype of the carrier. This may be the reason why, for example, one-third of female carriers of fragile-X syndrome are mildly affected (see below).

The major X-linked disorders are shown in Table 1.4.

1.2.3.1 Haemophilia

Haemophilia is an inability to form blood clots, resulting in prolonged bleeding upon injury and spontaneous internal bleeding. At a molecular

Table 1.4 Some examples of X-linked disorders.

Disorder	Frequency	Symptoms
Haemophilia A and B	1 in 10 000	Abnormally prolonged bleeding after trauma
Duchenne muscular dystrophy	1 in 4000 to 1 in 3000	Muscle wastage in teenage years
Fragile-X syndrome	1 in 1000	Mental retardation

level it results from a lack of either blood clotting factor VIII (haemophilia A, 80% of cases) or blood clotting factor IX (haemophilia B or Christmas disease). It is famous for affecting the Victorian royal family.

1.2.3.2 Duchenne muscular dystrophy

Duchenne muscular dystrophy (DMD) results in progressive muscle wasting or dystrophy, starting in the teenage years. It soon results in confinement to a wheelchair and affected boys normally die in their twenties. There is a milder form called Becker muscular dystrophy which maps to the same gene. DMD is notable for the size of the gene affected, 2.5 Mb, equivalent to 60% of the entire genome of *Escherichia coli* (see Chapter 5). However, after splicing the mRNA is 14 kb in size; thus 99% of the gene is in the form of introns and only 1% actually encodes amino acid sequence.

1.2.3.3 Fragile-X syndrome

Fragile-X syndrome results in mental subnormality. After Down's syndrome, it is the most common cause of mental subnormality in boys and is said to be responsible for the excess of boys in institutions for people with mental disabilities. It derives its name from a cytogenetic observation: when cells are cultured in medium that is deficient in precursors of DNA metabolism the X chromosome appears to have a break in metaphase spreads. At least five X-linked loci have been identified, FRXA–FRXE. FRXA has been shown to be caused by a trinucleotide repeat expansion of the type discussed above for MD and HD (see Chapter 5). Fragile-X syndrome mildly affects one-third of female carriers, so is partially dominant.

1.2.4 Complexity in single-gene disorders

There could be a temptation to regard analysis of monogenic disorders as straightforward, i.e. that the sole factor in their occurrence is the disease gene, whose inheritance follows simple and predictable patterns. In fact, this is often very far from the case. There are many factors that may modify the pattern of inheritance or symptoms of the disorder and these are discussed below. Such complexity provides a warning of the intricacies that may be encountered in the analysis of multifactorial disorders, which involve a combination of **polygenic** and environmental factors.

1.2.4.1 Genetic heterogeneity

This describes a situation where apparently clinically similar disorders are caused by mutations in different genes (non-allelic) or where mutations in the same gene result in clinically diverse conditions (allelic). Some examples of each are given in Table 1.5.

1.2.4.2 Penetrance

This refers to the frequency with which the disorder or phenotype is manifested in an individual who has inherited the disease allele. A mutation that does not inevitably cause the disorder is said to show incomplete penetrance. An example is split hand syndrome, a claw-like deformity of the hand caused by an autosomal dominant allele. Some individuals inherit the allele but have normally formed hands.

1.2.4.3 Expressivity

This describes differences in the severity of a disorder in individuals who have inherited the same disease alleles. In the case of sickle cell anaemia, in which all cases result from mutation affecting the same amino acid, symptoms may be sufficiently severe to cause childhood death or be so mild that the disease remains undiagnosed until middle age. One possible cause of variation is that the effect of the sickle cell mutation is modified by other genes. In the case of sickle cell anaemia, one such modifying gene has been mapped to the X chromosome.

1.2.4.4 Mosaicism

Mosaicism arises when not all cells in the body are genetically identical. This may come about through a mutation in early development and may result in either the **germline** or somatic cells being affected. If the germline is mosaic, gametes may arise from progenitor cells of different genetic constitutions. For example, in one line the progenitor cell may be heterozygous for an autosomal dominant mutation, while in another the progenitor cell may be completely normal. This will clearly disturb the proportion of gametes affected, expected to be 50% if the parent is heterozygous for an autosomal dominant allele. Note that the somatic cells of females heterozygous for a sex-linked mutation will necessarily be mosaic as a result of X-chromosome inactivation.

1.2.4.5 Phenocopy

Sometimes an environmental factor may result in a disorder with the same symptoms as an inherited disorder. For example, the infection of a mother during pregnancy with the rubella virus may result in a profoundly deaf baby, a condition that can also be caused by a number of different single-gene defects. This is known as **phenocopy**.

1.2.4.6 Environmental effects

Environmental factors may influence the penetrance or expressivity of a disease allele. A clear example of this is phenylketonuria, a disease that may be largely prevented by a low-phenylalanine diet.

Table 1.5 Some examples of genetic heterogeneity.

Allelic	Non-allelic
Muscular dystrophy	Profound deafness
Becker	Retinitis pigmentosa
Duchenne	Autosomal dominant
Cystic fibrosis	Autosomal recessive
Pancreatic sufficiency/deficiency	X-linked
CBAVD	Polycystic kidney disease

1.2.4.7 Anticipation

There are a number of diseases whose severity apparently increases with each succeeding generation. MD and fragile-X syndrome are two well-known examples. Often the disorder may be so mild in the parent that it has not been diagnosed before its occurrence in a child. This phenomenon is known as **anticipation**.

1.2.4.8 Genomic imprinting

Genomic imprinting is said to occur when the expression of an allele depends on the parent from which it was inherited. This has been revealed in a number of different situations. An example is the effect of parental origin of a deletion of chromosomal band 15q12. Individuals heterozygous for this deletion suffer from a different disease according to the parent from which it was inherited. If the deletion was inherited from the father, Prader–Willi syndrome results, characterised by mental retardation, hypotonia, gross obesity and hypogenitalism. If the deletion was inherited from the mother, Angelman syndrome results, characterised by mental and growth retardation, hyperactivity and inappropriate laughter. This suggests that the function of the remaining allele depends on its parental origin. The mechanism is thought to involve DNA methylation, but the details are currently unclear.

1.3 Multifactorial or complex disorders and traits

It is a common observation that members of the same family are likely to resemble each other more than they resemble the general population and are more likely to suffer from the same diseases. Because family members are likely to share the same environment as well as the same genes, it is not easy to disentangle the relative contributions of genetic and environmental factors that may determine what we are like and what diseases we are likely to suffer from. For the most part, susceptibility to common diseases and phenotypic characteristics do not follow simple patterns of inheritance, so neither is likely to be determined monogenically. Nevertheless, family, adoption and twin studies clearly show a strong genetic component. A list of common disorders in which a genetic component has been demonstrated is given in Table 1.6. This list encompasses such a broad range of diseases that a patient's genetic constitution may be considered a factor in the majority of medical conditions.

These diseases are often referred to as **multifactorial diseases** because there are both genetic and environmental factors responsible for the onset of the disease. The ways this may occur are discussed in more detail in Chapter 6. At this point it is sufficient to note that the interactions are likely to be complex because they involve the interactions of several genes (polygenes or oligogenes) with each other and with the environment. For this reason they are commonly referred to as **complex diseases**. Because of this complexity, it is unlikely that an individual's health prospects can ever be predicted in a deterministic fashion from genetic tests. However, genes that act as risk

Table 1.6 Examples of common disorders for which there is evidence of a genetic component.

Congenital disorders	Common diseases of later life	Psychiatric disorders
Neural tube	Rheumatoid arthritis	Manic depression
Spina bifida	Various cancers	Alcoholism
Anencephaly	Epilepsy	Schizophrenia
Congenital heart disease	Multiple sclerosis	Tourette's syndrome
Cleft lip and palate	Insulin-dependent	Dyslexia
Mental retardation	diabetes mellitus	Alzheimer's disease
	Non-insulin-dependent	
	diabetes mellitus	
	Peptic ulcer	
	Ischaemic heart disease	
	Hyperthyroidism	
	Gallstones	
	Migraine	
	Asthma and other allergies	

factors are being identified for a wide range of common diseases (see Chapter 6). These may well serve as the basis for tests to determine an individual's predisposition or susceptibility to common disorders, allowing early diagnosis and consequently more successful treatment if the disease should occur. Moreover, defining genetic variables simplifies the analysis of environmental factors. Thus genetic tests for predisposition to common diseases may allow at-risk individuals to change their lifestyle to avoid contracting the disease.

1.4 Chromosomal mutations

During gametogenesis the diploid complement of chromosomes is reduced through meiosis, so that each gamete receives one member of the 22 pairs of autosomes and one sex chromosome. During this process recombination takes place, resulting in the exchange of information on homologous chromosomes. Failure of this process results in gametes that do not have the correct complement of chromosomes; when such a gamete fuses with another the resulting zygote will have a chromosome imbalance. If the failure results in a whole chromosome failing to segregate, the event is called a **non-disjunction**. Non-disjunction can result in a gamete with an extra chromosome, in which case the zygote is said to be **trisomic**, or a gamete in which a chromosome may be missing, in which case the zygote will be **monosomic**.

Accidents at meiosis can also result in chromosome rearrangements, duplications and deletions. A reciprocal **translocation** is an exchange of segments between non-homologous chromosomes (Figure 1.5). If the pair of translocated chromosomes are inherited together there will be no change in the information content, apart from the breakpoint, which, if it occurs within a gene, may result in that gene being damaged. Such translocations

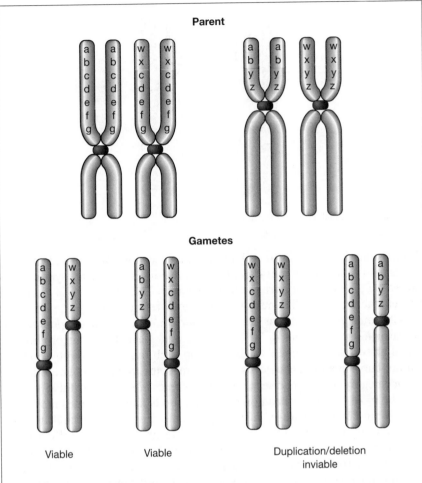

Parent

Gametes

Viable   Viable   Duplication/deletion inviable

Figure 1.5 A balanced translocation between non-homologous chromosomes. Each gamete receives one copy of each non-homologous chromosome. Only two of the combinations will result in the normal gene complement. The other two combinations result in duplication and deletions of genes, resulting in inviability or a mutant phenotype.

have been very valuable in mapping disease genes, because the position of the breakpoint may be mapped cytogenetically by examining banded chromosomes during metaphase in an affected individual (see Chapter 5). If the pair of translocated chromosomes are not inherited together in a gamete, then in the resulting zygote some segments of chromosomes may be missing and some present in triplicate. Such gametes are not likely to be viable, so normally both members of a pair of reciprocally translocated chromosomes must be inherited together. Such a situation is described as a **balanced translocation** (Figure 1.5). An **unbalanced translocation** is the non-reciprocal duplication of a chromosome segment (Figure 1.6). Inheritance of such a chromosome may lead to a chromosome imbalance; for example, a minority of Down's syndrome cases are familial. They result from trisomy for part of

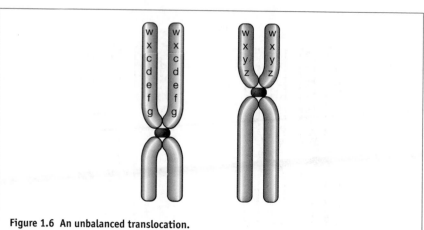

Figure 1.6 An unbalanced translocation.

chromosome 21 caused by an unbalanced translocation. Deletions of chromosomes lead to monosomy of chromosomal regions. *Cri-du-chat* syndrome is caused by a deletion of part of chromosome 5. This syndrome is identified by a characteristic cat-like mewing and mental retardation.

Chromosomal mutations are very common. One survey of spontaneous miscarriages showed that in 50% the fetus was affected by a major chromosomal abnormality. Some fetuses carrying such mutations survive to term but are consequently affected by the chromosome imbalance. These form a major category of genetic disease. Some examples are shown in Table 1.7.

The most common chromosomal mutation is Down's syndrome, caused by trisomy 21. The features of this syndrome are mental disability, characteristic broad facial features and short stature. This is usually a result of an extra copy of the whole chromosome 21, although it can be caused by a partial trisomy caused by a parent with a balanced translocation. The former result from accidents at meiosis and are therefore sporadic. Nevertheless, the frequency increases with maternal age. A young mother age 21 has only a 1 in 2000 chance of giving birth to a child affected by Down's syndrome, whereas the risk increases to 1 in 45 for mothers who are 45 years old. This is thought to be due to the fact that the mother's eggs

Table 1.7 Some examples of chromosomal mutations.

Condition	Frequency
Sex chromosomes	
45XO: Turner's syndrome	1 in 5000
47XXY: Klinefelter's syndrome	1 in 1000
Autosomes	
Trisomy 21: Down's syndrome	1 in 800 (maternal age-dependent)
Trisomy 18: Edward's syndrome	1 in 10 000
Balanced translocations	1 in 500

start to develop before birth and become arrested at prophase of meiosis I. Thus, at the time of ovulation an egg may be over 45 years old and the consequent deterioration results in the decreased fidelity of the meiotic process.

Another very common class of chromosomal mutations are those involving sex chromosomes. Monosomy for the X chromosome (45XO) results in Turner's syndrome, characterised by a sterile female phenotype with short stature and a web of skin between the neck and shoulders. Klinefelter's syndrome results from an XXY chromosome composition. This syndrome is characterised by a sterile male phenotype, and a tall and thin body form with breast development.

1.5 Mitochondrial mutations

Mitochondria provide 90% of cellular energy and thus the energy needed by organs, tissues and the body as a whole. Energy is generated in the mitochondria by the respiratory chain in a process known as **oxidative phosphorylation**. The respiratory chain consists of five protein complexes (complexes I–V) involving a total of 90 separate proteins. During oxidative phosphorylation electrons are passed along the chain from one complex to another. At the same time protons are pumped out of the mitochondrial matrix, generating a gradient across the inner mitochondrial membrane. The protons flow back into the matrix through complex V (ATP) synthase, which provides the energy for ATP production.

Each cell contains hundreds of mitochondria. Each contains between two and ten copies of a 16.6-kb circular DNA genome (mtDNA; Figure 1.7). In terms of size this corresponds to 0.0006% of the nuclear genome, but because there are approximately 10 000 mtDNA molecules per cell it amounts to about 1% of the total mass of cellular DNA. mtDNA encodes a number of essential functions, which are translated within the mitochondria using a mitochondrial-specific protein synthesis apparatus. The functions encoded by mtDNA are as follows:

- 13 respiratory chain subunits
    - seven subunits of complex I (NADH dehydrogenase)
    - three subunits of complex IV (cytochrome *c* oxidase)
    - two subunits of complex V (ATP synthase)
    - cytochrome *b* (a subunit of complex III)

- tRNA for each amino acid

- 12S and 16S rRNA for mitochondrial ribosomes.

The remaining functions are encoded by nuclear genes and synthesised in the cytoplasm before import into mitochondria.

Impairment of mitochondrial function leads to a decline in energy availability and results in clinical disorders. In principle, mutations to genes in both the nuclear and mitochondrial genomes could bring this about. Mutations to mtDNA have been shown to be the cause of a number of

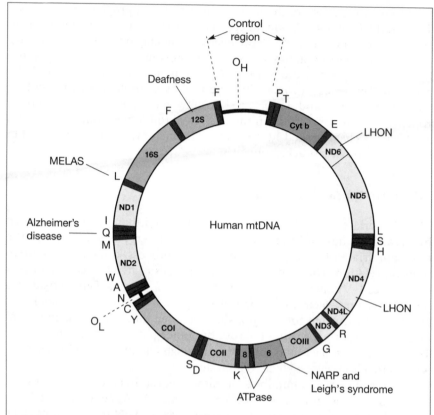

Figure 1.7 Map of the mitochondrial genome. The 22 tRNA genes are denoted by the single-letter amino acid code. 12S and 16S, 12S and 16S rRNA genes; COI–COIII, subunits of cytochrome *c* oxidase (complex IV); Cyt b, cytochrome *b*; LHON, Leber's hereditary optic neuropathy; MELAS, mitochondrial encephalomyopathy, lactic acidosis and stroke-like episodes; NARP, neurogenic muscle weakness, ataxia and retinitis pigmentosa; ND1–ND6, subunits of NADH dehydrogenase; ATPase6 and ATPase8, subunits of ATP synthase (complex V); O_H and O_L, origins of replication for the heavy and light strand respectively. The sequence of the control region is variable. It contains the D-loop, required for DNA replication.

disorders. In general these are multisystem disorders affecting the central nervous system, sight, hearing, heart and skeletal muscles, the kidneys and the endocrine glands (Table 1.8).

The part of the central nervous system most often affected is a region of the brain known as the basal ganglia. The basal ganglia are important for coordinated motion and its dysfunction results in ataxia. The effect of mitochondrial dysfunction on muscles is to cause mitochondrial myopathy, a general term describing the degeneration and loss of function of muscles associated with the presence of ragged red fibres: degenerating muscle fibres containing defective mitochondria that turn red in the presence of a specific stain. The exocrine gland most commonly affected is the pancreas, resulting in diabetes mellitus due to lack of insulin, and

Table 1.8 Diseases caused by mitochondrial mutations.

Disease	Acronym	Symptoms	Gene affected
Chronic progressive external ophthalmoplegia	CPEO	Mitochondrial myopathy and paralysis of eye muscles	Multiple gene loss because of deletion
Kearns–Sayre syndrome		CPEO plus ataxia, retinal deterioration, heart disease, hearing loss, diabetes and kidney failure	Multiple gene loss because of deletion
Leber's hereditary optic neuropathy	LHON	Blindness caused by damage to the optic nerve	Mutations in subunits of NADH dehydrogenase
Leigh's syndrome		Degeneration of basal ganglia leading to loss of motor and verbal skills	ATP synthase
Mitochondrial encephalomyopathy, lactic acidosis and stroke-like episodes	MELAS	Dysfunction of brain tissue causing dementia and seizures, mitochondrial myopathy and lactic acidosis	$tRNA^{Leu}$
Neurogenic muscle weakness, ataxia and retinitis pigmentosa	NARP	Muscle weakness, ataxia and blindness	Subunits of ATP synthase
Pearson's syndrome		Childhood bone marrow dysfunction, leading to multiple blood disorders and pancreatic failure	Multiple gene loss because of deletion

pancreatic exocrine deficiency, which is a failure to secrete the digestive enzymes that originate in the pancreas.

Mitochondria are maternally inherited, being passed on to progeny in the cytoplasm of the egg cell. Although the sperm acrosome contains mitochondria, they are not retained after fertilisation. Normally the 10 000 copies of the mitochondrial genome in each cell are identical and the cell is said to be **homoplasmic**. When a mutation occurs in one of the mitochondrial genomes there will be a mixture of different genome types in the cell and it is said to be **heteroplasmic**. The heteroplasmic state is short-lived and the descendants of a heteroplasmic individual become homoplasmic in a few generations. This is thought to occur because the oogonia contain a much smaller number of mitochondria (about 200) than other cell types, creating a situation similar to a **population bottleneck**, which reduces genetic diversity (see Section 9.2.4).

Mitochondrial disorders do not show regular Mendelian patterns of inheritance. In fact the pattern of inheritance can be complex. There are a number of interrelated reasons for this.

1. Mitochondria are maternally inherited so mitochondrial disorders can only be inherited from an individual's mother.
2. If a mother is heteroplasmic each of her children may receive different proportions of affected mitochondria. Therefore there will be a wide variety in the type and severity of symptoms in each child.
3. During development and the lifetime of an individual derived from a heteroplasmic zygote, the proportion of defective genomes may change. This may occur through chance sampling effects during segregation of mitochondria to daughter cells after cell division, or through the acquisition of new mutations to previously wild-type genomes. As a result, in an affected individual, there will be variation in the proportion of defective genomes, both spatially in different tissues and temporally throughout the individual's lifetime. The effect of mutation will lead to an increase with time in the proportion of defective genomes. This is probably the reason why mitochondrial disorders often only become manifest after a delay of several years and why some disorders become progressively more severe with age. A factor that will exacerbate this process is that interruption of the electron transport chain leads to the production of highly reactive free oxygen radicals, which are mutagenic. Thus a slight impairment of mitochondrial function, which initially is not severe enough to cause clinical symptoms, may nevertheless interrupt electron transport sufficiently to trigger this process and eventually result in the onset of a disorder.

Both large-scale deletions and mutations to specific genes can affect the mitochondrial genome.

Point mutations affecting NADH dehydrogenase cause LHON. This disease predominantly affects young men, resulting in blindness as a result of damage to the optic nerve and sometimes heart and neurological abnormalities.

Mutations to ATP synthase result in variable phenotypes. Severely affected individuals suffer from Leigh's syndrome, a devastating and often lethal childhood disorder in which the basal ganglia of the brain degenerate. Less severely affected individuals suffer from NARP, characterised by muscle weakness, ataxia and a form of blindness called retinitis pigmentosa (which can also be caused by a variety of chromosomal mutations). The difference in severity in different individuals with the same mutation reflects the proportion and tissue distribution of mutant mtDNAs in affected individuals. Individuals with Leigh's syndrome have high levels of mutant mtDNA in multiple tissues.

Point mutations affecting tRNA or rRNA have severe effects because they simultaneously reduce the ability to make different mitochondrial proteins. MELAS is caused by a mutation in the tRNA^{Leu} gene. Other mutations of this gene have been shown to result in a variety of severe symptoms, including mitochondrial myopathy, heart disease and diabetes mellitus. Indeed, mutations to tRNA^{Leu} are responsible for 1.5% of all incidences of diabetes mellitus. Mutations to tRNA^{Glu} have been found to be associated with 5% of patients with late-onset Alzheimer's disease. Mutations in the 12S rRNA gene also generally affect protein synthesis. They have been shown to be responsible for congenital deafness.

The effect of deletion in mtDNA (ΔmtDNA) is different in adults and children. Adults generally suffer from slowly progressive neurodegenerative disorders such as CPEO and Kearns–Sayre syndrome. Children suffer from severe disorders, such as Pearson's syndrome, that progress rapidly and affect multiple organs. CPEO is characterised by mitochondrial myopathy and paralysis of the eye muscles. Kearns–Sayre syndrome is characterised by CPEO plus retinal degeneration, heart disease, ataxia, deafness, diabetes mellitus and kidney failure. Pearson's syndrome causes a failure to make blood cells, resulting in severe anaemia and other blood disorders, pancreatic failure leading to diabetes mellitus and pancreatic exocrine dysfunction. Children who suffer from Pearson's syndrome generally die in the first few years of life. A few children who do survive generally develop Kearns–Sayre syndrome as they grow older.

The differences in clinical phenotypes between children and adults may result from differences in the proportion and tissue distributions of ΔmtDNA. A more widespread distribution with a higher proportion of ΔmtDNA will result in earlier onset with more severe symptoms. The tendency of mitochondrial deletions to worsen over time is probably caused by preferential replication of ΔmtDNA in non-dividing cells. The reasons for this are not clear at present. The origin of the deletions is also unclear, because they are rarely passed on from a mother with ΔmtDNA, presumably because an egg cell containing ΔmtDNAs could not survive. mtDNA with duplicated segments is often found to be associated with ΔmtDNA. A duplicated genome would not result in any impairment since all functions are still present. However, they may give rise to deletions as a result of complex intramolecular recombination events between the duplicated segments.

1.5.1 Mitochondria and ageing

Common symptoms of mitochondrial disorders, such as diabetes, dementia, muscle weakness, loss of sight and hearing, ataxia, etc., are also characteristic of the normal ageing process. This has led to the suggestion that accumulation of somatic mutations in mtDNA may contribute to the deterioration of body function with age. Many environmental toxins inhibit mitochondria and so may be a contributory factor. Evidence is accumulating in support of this hypothesis.

- The performance of the respiratory chain complex deteriorates with age in the brain, skeletal muscle, heart and liver.

- Rearrangements in mtDNA accumulate with age in these same tissues. ΔmtDNAs accumulate in skeletal muscle after age 40, consistent with other observations that show that ΔmtDNA is preferentially replicated in non-dividing tissue.

- Genetic predisposition to type II diabetes often tends to be maternally inherited, consistent with a mitochondrial component to the aetiology.

- Animals raised on restricted-calorie diets have longer lifespans. Animals fed on such diets produce fewer free oxygen radicals and accumulate less damage to mtDNA.

- Individuals who suffer from ischaemic heart disease experience temporary interruptions to the blood flow to the heart caused by athero-sclerotic plaques. During the resultant anoxia in the heart muscle, the respiratory chain is blocked; upon restoration of the blood supply (reperfusion) there is a burst of free oxygen radical production. The mitochondria in heart muscle of such patients contain a high level of mutant mtDNA, which may accelerate the onset of heart failure.

One intriguing possibility is that dietary supplements of antioxidants such as vitamin C may help to limit the production of free radicals in the mito-chondria and thus delay the ageing process.

1.6 The Human Genome Project

Most of the genes involved in common monogenic diseases have been cloned and the cDNAs sequenced. This has led to enormous advances in understanding the molecular nature of the diseases and in developing robust diagnostic tests for their detection. The genes involved in monogenic diseases will be discussed in Chapter 5 and the tests used in their detection will be described in Chapter 8. The focus of human genetics has now shifted to understanding complex or multifactorial diseases. As we shall see in Chapter 6 this is a very difficult task, undoubtedly far more difficult than for monogenic diseases. Even cloning the genes involved in mono-genic diseases proved very difficult and required large teams of scientists and large-scale international collaboration. From the time of the first efforts to clone genes involved in monogenic diseases it was appreciated that there was a need to map and then sequence the human genome. This led to a formal international collaboration called the Human Genome Project. Planning started in the late 1980s and work formally began in 1990. The impetus for the project initially came from the DOE and NIH in the United States. However, large-scale and important contributions came from laboratories all over the world, particularly the Genethon laboratory in Paris, funded by the French Muscular Dystrophy Association, and the

Sanger Centre in the UK, funded by the MRC and the Wellcome Trust for
Biomedical Research.

23

Summary

The work was carefully planned at a strategic level, with a number of
goals established in 1990. These were formally revised in 1993 and 1998.
Broadly speaking, the project was divided into two phases. The first, which
lasted from 1990 to 1998, was aimed at producing different types of map of
the human genome. The second, from 1998, was to derive the sequence
itself. The reason for planning the work like this was three-fold. First, the
maps would be needed before sequencing could begin. Second, in 1990
the technology for sequencing was inadequate for the task and would need
to be improved before sequencing the genome could be undertaken. Third,
maps of the human genome would be tremendously valuable in their own
right for identifying and isolating genes involved in disease. Moreover, the
technology for their construction was either available in 1990 or at least
imminent. A draft sequence was published in 2001 and a paper describing
the finished sequence of the human genome was published in 2004. The
process of annotation is now essentially complete – that is, identifying the
location of features such as genes, repetitive elements, genetic and physical
markers, etc. As we shall see in Chapter 4, genome databases and browsers
allow this information to be accessed online.

The physical and genetic characterisation of the human genome will
occupy the first part of this book. Chapter 2 will introduce the basic structure of the complex human genome. Chapter 3 will describe the different
types of map and how they are used. Chapter 4 will describe how the
sequencing was carried out and how it will be used.

1.7 Summary

- Genetic factors play a major role in determining lifetime health.
  Approximately 5% of babies born alive will suffer from a significant medical condition for which there is a genetic component. Susceptibility to
  many of the common diseases of later life is also influenced by our
  genetic constitution.

- There are four main classes of genetic diseases: (i) monogenic defects;
  (ii) multifactorial or complex disorders; (iii) chromosomal imbalances;
  and (iv) mitochondrial mutations.

- Monogenic disorders follow Mendelian patterns of inheritance, and
  are subdivided into autosomal recessive, autosomal dominant and
  sex-linked.

- In monogenic disorders, the pattern of inheritance and the manifesta-
  tion of the symptoms may be complicated by a variety of factors, such
  as genetic heterogeneity, penetrance, varying expressivity, mosaicism,
  anticipation, genomic imprinting and phenocopy.

- Multifactorial or complex disorders come about through the interaction
  between polygenic or oligogenic determinants and the environment.

- Chromosomal mutations arise through non-disjunction at meiosis, producing aneuploids. An extra copy of a chromosome results in trisomy, while loss of a chromosome results in monosomy. Trisomy 21 is responsible for Down's syndrome, one of the commonest genetic conditions. Other common chromosome abnormalities are balanced and unbalanced translocations, and deletions.

- Mitochondrial disorders are maternally inherited. A variety of factors lead to great complexity in the age of onset and severity of symptoms. Mitochondrial disorders affect many different organs simultaneously, including the central nervous system, skeletal muscle, heart, sight, hearing and endocrine function. Such disorders tend to become progressively worse as further mitochondrial mutations accumulate.

Further reading

General

House of Commons Science and Technology Committee (1995) *Third Report – Human Genetics: the Science and its Consequences. Volume 1. Report and Minutes of Proceedings.* HMSO, London.

Provides a general introduction to the ground covered by this book. As well as giving data for the incidence of various diseases, it provides an account of the views of leading clinicians, scientists and industrialists as to the influence of genetics on human health and the possibilities of the new developments to ameliorate the consequences of genetic defects. Finally, it provides a commentary on the legal, ethical and social impact of human genetics along with recommendations designed to ensure that scientific developments do not have a negative impact, issues discussed in Chapter 12 of this book.

WEATHERALL, D.J. (1991) *The New Genetics and Clinical Practice*, 3rd edn. Oxford Medical Publications, Oxford.

Chapter 1 provides a detailed review of the influence of genetic factors on human health. It is notable for the carefully researched data on the frequency of different genetic diseases and the extent of ethnic variation.

Monogenic disorders

INGRAM, V.M. (1957) Gene mutations in human haemoglobin: the chemical difference between normal and sickle cell anaemia. *Nature*, **180**, 326–328.

Online Mendelian Inheritance in Man (OMIM)
http://www.ncbi.nlm.nih. gov/entrez/query.fcgi?db=OMIM

This database started life as a catalogue assembled by Victor McCusick, one of the leading figures in the field of human genetics. It is now a website maintained by the National Center for Biotechnology Information (NCBI) in the United States.

Complex diseases

GHOSH, S. and COLLINS, F.S. (1996) The geneticist's approach to complex disease. *Annual Review of Medicine*, **47**, 333–353.

LANDER, E.S. and SCHORK, N.J. (1994) Genetic dissection of complex traits. *Science*, **265**, 2037–2048.

Inheritance of psychiatric disorders and personality traits

BOUCHARD, T.J. (1994) Genes, environment and personality. *Science*, **264**, 1700–1701.

MANN, C.C. (1994) Behavioural traits in transition. *Science*, **264**, 1686–1689.

PLOMIN, R., OWEN, M.J. and McGUFFIN, P. (1994) The genetic basis of complex human behaviours. *Science*, **264**, 1733–1739.

ROSE, S.J., LEWONTIN, R.C. and KAMIN, L.J. (1990) *Not in our Genes: Biology, Ideology and Human Nature*. Penguin, Harmondsworth.

An outspoken critique of the evidence that underpins the claims that genetic factors control behaviour and the incidence of psychiatric disorders.

Mitochondrial disorders

LARSSON, N.G. and CLAYTON, D.A. (1995) Molecular aspects of human mitochondrial disorders. *Annual Review of Genetics*, **29**, 151–178.

WALLACE, D.C. (1997) Mitochondrial DNA in aging and disease. *Scientific American*, **276**, 22–29.

CHAPTER TWO

An introduction to the structure of the human genome

Key topics

2.1 Introduction

The human **genome** is the term used to describe the sum total of DNA molecules found within every cell of the human body except red blood cells. Nearly all the genome is found in the nucleus; however, the mitochondria also contain essential genetic information. The haploid genome, the DNA found in a gamete, is half the complement of a somatic cell. The human genome consists of 3200 million base pairs of DNA (3.2×10^9 bp or 3200 Mb).

This DNA contains all the information required for a zygote to develop into an adult human. The information is in the form of genes. Each gene consists of a length of DNA that performs some function, usually specifying the amino acid sequence of a protein. A surprising aspect of the human genome is that coding sequences form only 1.5% of the total. Some of the rest of the DNA is non-coding but has a functional role in regulating and promoting gene expression or a structural role in chromosome integrity and segregation of chromosomes at nuclear division. A large fraction of the genome does not appear to have a function; if it does, it is a function that does not depend on the sequence of bases within the DNA. Much of this DNA consists of a menagerie of repetitive and mobile elements that may be parasitic or 'selfish' in origin. The genes themselves are not continuous stretches of DNA: the **coding sequences (exons)** are interrupted by non-coding sequences (**introns**). Sometimes the exons form only small patches in long stretches of introns. Indeed, as discussed below, in some cases only 1% or less of the total DNA that forms a gene may be coding sequence.

Thus the human genome is large and complicated in its organisation. Before gene cloning techniques were developed, it was impossible to study individual genes. Even with gene cloning, it is still often very difficult to find and clone human genes. The small fraction of the genome that forms genes and the technical complications introduced by the repetitive and mobile elements exacerbates the problem further.

In the past few years there has been spectacular progress in characterising the genome, including mapping physical and genetic landmarks and determining the entire nucleotide sequence. This book is an account of how the maps and sequence of the human genome were obtained and how they will be exploited in research into basic human biology and in medicine. Before this account can begin, it is necessary to describe the general features of the genome and how the DNA is packaged into chromosomes. This account is only intended to be a summary of material treated in much more detail elsewhere. You should refer to one of several excellent text books listed at the end of the chapter for more detail. The facts and figures given in this chapter come mostly from the analysis of the draft human genome sequence, considered in detail in Chapter 4.

2.2 Sequence architecture of the human genome

In the 1960s a simple experiment measuring the rate of reassociation of denatured DNA led to the surprising conclusion that a large fraction of mammalian DNA consisted of repetitive sequences (see Further reading). As a result of these experiments, human DNA may be categorised as follows:

1. **Single sequence DNA** or low-copy DNA (about 45% of the total). This class contains the sequences that form genes, as would be expected

from the Mendelian laws of genetics. However, coding sequences account for only about 1.5% of the total genome. The remainder of this class is DNA from introns or from spacer DNA, sequences of no apparent function that separate genes.

2. **Intermediate repeated DNA**, present at between 10^2 and 10^5 copies per genome (about 45% of the total).

3. **Highly repetitive DNA**, present at up to 10^6 copies per genome (about 10% of the total).

2.2.1 Genes

A functional definition of a **gene** is a DNA sequence that contributes to the phenotype of an organism in a way that depends on its sequence. A change in the sequence may affect its function and may have phenotypic consequences. Most genes encode proteins. However, sequences that encode functional RNA molecules are also classified as genes, as their function depends on their sequence.

2.2.1.1 Gene expression

Transcription in humans, as in all other eukaryote cells, is catalysed by three types of RNA polymerase, designated RNA polymerase I, II and III (or pol I, pol II and pol III respectively). RNA polymerase I transcribes rRNA. RNA polymerase II transcribes nuclear genes that encode proteins. RNA polymerase III transcribes tRNA genes, and a small number of functional RNA molecules such as 5S RNA and nuclear RNAs that mediate splicing (see below). All three RNA polymerases require the action of transcription factors to bind DNA and promote transcription.

The structure of the **upstream** region of a typical gene transcribed by pol II is shown in Figure 2.1. A typical gene contains a **promoter** region, which extends for about 200 bp upstream of the transcription start site at nucleotide +1. In addition it may contain one or more of the following elements: **enhancers**, **response elements** and **silencers**. The promoter contains all of the elements required for a basal level of transcription. These elements are short conserved sequences that bind different **transcription factors**. Transcription factors are said to be *trans*-acting because they are encoded by another gene that must be translated in the cytoplasm and imported into the nucleus. The sequence elements on the DNA to which they bind are said to be *cis*-acting because they control the transcription of a gene that lies immediately adjacent on the same DNA molecule.

Most human promoters contain an element called a TATA box centred around position –30 relative to the transcription start site. The TATA box binds **general transcription factors**, designated TFIIX, where the roman numeral denotes the RNA polymerase involved and X denotes the particular protein. In pol II transcription the process is initiated by the binding of TFIID mediated by another protein called **TATA-binding protein** (TBP). This is followed by the assembly of an initiation complex with pol II and the other general transcription factors.

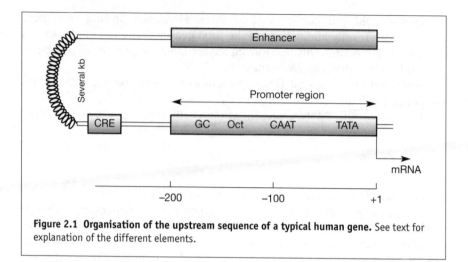

Figure 2.1 Organisation of the upstream sequence of a typical human gene. See text for explanation of the different elements.

The TATA box is responsible for correctly locating the site at which transcription starts. Three other commonly found elements within the promoter affect the efficiency of transcription:

- **GC box** (GGGCGG), which binds the transcription factor SP1.
- **CAAT box** (GGCCAATCT), which binds the transcription factors CTF and NF1.
- **Oct box** (ATTTGCAT), which binds the transcription factors Oct-1 and Oct-2.

The CAAT box is normally found around position –80 relative to the transcription start site. The spacing of the GC box and Oct box may vary. Active promoters contain at least one of these elements but do not necessarily contain all three. As well as these elements that bind ubiquitous transcription factors, tissue-specific elements may also be found within the promoter region.

Enhancers are regions of DNA that operate to stimulate the basal level of transcription from its promoter. They are operationally distinguished from promoters by three criteria:

- they may be located a considerable distance from the transcription start site;
- their action is not dependent on their location: they may be located **upstream** or **downstream** of the gene they control;
- their action is not dependent on orientation.

Enhancers usually contain multiple sites where transcription factors bind to stimulate transcription. These may include sites for the same ubiquitous transcription factors found within the promoter as well as sites for the binding of specific transcription factors. Enhancers probably act at a distance through the formation of DNA loops, which bring the transcription factors bound to the enhancer region into close proximity with the basal transcription complex.

Response elements induce genes in response to particular signals. Examples include cyclic AMP response element (CRE), serum response element (SRE) and heat-shock response element (HRE). Response elements may be found within the promoter region, closely upstream of the promoter or in more distant enhancers.

Transcription control is generally positive in eukaryotes: transcription is stimulated by the presence of transcription factors. However, there are examples of negative control where genes are turned off by the binding of proteins to elements known as **silencers**. In yeast, where this phenomenon has been well documented, the same transcription factor can stimulate transcription at one gene and silence transcription at another. The action of silencers is less well characterised in humans.

The overall level of transcription of a gene is the outcome of the different influences exerted by the promoter and enhancers. Some genes, termed **housekeeping genes**, are expressed in most tissues most of the time, and are responsible for functions likely to be necessary in any cell. Such genes may be transcribed from a promoter alone if it contains sufficient elements for the binding of ubiquitous transcription factors such as SPI1. Other genes, whose transcription is tissue specific or depends on the presence of a specific signal, require the action of enhancers and response elements to stimulate the basal transcription from their promoters. There is great scope for sophisticated fine tuning of transcription arising from the interaction of many transcription factors to sites within the promoter and enhancers.

As soon as transcription starts, the 5′ end of the nascent mRNA molecule is covalently modified to form a structure known as a **cap**. This is formed by the addition of 7-methylguanosine to the 5′ nucleotide of the nascent mRNA in a 5′–5′ triphosphate linkage. This effectively blocks the 5′ end of the molecule because the 7-methylguanosine is in the opposite orientation to the rest of the molecule. The cap is essential for initiation of translation at the first AUG codon in the mRNA. The appearance of sequence AAUAAA in the nascent RNA molecule causes transcription to terminate a few hundred base pairs downstream; the RNA molecule is then cleaved about 20 bp downstream of the AAUAAA signal. The 3′ end is also covalently modified by the addition of 100–200 adenine residues to form a structure known as the poly-A tail.

The part of the gene that encodes the protein contains sequences called **introns**, which interrupt the protein-coding regions known as **exons** (see Figure 2.2). The **primary RNA transcript** contains sequences derived from the introns. These are removed by a process known as **splicing** before the transcript leaves the nucleus. Splicing occurs through a cyclical process that occurs in a structure called a spliceosome. The spliceosome consists of small nuclear RNA molecules complexed to proteins called snRNPs (small nuclear ribonucleoprotein particles). These are sometimes colloquially known as 'snurps'. Splicing requires the concerted action of snRNPs and specific sequences at the junctions of introns and exons called **splice donor** and **splice acceptor** sites. The splice donor site is sometimes known as the 5′ splice site; it consists of the first two nucleotides of the intron, which is always the dinucleotide GT. The splice acceptor site is sometimes known as the 3′ acceptor site; it consists of the last two nucleotides of the intron, which is always the dinucleotide AG. Alteration of these sequences will prevent splicing and will have phenotypic consequences. In addition, there is a sequence in the intron known as the branch site, 18–40 nucleotides upstream of the splice acceptor site, where a splicing intermediate called a **lariat** is formed. However, in humans this sequence is not highly conserved and, if altered by mutation, splicing can proceed using other related sequences in the vicinity.

RNA processing is a point at which gene expression can be modified. There are now many examples of genes that are differentially spliced and give rise to sets of different polypeptides that originate from the same primary transcript.

2.2.1.2 CpG islands

DNA can be modified by methylation of cytosine to form 5-methylcytosine. This normally occurs at the dinucleotide CpG, i.e. a C residue followed in the 3′ direction by a G residue on the same strand of DNA. The dinucleotide CpG occurs less frequently than would be expected by chance from the base composition of DNA. This is a consequence of DNA repair mechanisms. Accidental deamination of cytosine produces uracil, whereas deamination of 5-methylcytosine produces thymine. Because uracil is not found in DNA, it is efficiently excised and replaced with cytosine by an enzyme called glycolase. However, thymine, a natural constituent of DNA, is now mispaired with a G residue. Although this will also be repaired, the process is not perfectly efficient and over a long period has led to a gradual reduction in the frequency of the CpG dinucleotide. DNA methylation is generally associated with repression of transcription.

So CpG is relatively rare in the human genome and where it occurs it will normally be methylated. However, so-called **CpG islands** exist where the frequency of CpG is greatly elevated, approaching that expected from the percentage (G+C) base composition of the human genome, and the CpG dinucleotides are hypomethylated or not methylated at all. These CpG islands are about 1 kb in length and tend to extend over promoters of expressed genes; about 56% of human genes are estimated to be associated with such sequences. They can be recognised by restriction enzymes that contain CpG in their target sequence and which will thus cut less frequently than expected because of the relative rarity of the CpG dinucleotide and because methylation inhibits the action of the restriction enzyme. Sites where these enzymes cut DNA will be clustered in CpG islands.

2.2.1.3 The number, size and spacing of human genes

A question that has provoked much interest over the years is how many genes there are in total in the human genome. Although, as we shall see in Chapter 4, there is still some uncertainty about the final number, the completion of the draft human genome sequence has led to the conclusion that there are about 30 000–35 000 genes in total. This figure is lower than previous estimates and was one of the most surprising conclusions of the genome project. The draft genome sequence allows some of the properties of genes to be enumerated (these are summarised in Table 4.3 on page 124). Human genes show enormous variation in overall size and the size and number of introns. Some genes, such as the histone genes, do not contain introns. The α-globin gene is 0.8 kb in size and contain three introns, which account for 30% of the total genomic locus. Other genes are extremely large. For example, the structures of the α-globin and the blood clotting factor VIII genes are compared in Figure 2.2. The dystrophin gene (the gene defective in Duchenne muscular dystrophy) is 2.4 Mb in size and contains 79 introns, which account for 99.4% of the total genomic locus. The mean size of genes is 27 kb, including introns and exons. The whole of each gene is transcribed, but the resulting RNA molecule is processed in the nucleus to produce mRNA that is translated into protein in the cytoplasm. Only that portion of the gene that remains in the translated portions of its mRNA can be counted as coding sequence, the mean size of which is 1.34 kb. Thus on average only 5% of each gene is coding sequence. Of the total genome, 33% is transcribed but only 1.5% is coding sequence. The density of genes varies in each chromosome from 9 to 14 genes/Mb or one gene every 71–111 kb.

2.2.1.4 Gene families
Between 25% and 50% of protein-coding sequences in the genome are unique. The remainder belong to families of similar or related genes. Some of these genes are dispersed through the genome, while others are present in clusters of related genes.

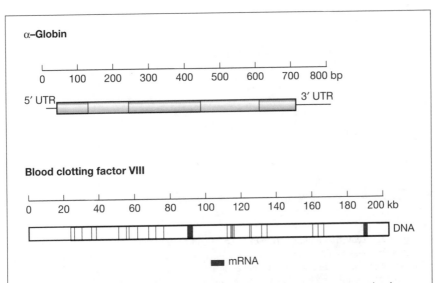

Figure 2.2 Human genes vary in size and intron content. The figure compares the size and intron content of the genes for α-globin and blood clotting factor VIII. In each case the coding regions are shown in solid blue boxes and the introns as open boxes. Note the difference in the scale shown above each gene. In the case of α-globin the 5′ and 3′ untranslated regions of the mRNA are shown (5′ UTR and 3′ UTR); these are normally classed as exons. The blood clotting factor VIII gene occupies 186 kb of the human X chromosome. It contains 26 exons ranging in size from 69 bp to 3106 bp and introns as large as 32 kb. The mRNA is only 9 kb in size and is shown to scale. It comprises a 7053-nucleotide coding sequence and a 3′ UTR of 1806 nucleotides.

Gene families are thought to have arisen by a process of gene duplication and divergence from an ancestral gene. Figure 2.3 shows how the evolution of the different proteins that make up haemoglobin is thought to have occurred. Haemoglobin is the oxygen-carrying molecule in blood, comprising two β-globin and two α-globin protein subunits, each complexed to a molecule of haem. The different members of the globin gene families are expressed at different times during fetal, embryonic and adult stages of development (see Chapter 5).

As well as whole genes it is very common to find proteins of different overall function containing similar protein **domains**. For example, many different proteins contain nucleotide-binding domains but show no similarity throughout the rest of the protein.

Because there are multiple copies of genes in a family, sometimes loss of function in one of them may be tolerated. This has led to the evolution of **pseudogenes**, which are sequences recognisably similar to functional genes but which have accumulated nonsense and frameshift mutations that will prevent them from functioning. They may be viewed as the decaying relics of genes that were once functional. For example, the β-globin gene cluster contains two pseudogenes (Figure 2.3).

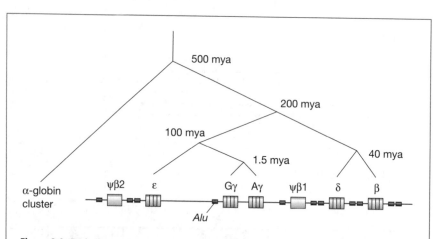

Figure 2.3 Evolution and structure of the β-globin gene family. About 500 million years ago (mya) gene duplication of a single globin-like gene produced the two genes that diverged to give rise to the ancestral genes of the α- and β-globin gene clusters. Initially the α and β genes were found as tandem repeats, as is still the case in modern amphibians. A translocation event separated the two genes to different chromosomes and this was followed by further rounds of gene duplication and divergence. ψβ1 and ψβ2 are pseudogenes. The positions of eleven *Alu* elements are shown by the solid boxes.

2.2.1.5 Non-protein-coding genes

Not all genes encode proteins. The bulk of cellular RNA consists of rRNA and tRNA required for protein translation. There are also a number of other RNA molecules that function directly within the cell. The 7SL RNA molecule forms part of the signal recognition particle required for the translocation of proteins across the endoplasmic reticulum. This is the first step in the secretory pathway that will result in the protein being exported from the cell. Small nuclear RNA molecules form complexes with protein to mediate the splicing of protein-coding RNA transcripts (see above).

2.2.1.6 Tandem repeat arrays

The rRNA molecules consist of three separate species: 28S, 18S and 5.8S. They are transcribed from a single transcription unit by a dedicated RNA polymerase known as RNA polymerase I. A single 45S transcript is produced that is processed to form the separate species. tRNA genes are also transcribed by a dedicated polymerase known as RNA polymerase III. (Protein-coding genes are transcribed by RNA polymerase II.)

The transcription units for rRNA are repeated 150–200 times, so that each copy lies immediately adjacent to the next, and are arranged so that the end of one unit abuts the start of the next. Such an arrangement is known as a **tandem repeat array** (Figure 2.4). The genes encoding the five histone proteins (see below) are found in similar tandem repeat arrays. In contrast to gene families, the sequences of these genes in tandem repeat arrays are all identical or nearly identical. There are 500 tRNA genes dispersed around the genome.

rRNA and tRNA constitute about 90% of the total RNA within the cell. Histones are also present in large numbers as they are bound to DNA in a stoichiometric fashion (see below). When the DNA is replicated the number of histones must also double, so large quantities have to be synthesised in a short period of the cell cycle. A single copy of the histone, rRNA and tRNA genes would not be capable of synthesising enough product. Only by having several hundred copies can the rate of synthesis keep up with demand.

2.2.1.7 Spacer regions

Genes are separated from each other by long tracts of DNA known as **spacer regions**. The sequence of spacer regions changes much more rapidly in evolution than the sequence of genes, indicating that there is less selection

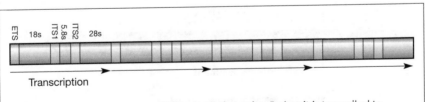

Figure 2.4 Tandem repeat array rRNA transcription units. Each unit is transcribed to produce a 45S transcript, which contains the ETS (external transcribed spacer unit), 18S, ITS1 (internal transcribed spacer 1), 5.8S, ITS2 and 28S sequences. This transcript is processed in a number of sequential steps to release the 18S, 5.8S and 28S RNA molecules.

against changes in its sequence. Thus it is thought that spacer DNA does not have a function that depends on its sequence. It is possible that the physical separation of genes is important in some way, so spacer DNA may have a sequence-independent function. The long tracts of DNA found in introns may be regarded as a particular form of spacer DNA. An example of the physical separation of genes may be seen in the region that contains the five members of the β-globin gene cluster. Each gene is 1.6 kb in size including two introns, yet the cluster is spread over 60 kb of chromosome 11 (Figure 2.3).

2.2.2 Intermediate repeated sequences

Much of the intermediate class of DNA consists of a small number of families, each consisting of sequences that are similar but not identical to each other. Most of these are **transposons**, so called because new copies are generated in a new location by a process of **transposition**. When this happens, one copy is found at the original location, and one copy at the new location, thus duplicating the original DNA sequence. Because the removal of the extra copies is a slow process on an evolutionary time-scale, they will increase in number. In humans they have come to occupy 45% of the genome. This has led to the concept of **selfish** or **parasitic DNA** because they do not contribute to the phenotype of the organism but evolve only to increase in number. However, some of these **mobile genetic elements** have been co-opted to have a role (see Section 4.9.1.3). The signature of mobile genetic elements is a short direct repeat sequence either side of the point of insertion into the host chromosome. There are four different classes of repeated DNA elements:

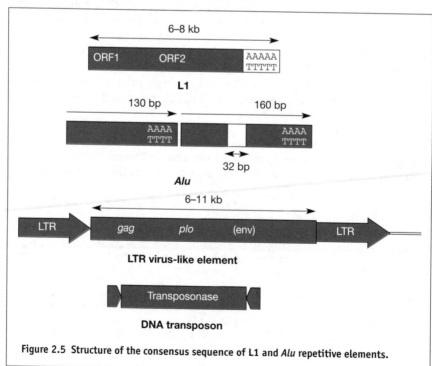

Figure 2.5 Structure of the consensus sequence of L1 and *Alu* repetitive elements.

- long interspersed elements (LINEs)
- short interspersed elements (SINEs)
- LTR retrovirus-like elements
- DNA transposons

The structures of these elements are illustrated in Figure 2.5.

2.2.2.1 LINEs

Complete **long interspersed elements** (LINEs) are about 6–8 kb in size and contain a promoter for RNA polymerase II and two **open reading frames** (ORFs). One of the ORFs encodes a protein with similarities to **reverse transcriptase** found in **retroviruses** that synthesise a DNA copy of an RNA template. The other ORF encodes an endonuclease. An AT-rich region is located near the 3′ end of the element. The propagation of LINE elements occurs by a process called **retrotransposition**, in which the LINE mRNA molecule serves as a template for the reverse transcriptase it encodes and the resulting DNA copy is inserted into a new chromosomal site (Figure 2.6). The reverse transcription of LINE elements usually fails to proceed to completion. Because the LINE mRNA template is copied from its 3′ end, most LINEs are truncated at the 5′ end and average only 1 kb in size. There are four distantly related LINE families in the genome (LINE1–4), together occupying over 20% of the genome. The most common is LINE1, with over 500 000 copies, representing nearly 17% of the total sequence.

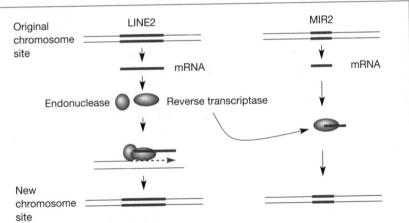

Figure 2.6 Propagation of LINEs and SINEs by retrotransposition. The retrotransposition of a LINE element is shown on the left. After transcription and translation, the endonuclease and reverse transcriptase form a complex with the LINE mRNA, which migrates to the nucleus. The endonuclease nicks chromosomal DNA and the free 3′ OH end generated is extended by the reverse transcriptase, using the LINE mRNA as a template. As a result, a new copy of the LINE mRNA is inserted into the genome. The reverse transcriptase of each LINE element binds to a sequence, particular to each element, at the 3′ end of the LINE mRNA. SINEs are thought to propagate by also having the recognition sequence of a reverse transcriptase encoded by a LINE in the same genome. In this case, the human MIR element has the recognition sequence for the LINE2 reverse transcriptase, which is thus able to copy it into DNA. Priming of the reverse transcription may occur by a hairpin loop formed by the poly-A tail of the mRNA and the poly-T region in the second repeat (see Figure 2.5).

2.2.2.2 SINEs

Short interspersed elements (**SINEs**) are 300–400 bp in size and are found in most metazoan organisms. SINE elements do not encode any proteins and are not able to transpose autonomously. Most SINE elements are similar at their 3′ end to the sequence recognised by the reverse transcriptase encoded by a LINE in the same genome. This allows them to propagate by retrotransposition (Figure 2.6). SINEs were originally derived from genes that are transcribed by RNA polymerase III (pol III), normally tRNA genes. The pol III promoter is retained in SINEs and is responsible for their transcription.

There are three SINE families in the human genome – *Alu*, MIR and MIR3. Together they occupy 13% of the genome. The **Alu** element derives its name from the restriction enzyme *Alu*1, as most copies contain a target sequence for this enzyme. It is the most common SINE in the human genome, with about 10^6 copies, representing 10% of the genome. This means that on average there will be an *Alu* element every 3 kb, although, as we shall see in Chapter 4, the density is not uniform. It is the only SINE that is still active in the genome. The *Alu* element is unusual because its sequence shows that it was not derived from a tRNA gene, but from another pol III-transcribed gene, 7SL RNA, which forms part of the signal recognition complex required for the translocation of proteins into the endoplasmic reticulum during secretion. Moreover, unlike most SINEs, its 3′ end is not similar to the recognition sequence of a reverse transcriptase encoded by a LINE, so which reverse transcriptase is responsible for its propagation is still uncertain. The *Alu* elements are confined to the primate line, but there are other mammalian SINEs derived from 7SL RNA, such as the rodent B1 element. In contrast, MIRs are found in all mammalian lines, hence their name: mammalian-wide interspersed elements. The 3′ end of the MIR3 transcript contains the 50 bp recognition sequence of the reverse transcriptase encoded by LINE2. Neither LINE2 nor MIR3 is still active. It is thought that when the last active LINE2 element disappeared from the genome 80–100 million years ago, transposition of MIR3 also ceased (see Section 4.9.1.3).

2.2.2.3 LTR retrotransposons

LTR **retrotransposons** (LTRs) derive their name from **long terminal repeats** at each end of the element. Between the LTRs are *gag* and *pol* genes that encode a reverse transcriptase, protease, RNAse H and integrase. These proteins are sufficient to programme the autonomous transposition of the element. The LTRs contain all the necessary regulatory elements to ensure transcription of the *gag* and *pol* genes. The resulting transcript is reverse transcribed in the cytoplasm using a tRNA primer. LTRs are very similar to retroviruses, and it is thought that retroviruses arose from LTRs by the acquisition of an *env* gene. Recombination often occurs between the LTRs, resulting in the excision of the intervening sequence and leaving a single inactive LTR in the genome. A variety of LTR elements are found in eukaryotes. Only endogenous retroviruses (ERV) are found in the

Together LTR-derived sequences occupy 8% of the genome, but 85% of this consists of isolated LTRs. Nearly all LTR transposons in the mammalian genome are now inactive (see Section 4.9.1.3).

2.2.2.4 DNA transposons

DNA transposons resemble bacterial transposons. They have terminal inverted repeats and encode a transposonase. Only an intact element can encode an active transposonase. However, the transposonase can act in the nucleus on deleted and mutated copies of the element. This results in the gradual accumulation of inactive elements and erodes the efficiency of the transposition process. For this reason DNA transposons tend to be relatively short-lived unless they can move by horizontal transfer to virgin genomes, which does happen. There are seven major classes of DNA transposons in the human genome, occupying less than 3% of the total sequence. All are thought to have been inactive for at least 50 million years (see Chapter 4).

2.2.2.5 Retrotransposons

The reverse transcriptases of the mobile elements described above may sometimes act on the mRNA sequences of other genes, resulting in the transposition of a copy to a new location. Because the reverse transcriptase uses mRNA as a template, which has been processed to remove introns, the new copy contains only the exonic sequences from the source gene. Usually these copies become inactivated by mutation and are known as **processed pseudogenes**. However, the draft sequence of the human genome shows that there are about 300 functioning genes that probably originated by retrotransposition (see Section 4.9.1.3).

2.2.3 Highly repetitive DNA

Highly repetitive DNA consists of short sequences repeated up to a million times in the genome. Because the average base composition of the repeated sequence may be different from the average of the rest of the genome, it often has a different density. Thus when total human DNA is fractionated by buoyant density ultracentrifugation, the highly repetitive DNA may form separate or satellite bands to the main peak. For this reason highly repetitive sequences are often known as **satellite DNA**. Most satellite DNA is found in long tandem arrays around the centromere, in subtelomeric regions and in the heterochromatic short arms of acrocentric chromosomes and most of the Y chromosome (see below for a description of chromosome structure). An important example of a satellite DNA sequence is α satellite DNA. The α satellite is a highly repetitive sequence that has an essential function in the centromeres of chromosomes (Box 2.1). Centromeres are the point where chromosomes attach to the spindle at nuclear division, ensuring the proper segregation of chromosomes to daughter cells.

BOX 2.1: α SATELLITE DNA

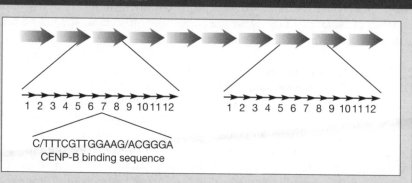

C/TTTCGTTGGAAG/ACGGGA
CENP-B binding sequence

Each centromere contains a tandem array of α satellite repeats that extend for millions of base pairs, uninterrupted by any other sequence. Each array is organised in a hierarchy of higher-order repeats. These higher-order repeats are represented by the large arrows in the figure. Each higher-order repeat contains a number of monomers, which varies between 4 and 32 depending on the particular chromosome; these are represented by the smaller arrows numbered 1–12 in the figure. Some, but not all, of these monomers contain a 17-bp binding site for the centromere-specific DNA-binding protein CENP-B. The monomers are not identical to each other; indeed they can show up to 20% sequence divergence. They may be no more similar to each other than they are to the α satellite in other primates. In any one array, each of the higher-order repeats (large arrows) are virtually identical (<2% sequence divergence). They contain the same monomer subunits repeated in the same order. The organisation and sequence of the monomers is particular to each chromosome. The number of higher-order repeats varies from 100 to 5000 on different chromosomes, giving a range of overall array sizes from 0.2 Mb to 10 Mb. Altogether, the α satellites represent several per cent of the total genome.

The repeat structure of the α satellite sequences makes them extremely difficult to clone in bacteria, because they are continually rearranged by recombination. Nevertheless this has recently been achieved, and it has been shown that when introduced into cells in culture they will function as centromeres. They have been used to construct artificial human chromosomes.

2.2.3.1 Minisatellites and microsatellites

Some types of satellite DNA are interspersed throughout the whole of the genome. There are two such classes, known as **minisatellites** and **microsatellites**. Both types have proved extremely important in the construction of maps of the human genome, which is discussed in Chapter 3. Minisatellites were also the original basis of **genetic fingerprinting** used in forensic science.

Minisatellites consist of sequences between 10 bp and 100 bp long repeated in tandem arrays that vary in size from 0.5 kb to 40 kb. They tend to occur near telomeres although they have been found elsewhere. Some, but not all, minisatellites show variation in repeat number so they are sometimes referred to as **variable number tandem repeats** (VNTRs). An individual who carries two different alleles that vary in size is said to be **heterozygous**. A locus that commonly exists in different forms is said to be **polymorphic**. As we shall see in Chapter 3, polymorphic loci such as minisatellites may be used as markers to construct genetic maps, although their tendency to be clustered near telomeres limits their use as a genomic mapping marker. Some minisatellite loci are hypervariable; for example, alleles of a locus called D1S8 contain between 120 and 1000 repeats of a 29-bp sequence, resulting in a variation in size from 3.3 kb to 29 kb. As well as variation in repeat number, the sequence of the repeat unit itself can vary in different members of the repeat array, so that loci that are monomorphic for length may still be highly polymorphic in structure. The variability of minisatellite loci forms the basis of genetic fingerprinting used in forensic science and paternity testing (see Chapter 11).

Microsatellites consist of tandem repeats of units two to four nucleotides in length. They are also known as simple sequence tandem repeats (SSTRs). They are found at all locations within the genome, even within protein-coding sequences. Like minisatellites, the number of repeats in each microsatellite varies, changing its overall size. Thus microsatellites are polymorphic and they have proved to be valuable genetic markers. They have been more useful in this respect than minisatellites because their genomic distribution is more uniform. The most commonly used microsatellite for this purpose has the structure $(CA)_n$. Tetrameric STRs form the basis of the current method of DNA profiling for forensic casework (see Chapter 11).

2.2.4 Telomeres

At the end of each chromatid is a structure known as the **telomere**. The telomere consists of a large and variable number of tandem repeats of the sequence TTAGGG and has a protruding 3′ end. This 3′ end does not have the usual properties of single-stranded DNA. One possibility is that the free 3′ end folds back on itself to form a hairpin structure through unusual base-pairing G residues. Other more elaborate structures have been proposed. The telomere is synthesised by an RNA-containing enzyme called **telomerase**. The RNA molecule in telomerase is used as a template to extend the free 3-OH end of the chromosome.

The telomere serves two essential functions.

1. Free ends of chromosomes produced by breakage are highly unstable and fuse with a high frequency to other chromosome ends. The telomere stops this happening to the ends of normal chromosomes by binding specific proteins to form a protective cap.
2. Telomeres solve the problem of replicating linear DNA molecules. As explained in Figure 2.7, the properties of the replication fork mean that the 3′ end of a DNA molecule will be successively eroded with each round of replication. If the 3′ end consists of a stretch of telomeric repeats, the erosion will affect these repeats rather than any essential functions in chromosomal DNA.

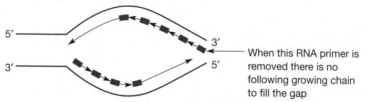

Figure 2.7 The problem of DNA replication at a chromosome end. A bidirectional replication origin near the end of a DNA molecule. Synthesis of each new DNA strand (shown in blue) proceeds in a 5′ to 3′ direction, extending an RNA primer (shown as a box) that is subsequently degraded. The resulting gap is filled by extension of the following chain. One strand of DNA can be synthesised continuously. The other strand, called the lagging strand, must be replicated in short sections, called Okazaki fragments, as the replication bubble grows. When the bubble reaches the end of the chromosome, an RNA primer can be made for lagging strand synthesis; but when it is degraded, there is no following strand to fill the gap. Thus, that strand of the DNA molecule will be shortened by the length of the RNA primer.

The telomere may play an important part in both the ageing process and cancer. Human cells are **mortal** – in tissue culture they can only go through about 50 divisions before they stop dividing and die. This number of generations is approximately the maximum number of generations in any cell lineage from fertilisation to ageing and death, so cell mortality *in vitro* probably reflects the normal ageing process. As a cell ages its telomere becomes shorter. It is possible that cell mortality occurs when the telomere becomes too short to function efficiently and vital sequences in the chromosome are damaged. Telomerase is active in early embryonic cells but is turned off at later times and normal mortal cells do not have telomerase activity. Thus there is an in-built clock that counts cell generations and will eventually lead to cell death. This could be responsible for the normal ageing process. It is also a defence against cancers, because if cells escape other mechanisms controlling their proliferation, eroded telomeres will eventually cause their death. However, just as cancer cells have escaped other controls over their proliferation so they also eventually escape this control. One characteristic of cancer cells is that they are **immortal** and their telomeres do not shorten. Telomerase is not active in normal mortal cells, but it is active in at least some cancer cells. Forcing the expression of telomerase in human tissue culture cells that are mortal greatly increases the number of generations before division stops. Telomerase thus presents a target for anti-cancer drugs. Conversely, forced telomerase expression or delivery of telomerase to cells could theoretically provide a way of overcoming ageing. However, even if this were technically possible, and apart from the obvious ethical considerations, it would remove an important natural defence against cancer.

2.3 Structure of chromosomes

The 3200 Mb of DNA that constitutes the human genome is divided between 23 pairs of chromosomes; 22 of these chromosomes are found in both males and females and are known as **autosomes**. One pair of chromosomes is different in each sex and these are known as the sex chromosomes. Females have two copies of the **X chromosome**, while males have one X chromosome and one Y chromosome.

Further detail may be revealed by **banding**. In this process the metaphase spread is subject to light digestion with an enzyme such as trypsin, which breaks down proteins. It is then treated with **Giemsa**, a dye that binds DNA. Giemsa binds with different intensities at different points along the chromosome, producing a series of bands called **G-bands**. G-bands correspond to regions of the chromosome that have a lower than average proportion of GC base pairs and contain fewer genes (Section 4.9.1.1). The banding pattern is reproducible and is characteristic of each chromosome. Thus each chromosome, and even each part of a chromosome, may be recognised by its banding pattern (Figure 2.8). This provides the basis of a mapping procedure and allows changes to chromosomes to be recognised (Figure 2.9). Treatment of the metaphase spread with the fluorescent dye **quinacrine** produces a different type of banding pattern known as **Q-bands**.

At metaphase the chromosomes are **bivalent**. The two **chromatids** are joined at a constriction called the **centromere**. A protein structure called the **kinetochore** is located at the centromere and is the point at which the chromosome is attached to the mitotic spindle. The centromere divides the chromosome into two arms, the larger of which is called the **q arm** and the smaller the **p arm**. If the centromere is centrally placed the chromosome is said to be **metacentric**; where it is nearer one end the chromosome is said to be **submetacentric**. Where the centromere is near one end the chromosome is said to be **acrocentric** and when the centromere is at the very end the chromosome is said to be **telocentric**.

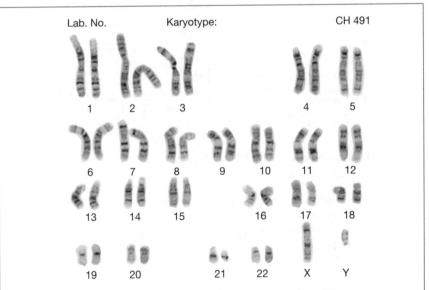

Figure 2.8 Karyotype of a human male. In the metaphase spread the different chromosomes are randomly arranged. The metaphase spread may be photographed, and the individual chromosomes cut out and arranged so that each chromosome is paired with its homolog. It can be seen that chromosomes vary in size. By convention the largest pair is called chromosome 1 and the smallest chromosome 22, although it has subsequently been shown that chromosome 21 is in fact smaller than chromosome 22. (From Maltby, E., Langhill Centre for Human Genetics, Sheffield, UK.)

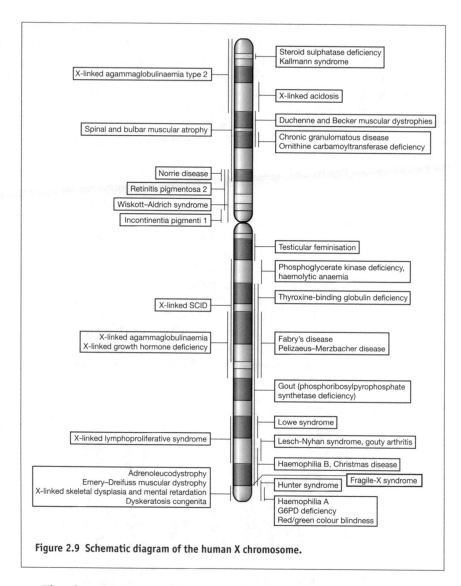

Figure 2.9 Schematic diagram of the human X chromosome.

The chromosome number, the chromosome arm and the G-bands are used together to specify the location of genetic elements. Figure 2.9 shows a map of the X chromosome by way of illustration. Careful microscopy leads to the subdivision bands. For example, the Duchenne muscular dystrophy locus is said to be located at Xp21.2, i.e. a subdivision of band Xp21.

2.3.1 Fluorescence *in situ* hybridisation

Chromosomes attached to a microscope slide in a metaphase spread can be hybridised with complementary sequences in probes, allowing the chromosomal origin of the probe to be identified. This process is known as ***in situ* hybridisation**. Originally it was carried out using ^{3}H-labelled DNA. This was technically demanding and needed the use of autoradiograms, which needed long development times. The procedure was revolutionised by the use of DNA probes that were labelled using different coloured **fluorophores**. These were visualised using a fluorescence microscope that excited fluorescence using the particular excitation wavelength of the fluorophore. The use of fluorescently labelled DNA probes in this way is called **fluorescence *in situ* hybridisation** (FISH). Using different coloured fluorophores it is possible to visualise multiple regions. FISH is an important tool in physically mapping the human genome because it allows the chromosomal origin of any cloned piece of DNA to be determined. Plate 1 shows an example of this where a probe for the elastin locus hybridises to chromosome 7. The resolution of this technique is several megabases because chromosomes are in a highly condensed state in metaphase. Recently, hybridisation to interphase chromosomes or to stretched chromosome fibres has allowed mapping where the resolution is 50 kb or higher.

FISH has also proved to be a highly effective tool for karyotyping. Chromosome-specific probes can be used as 'chromosome paints' that allow the rapid identification of chromosomes without the careful characterisation of G-banded karyotypes. Plate 2 shows how this can be used for sex determination of a fetus using whole, uncultured cells from amniotic fluid. This is a powerful technique where there is a risk of a sex-linked disease. Plate 3 shows how chromosomes in a metaphase spread may be visualised using a chromosome paint. Chromosome paints are used extensively in the analysis of cancer cell karyotypes, which can show characteristic rearrangements during the course of the disease.

2.3.2 Packaging DNA into chromosomes

A single DNA molecule runs the length of each chromosome. This molecule is about 279 Mb in size in the largest of the chromosomes (chromosome 1). Physically the DNA in chromosome 1 is about 8 cm long, yet it is packaged at mitotic metaphase into a chromosome only 8 μm in length, a compaction ratio of about 10 000. How is this done?

The DNA is wound around itself in a hierarchy of coils and supercoils. At the deepest level, the DNA is wound around an octamer composed of two molecules each of the four different basic proteins called **histones**

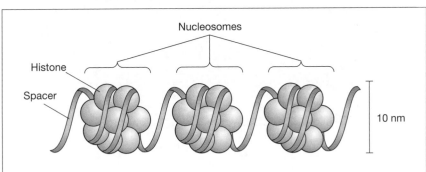

Figure 2.10 Structure of nucleosomes. DNA is wound around an octamer consisting of two copies each of four histones: H2A, H2B, H3 and H4.

H2A, H2B, H3 and H4 (Figure 2.10). This basic unit is called a **nucleosome** and in the electron microscope the DNA has the appearance of beads on a string. Each nucleosome occupies about 200 bp of DNA and is linked to the next by a short stretch of naked DNA. The register of DNA in nucleosomes may be very important in the control of gene expression.

DNA wound up in nucleosomes is coiled again to form a fibre 30 nm in diameter (Figure 2.11). The 30-nm fibre is stabilised by a fifth histone called histone H1. Finally, at mitosis, the 30-nm fibre winds around a scaffold of protein to form a fibre 600 nm in diameter (Figure 2.12). Various non-histone proteins are also associated with the DNA. The whole DNA–RNA–protein complex is known as **chromatin**.

During interphase, chromosomes are not so tightly condensed as metaphase chromosomes but the DNA must still be highly packaged. The transcription machinery must gain access to the DNA. Simple cartoons that depict transcription as an RNA polymerase moving along a DNA molecule like a train on a railway track are distorting reality to the point where it is

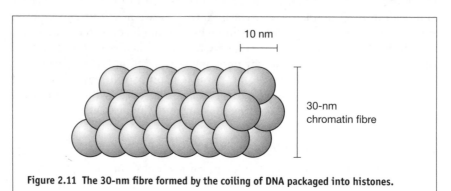

Figure 2.11 The 30-nm fibre formed by the coiling of DNA packaged into histones.

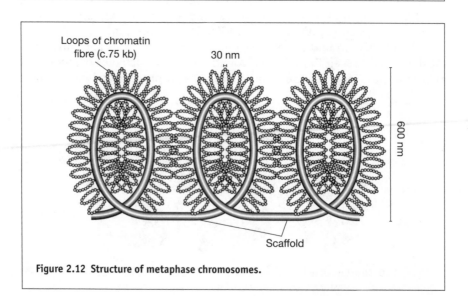

Figure 2.12 Structure of metaphase chromosomes.

A

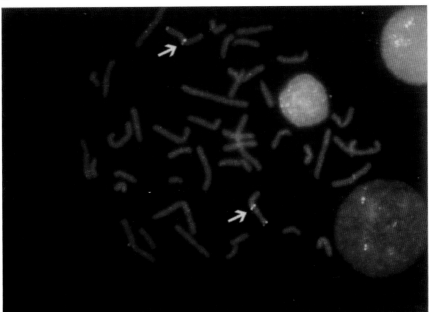

Plate 1 A gene probe for the elastin locus on the long arm of chromosome 7 is used to identify a deletion of the locus in a patient with Williams syndrome. (A) The normal karyotype. The two copies of chromosome 7 are arrowed. The signal near the telomere is a chromosome-specific probe to identify the chromosome; the signal near the centromere is the elastin locus. (B) One of the two copies of chromosome 7 (arrowed) lacks the signal near the centromere, showing the presence of a deletion. Williams syndrome affects about 1 in 10 000 of the population. It is characterised by an elfin-like appearance, hyperactive 'cocktail party' personality and multisystem disorders including multiple cardiac abnormalities, childhood hyperglycaemia, growth retardation, learning difficulties and neuromuscular disorders. It is caused by hemizygosity of the elastin locus. Elastin is a connective tissue protein that is a major component of elastic fibres in blood vessels, ligaments, etc.

B

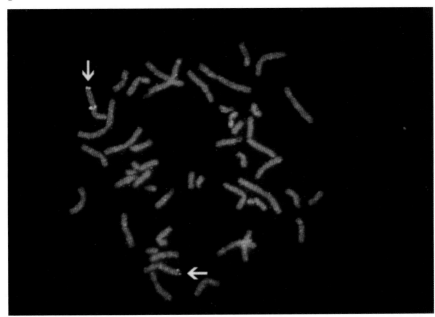

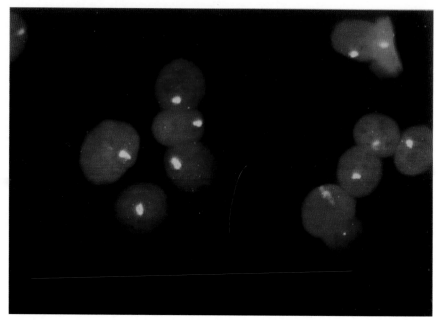

Plate 2 Probes specific for X and Y chromosones can be used to identify the sex of cells with interphase nuclei. This example shows a male foetus identified by the use of a Y-specific probe on uncultured cells from amniotic fluid.

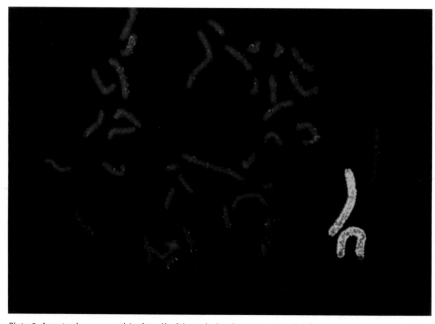

Plate 3 A metaphase spread 'painted' with a whole-chromosome paint for chromosome 2.

no longer a valid model. It is more likely that there are about 200 RNA factories within the nucleus of the human cell. The DNA is moved through these factories rather than the other way round. Perhaps a more accurate analogy is the way a tape in an audio cassette is fed through the playing head. When the tape is not being played it must be stored in a highly packaged fashion. What happens when the tape is not properly wound inside the cassette is something we are all familiar with!

2.3.3 Heterochromatin and euchromatin

Chromatin exists in two states, known as **euchromatin** and **heterochromatin**. Euchromatin is fully decondensed in interphase and is available for transcription. Heterochromatin differs in the following ways from euchromatin:

- Heterochromatin remains condensed in interphase and is stained with DNA-binding dyes such as Feulgen. In metaphase chromosomes it can be recognised by staining with a combination of distamycin and DAPI.
- It is transcriptionally inactive.
- It is late replicating.

Heterochromatin is found in the following regions:

- Regions surrounding centromeres.
- Subtelomeric regions.
- The short arms of acrocentric chromosomes.
- The non-expressed portion of the Y chromosome.
- Short regions along the length of chromosome arms (interstitial) that are otherwise euchromatic.

These regions are always in the heterochromatic state. In addition, the inactive X chromosome in females is heterochromatic. This is known as facultative heterochromatin because the X chromosome is active in males and the choice of which is inactivated in females is random. Much of the obligate heterochromatin consists of highly repetitive simple sequence DNA or satellite DNA.

2.4 Summary

- The haploid human genome consists of 3200 million base pairs (3200 Mb) of DNA.
- It consists of different sequence classes:
  - Single sequence or low-copy DNA.
  - Intermediate repeated DNA sequences consisting of families of DNA sequences repeated between 10^2 and 10^5 times per genome.
  - Highly repetitive or satellite DNA, repeated 10^6 times per genome.

- Coding sequences represent only about 1.5% of the genome.

- Not all genes encode proteins – some encode functional RNA molecules.

- There are multiple copies of genes encoding rRNA and histones. These are arranged in long tandem repeat arrays.

- There are two different classes of intermediate repeat sequences called SINEs and LINEs. SINEs and LINEs are mobile genetic elements that may be selfish or parasitic because they undergo selection to increase in number but do not contribute to the phenotype.

- The most common LINE is called L1. It encodes a reverse transcriptase that mediates transposition to a new location by a process known as retrotransposition.

- The most common SINE is a family of sequences called *Alu* which, like the L1 family, has also spread through the genome by a process of retrotransposition.

- Most highly repetitive DNA is found in long tandem repeats at centromeres. The sequences bind proteins of the kinetochore where the spindles are attached to chromosomes at mitosis and meiosis.

- Minisatellites consist of tandem repeats of units 10–100 bp in size. The number of repeats and the sequence of the repeat units themselves are both polymorphic, making them useful genetic markers. They form the basis of genetic fingerprinting.

- Microsatellites are interspersed throughout the genome. They consist of tandem repeats of units 2–4 bp in size. They are also widely used as genetic markers.

- The number, size and appearance of chromosomes is called the karyotype. There are 22 autosomes and one pair of sex chromosomes.

- G-bands are used to construct cytogenetic maps of chromosomes, which can be used to specify the chromosomal location of genes.

- The telomere is a special structure that protects the ends of chromosomes. It may be important in cancer and ageing.

- The DNA molecules in chromosomes are physically long and so have to be tightly packaged.

- The DNA is wound around histones to form nucleosomes.

- The nucleosomes are coiled to make a fibre 30 nm in diameter.

- At mitosis and meiosis the 30-nm fibre is coiled around a scaffold of proteins to produce a 700-nm fibre.

- Chromatin exists in two states, euchromatin and heterochromatin. Euchromatin is decondensed in interphase and may be transcriptionally active. Heterochromatin is condensed in interphase and is transcriptionally inactive.

Further reading

General

BROWN, T.A. (2007) *Genomes 3*. Garland Science, New York.

LEWIN, B. (2007) *Genes IX*. Jones and Bartlett Publishers, London.

LODISH, H., BERK, A., KAISER, C. *et al.* (2007) *Molecular Cell Biology*, 6th edn. W.H. Freeman, New York.

Reassociation experiments

BRITTEN, R.J. and KOHNE, D.E. (1968) Repeated sequences in DNA. *Science*, **161**, 529–540.

Selfish DNA

DOOLITTLE, W.F. and SAPIENZA, C. (1980) Selfish genes, the phenotype paradigm and genome evolution. *Nature*, **284**, 601–603.

ORGEL, L.E. and CRICK, H.C. (1980) Selfish DNA: the ultimate parasite. *Nature*, **284**, 604–607.

Transcription

LOO, S. and RINE, J. (1995) Silencing and heritable domains of gene expression. *Annual Review of Cell Biology*, **11**, 519–548.

MANIATIS, T., GOODBOURN, S. and FISCHER, J.A. (1987) Regulation of inducible and tissue-specific gene expression. *Science*, **236**, 1237–1245.

MITCHELL, P. and TIJAN, R. (1989) Transcriptional regulation in mammalian cells by sequence specific DNA binding proteins. *Science*, **245**, 371–378.

PABO, C.T. and SAUER, R.T. (1992) Transcription factors: structural families and principles of DNA recognition. *Annual Review of Biochemistry*, **61**, 1053–1095.

ZAWEL, L. and REINBERG, D. (1993) Initiation of transcription by RNA polymerase II: a multi-step process. *Progress in Nucleic Acid Research and Molecular Biology*, **44**, 67–108.

Splicing

KRAMER, A. (1996) The structure and function of proteins involved in mammalian pre-mRNA splicing. *Annual Review of Biochemistry*, **65**, 367–410.

PADGETT, R.A., GRABOWSKI, P.J., KONARSKA, M.M. *et al.* (1986) Splicing of messenger RNA precursors. *Annual Review of Biochemistry*, **55**, 1119–1150.

α Satellite sequences, centromeres and human artificial chromosomes

HARRINGTON, J.J., VAN BOKKELEN, G., MAYS, R.W. *et al.* (1997) Formation of *de novo* centromeres and construction of first generation human artificial microchromosomes. *Nature Genetics*, **15**, 345–355.

SCHULMAN, I. and BLOOM, K.S. (1991) Centromeres: an integrated protein/DNA complex required for chromosome movement. *Annual Review of Cell Biology*, **7**, 311–336.

WILLARD, H.F. (1991) Evolution of alpha satellite. *Current Opinion in Genetics and Development*, **1**, 509–514.

LINEs

HATTORI, M., KUHARA, S., TAKENAKA, O. and SAKAKI, Y. (1986) L1 family of repetitive elements in primates may be derived from a sequence encoding a reverse-transcriptase related protein. *Nature*, **321**, 625–627.

Retrotransposons

ROGERS, J. (1986) The origin of retroposons. *Nature*, **319**, 725.

WAGNER, M.A. (1986) Consideration of the origin of processed pseudogenes. *Trends in Genetics*, **2**, 134–136.

Alu elements

BRITTEN, R.J. (1996) Evolution of *Alu* retroposons. In *Human Genome Evolution*, Jackson, M., Strachan, T. and Dover, G. (eds), pp. 211–228. Bios Scientific Publishers, Oxford.

OKADA, N. (1991) SINEs. *Current Opinion in Genetics and Development*, **1**, 498–504.

ULLU, E. and TSCHUDI, C. (1984) *Alu* sequences are processed 7SL RNA genes. *Nature*, **312**, 171–172.

Minisatellites

ARMOUR, J.A.L. (1996) Tandemly repeated minisatellites: generating human genetic diversity via recombinational mechanisms. In *Human Genome Evolution*, Jackson, M., Strachan, T. and Dover, G. (eds), pp. 171–190. Bios Scientific Publishers, Oxford.

JEFFREYS, A.J. (1987) Highly variable minisatellites and DNA fingerprints. *Transactions of the Biochemical Society*, **15**, 309–316.

JEFFREYS A.J., WILSON, V. and THEIN, S.L. (1985) Hypervariable minisatellite regions in human DNA. *Nature*, **314**, 67–73.

Microsatellites

HANCOCK, J.M. (1996) Microsatellites and other simple sequences in the evolution of the human genome. In *Human Genome Evolution*, Jackson, M., Strachan, T. and Dover, G. (eds), pp. 191–210. Bios Scientific Publishers, Oxford.

WEBER, J.L. and MAY, P.E. (1989) Abundant class of human polymorphisms that can be typed using the polymerase chain reaction. *American Journal of Human Genetics*, **44**, 388–396.

Telomeres

ZAKIAN, V. (1995) Telomeres: beginning to understand the end. *Science*, **270**, 1601–1607.

Chromatin structure

FELSENFELD, G. (1992) Chromatin as an essential part of the transcriptional mechanism. *Nature*, **355**, 219–224.

FELSENFELD, G. and MCGHEE, J.D. (1986) Structure of the 30 nm fibre. *Cell*, **44**, 375–377.

TRAVERS, A.A. and KLUG, A. (1987) The bending of DNA in nucleosomes and its wider implications. *Philosophical Transactions of the Royal Society of London Series B*, **317**, 537–561.

CHAPTER THREE

Mapping the human genome

Key topics

- Importance of genomic maps
- Sequence tagged sites
- Genetic maps
  - Key concepts, LOD scores, informative meioses, phase, haplotype
  - RFLP, minisatellite and microsatellite mapping markers
  - The Genthon map
- Physical maps
  - Clone maps
  - Radiation hybrid maps
  - STS content maps
  - FISH
  - Long-range restriction maps
- Expression maps
- Integration of genetic, physical and expression maps

3.1 Introduction

Maps allow us to find and isolate genes when we have no other information about them apart from their location. Detailed maps provide the only means by which the genes that contribute to disease susceptibility can be identified and ultimately characterised. In the past few years, genes responsible for many of the major monogenic disorders have been isolated by such positional cloning (Figure 3.1). Furthermore, the focus of research in human genetics has now shifted to the analysis of complex diseases. Maps play a central role in the search for these genes.

Maps of the human genome can take different forms. Genetic maps are based on recombination frequencies between genetic markers at meiosis.

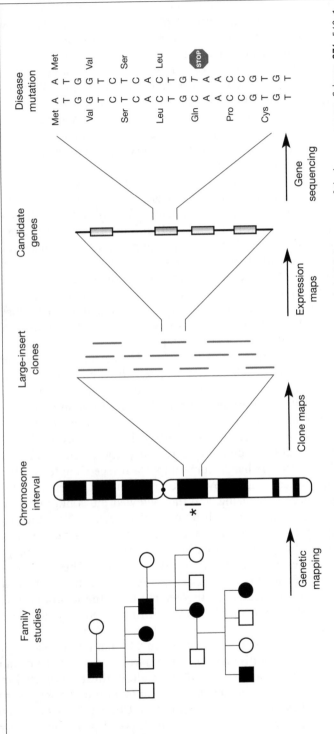

Figure 3.1 Different types of map allow genes to be identified quickly. (From Schuler, G.D. *et al.* (1996) A gene map of the human genome, *Science*, **274**, 540–1. Reprinted with permission from AAAS.)

Genetic markers are features of the human genome that are variable in different individuals. They may control phenotypic characteristics, which includes inherited diseases, or they may be a variable feature of the genotype that can be observed but that has no phenotypic consequence. Genetic markers freely recombine at meiosis unless they are located in the same region of the genome, in which case they are said to be linked. Linked markers tend to be co-inherited, and the closer they are the higher the frequency of co-inheritance. Once a detailed genetic map has been constructed using a reference set of neutral markers, a gene responsible for a single gene disorder can be mapped by observing co-inheritance between the disease and a reference marker.

The case of complex diseases is more difficult, because there is no single allele that will co-segregate with the disease. Here a different type of map, based on single nucleotide polymorphisms (SNPs), has been important, which charts the location of regions called haplotype blocks. These are chromosomal regions where particular combinations of alleles at closely liked loci, called a haplotype, occur more often than expected by chance. The HapMap project charted the location of all the different blocks using **single nucleotide polymorphisms (SNPs)** and identified a set of representative SNPs that could be used to identify which version of each block had been inherited. We shall consider this type of map and its application to identifying complex disease alleles in Chapter 6.

Physical maps describe the location of DNA sequences and other features of the genome, which do not necessarily vary between different individuals. They are constructed by plotting the location of physical mapping markers, such as **sequence tagged sites**. Expression maps plot the origin of RNA molecules. Originally these were concerned with the origin sequences found in mRNA, which identify the small proportion of the genome that encodes proteins. More recently, the origin of other RNA species, such as miRNAs has become important (see Chapter 7). In their original form, physical maps allowed clone libraries of the human genome to be ordered before sequencing, and then allowed the contigs to be joined together to make a continuous chromosomal sequence. Now the genome has been sequenced and fully annotated, a detailed physical map is available describing the physical location and structure of human genes, expressed sequences and many other genomic features. These are stored in databases such as **Ensembl**, which also cross-references the genetic map. So once a gene has been located on the genetic map, the location of all the nearby genes is readily available along with many clues as to the likely function of each.

In this chapter we shall consider how the different maps were originally constructed, including a discussion of the particular problems of human linkage mapping. In Chapter 4 we shall see how these were used in the sequencing of the human genome and how annotation of the sequence provides a detailed physical map of the genome. In Chapter 5 we shall see how these resources can be used to map and physically characterise genes affected in single gene disorders. Chapter 6 describes how the HapMap project has played a key role in mapping alleles contributing to the risk of complex disease.

3.2 Sequence tagged sites: a common currency for the different types of map

Mapping markers are the landmarks whose positions are plotted to construct the different types of map. As discussed above, it is important that the different types of map can be aligned with each other. However, markers that can be used for one type of mapping cannot necessarily be used for another. For example, genetic markers must be based on features that are naturally polymorphic. A disease segregating in a family is a genetic marker because some members are affected and others are not. However, unless the disease gene has been cloned, it cannot be used as a physical marker. A physical marker could be a sequence of DNA whose presence in a clone or location on a chromosome can be determined. However, if the sequence is the same in all individuals it cannot be used as a genetic marker.

There is clearly a need for a marker that can be used in the different types of map. The **polymerase chain reaction** (PCR) is used to generate a type of physical landmark called an STS that meets this requirement. PCR amplifies DNA using synthetic oligonucleotides as primers (Box 3.1). An STS is a stretch of DNA about 300 bp in length. As its name implies, the sequence of the STS is used to tag the larger DNA molecule from which it was derived so that the DNA region can be identified wherever it occurs. Figure 3.2 illustrates how STS markers may be generated and used to order a series of clones. The nucleotide sequence of the STS is used to specify the sequence of two synthetic oligonucleotides that will bind in opposite orientations at either end of the STS. Thus they will amplify the sequence if they are used as primers in a PCR reaction. PCR using these primers can therefore be used as a test to discover whether the sequence is present in a particular DNA sample.

Originally the concept of an STS marker was developed for physical mapping because it was based on recognition of a sequence. However, it was subsequently realised that an STS could be polymorphic, so long as the primers were designed to anneal to constant regions flanking some form of length polymorphism. We shall see below how microsatellites can be used to provide polymorphic STSs that are used for genetic mapping. STSs can also be generated from cDNA molecules and can be used to tag sequences that are expressed (**expressed sequence tag** or EST). The STS therefore provides a type of marker that is compatible with the three major map formats and so allows them to be aligned with respect to each other.

Apart from the fact that they provide a common marker for different types of map, there are four other important advantages of the STS methodology:

BOX 3.1: THE PCR REACTION

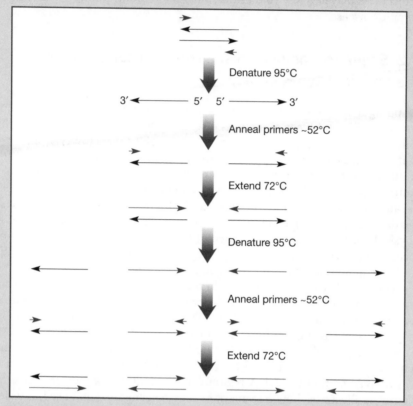

The PCR reaction is a form of *in vitro* cloning that can be used to amplify very small quantities of DNA. It depends on DNA polymerases from thermophilic bacteria, which can withstand temperatures of 95°C and have a temperature optimum of 72°C. DNA polymerases can only synthesise new DNA by extending the 3' end of a pre-existing primer. This is exploited to target new synthesis to a region between sites where two primers bind. In each cycle, the template DNA is denatured by heating to 95°C, annealed to the primer at a temperature of about 52°C (see below) and then incubated at 72°C for the polymerase to extend the primer using the target DNA as a template. In the figure the target DNA is shown in black and newly synthesised DNA is shown in blue; the arrows on each DNA strand point toward the 3' end. Two complete cycles are represented, and it can be seen that in each the amount of target DNA is doubled. A total of 30 cycles means that amounts of DNA as small as a single molecule can be amplified to produce microgram quantities of DNA. The reaction is carried out in a thermocycler that can be programmed to carry out the cycles as specified by the user. The primers can be readily synthesised using automated technology. The length of the primers is usually about 20 bases; the exact length and its base composition determines the optimum annealing temperature. The annealing temperature and ionic composition of the buffer are adjusted to ensure maximum stringency.

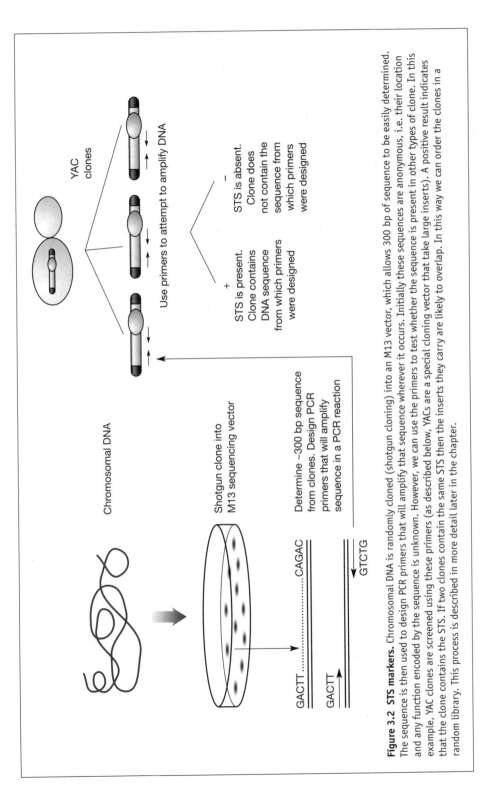

Figure 3.2 STS markers. Chromosomal DNA is randomly cloned (shotgun cloning) into an M13 vector, which allows 300 bp of sequence to be easily determined. The sequence is then used to design PCR primers that will amplify that sequence wherever it occurs. Initially these sequences are anonymous, i.e. their location and any function encoded by the sequence is unknown. However, we can use the primers to test whether the sequence is present in other types of clone. In this example, YAC clones are screened using these primers (as described below, YACs are a special cloning vector that take large inserts). A positive result indicates that the clone contains DNA sequence from which primers were designed. If two clones contain the same STS then the inserts they carry are likely to overlap. In this way we can order the clones in a random library. This process is described in more detail later in the chapter.

1. They allow different sources of DNA fragments to be examined for the presence of a common sequence; for example, two clones may be compared to detect any overlap between them. The sources of DNA may be totally different: we discuss below how YAC clones and radiation hybrid clones are forms of physical mapping that both use STS markers. Because of this the two types of map can be aligned with each other.

2. If a researcher in one laboratory wishes to examine whether an STS is present in clones on which they are working, they do not have to physically obtain a clone from another laboratory that produced the STS. Instead the sequence of the primers is used to synthesise the oligonucleotides using a widely available technology for automatic synthesis. There is no need to physically maintain and distribute stocks of DNA. This solves an important logistical problem of maintaining and distributing clone libraries.

3. The PCR procedure requires both primers to bind at the correct distance apart *and* in the correct orientation. This reduces the number of false positives that can arise using other methods to compare DNA molecules, such as hybridisation.

4. The PCR test lends itself to automation, allowing very large numbers of tests to be performed. Since the human genome is so large, this is an important practical consideration.

3.3 Genetic maps

3.3.1 Introduction

Genetic maps are based on the order of genetic mapping markers and the genetic distance between them measured by the recombination frequency. There are a number of concepts connected with genetic maps that we need to consider before we discuss the construction of a comprehensive map of the human genome.

3.3.1.1 Map units

All maps must have a unit of scale. The unit of human genetic maps is the **centimorgan** (cM). 1 cM is defined as a recombination fraction of 0.01, i.e. 1 in 100 gametes will be recombinant and the remaining 99 will have the parental configuration. Physically, 1 cM corresponds to between 0.7 Mb and 1 Mb of DNA sequence, but there is no invariant relationship between physical and genetic distances (see Figure 3.9).

3.3.1.2 Phase

Phase specifies whether particular alleles at adjacent loci are on the same (*cis*) or different (*trans*) chromosomes. For example, in the informative meiosis depicted in Figure 3.3, alleles A and 1 are in *cis* while alleles A and 2 are in *trans*. Clearly it is necessary to know the phase in order to know which outcomes of the meiosis represent recombinant chromosomes and

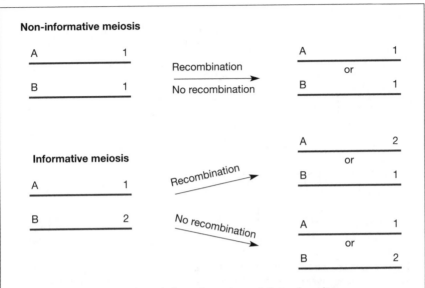

Figure 3.3 Informative and non-informative meioses. *Left:* two homologous chromosomes in a diploid cell about to undergo meiosis. Two adjacent loci, each with two alleles, are located on these chromosomes. *Right:* the structure of haploid gametes that result from the meiosis is shown. *Top:* the meiosis is uninformative because one of the loci is homozygous; after meiosis the gametes will have chromosomes that are identical whether or not a recombination event has occurred. *Bottom:* recombination results in chromosomes that can be distinguished from non-recombinant chromosomes; the meiosis is therefore said to be informative.

which represent parental chromosomes. Figure 3.4 shows an example of how the phase can be established by examining three or more generations.

3.3.1.3 Informative and non-informative meioses

To measure the recombination frequency between two markers at meiosis, it is necessary that the meiosis is informative, i.e. it is possible to distinguish between parental and recombinant chromosomes. This requires that both markers on a chromosome are heterozygous. Consider the three-generation family in Figure 3.4. The meioses in individual II1 are informative because we can tell by the genotype of the children in generation III whether recombination has taken place between the two markers. However, the meioses in II2 are uninformative as gametes with the same haplotype are produced whether or not recombination has taken place.

3.3.1.4 Haplotype

A **haplotype** is a set of closely linked alleles that tend to be inherited together at meiosis, i.e. not separated by recombination ('haplotype' is a contraction of 'haploid genotype'). In the previous example there are two parental haplotypes, A1 and B2. Haplotypes may consist of many more than two alleles. Generally, alleles making up a haplotype will be inherited

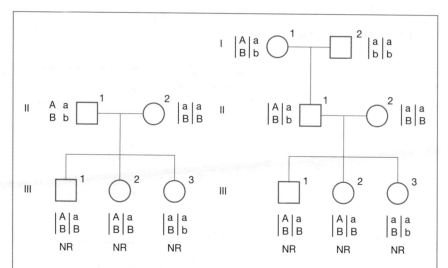

Figure 3.4 Phase can be worked out in a three-generation family. The figure shows the segregation of two linked loci each with two alleles (A/a and B/b). In the two-generation family on the left, the phase of the alleles in individual II-1 is unknown; therefore it cannot be established whether the progeny in generation III have inherited recombinant or parental chromosomes from II-1. The genotype of the grandparents allows the phase of the alleles in II-2 to be established. This reveals whether individuals in generation III have inherited parental or recombinant chromosomes. The vertical lines indicate the phase of the alleles where it is known.

as a block because their close proximity makes it unlikely they will be separated by recombination.

3.3.1.5 CEPH families

Recent genetic maps have been constructed using a set of reference families. These consist of samples collected from Mormon families living in Utah, USA, and French–Venezuelan families. Mormon families are typically much larger than normal and the keeping of detailed family records is part of the Mormon culture. The samples were used to establish permanent cell lines, which are maintained by the Centre d'Étude Polymorphism Humain (CEPH) in Paris. Each family consists of three generations with four grandparents, two parents and at least six children. Three generations allows the phase of markers in the parents to be established. The segregation of markers in the six children allows the results of meiosis in the parents to be determined. These cell lines provide an invaluable source of reference pedigrees, which are used in many different sorts of investigation into the human genome.

3.3.1.6 LOD score analysis

In experimental organisms, a large number of progeny from a genetic cross may be examined to measure recombination frequency accurately. Clearly this is not possible in human families. This problem is overcome by using a special form of mathematical analysis known as log ratio of odds or **LOD score** analysis (Box 3.2). The result of LOD score analysis is a numerical value that measures the ratio of odds that two loci are linked at given recombination fraction θ, compared with the chances that they are unlinked ($\theta = 0.5$). The threshold for declaring linkage is an LOD score of 3 or more. It is unlikely that an LOD score of 3 will be obtained from one family. However, the mathematical properties of the test allow data from a number of families to be combined.

BOX 3.2: LOD SCORE ANALYSIS

An LOD score is defined as the logarithm of the odds (Z) of observing the outcome of a meiosis if two loci are linked with a recombination fraction θ, compared with observing outcome if they are unlinked ($\theta = 0.5$). Thus

$$Z_x = \frac{\text{Probability of observed result if } \theta = x}{\text{Probability of observed result if } \theta = 0.5}$$

$$LOD_{\theta=x} = \log_{10} Z_x$$

The probability of each gamete being recombinant $= \theta$.
Probability of each gamete being non-recombinant $= 1 - \theta$.

$$\text{Thus in a family, the LOD score} = \frac{(1-\theta)^P . \theta^R}{0.5^{R+P}}$$

where R = number the number of children inheriting recombinant gametes and P = number of children inheriting parental gametes from the meiosis in question.

The calculation is then repeated using different values of θ. The value of θ when the LOD is at a maximum is called the maximum likelihood score (MLS).

The reason that the Z value is expressed as a logarithm is that the data from multiple families can be combined as the product of their individual odds. The $\log_{10}$ of the product of two numbers is the sum of their individual $\log_{10}$ values. In other words, the combined LOD score from multiple families can be conveniently calculated as the sum of their individual LOD scores. This is the strength of LOD score analysis: it allows the data from individual families to be combined easily, whereas on their own the data from each family consists of numbers that are too small to produce a statistically significant result.

The simplest application of LOD score analysis compares two markers, but it may be adapted to analyse multiple loci. This is known as multipoint LOD score analysis.

Statistical significance

The calculation of LOD score analysis is based on Bayes' theorem, which states that:

Continued ▶

Posterior odds = prior odds × conditional odds

Based on empirical observation, the odds that any two random loci are linked is about 1 in 50. This is the prior probability. The odds that they are linked based on observation, which is measured by the LOD score, are the conditional odds. The overall odds that they are linked are the posterior odds.

Conventionally, for a result to be judged statistically significant the odds of it being true should be equal to or greater than 20:1. Therefore a LOD score of 3 is required (i.e. $Z = 1000:1$) for statistical significance because:

Posterior probability = prior probability x conditional probability
$$= 1:50 \times 1000:1 = 20:1$$

A worked example

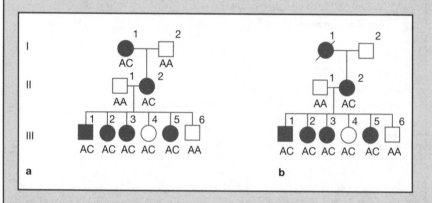

a

b

Figure **a** shows a pedigree where an autosomal dominant disease and a biallelic mapping marker with alleles A or C are segregating. We can tell from the three-generation pedigree that allele C is on the same chromosome as the disease allele. In generation III, five children received a non-recombinant chromosome from their mother II2 and one child (III4) received a recombinant chromosome.

$$\text{LOD score } \theta_{=0} = \text{Log}_{10}\frac{(1-0)^5 \times 0^1}{0.5^6} = -\infty$$

This negative LOD, which is greater than 3 score eliminates the possibility that the marker and disease gene are linked with a recombination fraction = 0. This is intuitively obvious, since if there has been a recombination event $\theta \neq 0$ by definition.

$$\text{LOD score } \theta = 0.05 = \text{Log}_{10}\frac{(1-0.05)^5 \times 0.05^1}{0.5^6} = 0.39$$

We now repeat this calculation for different values of θ, producing the results shown in the table below.

θ	0	0.05	0.1	0.2	0.3	0.4	0.5
LOD score phase known	$-\infty$	0.39	0.57	0.62	0.51	0.30	0
LOD score phase unknown	$-\infty$	0.09	0.27	0.32	0.22	0.08	0

Thus the MLS is generated when $\theta = 0.2$, suggesting that this is the most likely recombination fraction between the mapping marker and the disease. It appears likely that the mapping markers are linked, but the LOD score of 0.62 falls short of the threshold of significance of 3.0. However, if several other pedigrees gave a similar value, the combined LOD score could exceed 3 and linkage could be declared.

In fact, the pedigree is part of a larger pedigree where Huntington's disease (HD) was segregating and was actually used in initial mapping of the HD gene. However, for the purposes of illustration, the genotypes of the grandparents (I1 and 12) in the pedigree shown in Figure **a** were assumed because in reality their genotypes were not known (Figure **b**). The phase could have been HD allele and allele C on the same chromosome (phase 1) or HD allele and allele A on the same chromosome, in which case there were five recombinant children and one parental (phase 2). However, it is still possible to use LOD score analysis by assuming that in the parent II2 each phase was equally likely:

$$\text{LOD score} = \frac{1}{2} \frac{(1-\theta)^R . \theta^P}{0.5^{R+P}} \text{ (phase 1)} + \frac{1}{2} \frac{(1-\theta)^R . \theta^P}{0.5^{R+P}} \text{ (phase 2)}$$

For $\theta = 0.05$

$$\text{Log}_{10} \frac{1}{2} \frac{(1-0.05)^5 . 0.05^1}{0.5^6} + \frac{1}{2} \frac{(1-0.05)^1 . 0.05^5}{0.5^6} = 0.09$$

Again, the calculation is iterated for different values of θ, and the results are shown in the previous table. The result still suggests linkage with an MLS value when $\theta = 0.2$. However, the LOD score is lower than when the phase was known, representing a loss of power because of the missing information.

The fact that the pedigree is actually Huntington's disease raises another problem. We assumed that individual III4 was recombinant because she appears unaffected. However, Huntington's disease only manifests itself later in life. Perhaps she will develop the disease at a later date and so the LOD score calculation will be based on false information. The calculation of LOD score will have to take account of this by incorporating a parameter for age-related penetrance, i.e. incorporating the probability that she is genuinely unaffected based on her age.

A related parameter to the LOD score is the **maximum likelihood score (MLS)**. This is the LOD score for the most likely of a series of alternatives. For example, an MLS score can be calculated for the most likely recombination fraction (θ) between two markers. This is done by iterating the LOD score calculation assuming a different value of θ each time. The MLS obtained gives the most likely value of θ. MLS can also be calculated when other parameters of a genetic model are varied, i.e. degree of penetrance. In practice, LOD scores and MLS values are derived by computer analysis of the observed results. One popular program for this purpose is called LIPED.

3.3.2 Mapping markers

The first maps of the human genome were produced in the 1970s and were based on linkage between markers such as enzyme polymorphisms, disease genes, blood groups and other phenotypic traits that were monogenic in origin. Linkage between such markers was rare, so the resolution of the maps was low and large areas of the genome were not represented. Progress in the construction of human genetic maps was blocked because of a lack of usable markers in relation to the size of the genome.

3.3.2.1 Restriction fragment length polymorphism

The breakthrough came in 1980 with the realisation that many sequences in the genome were polymorphic in a way that did not have phenotypic effects. Such markers are called **neutral molecular polymorphisms**. The first such marker to be used is called a **restriction fragment length polymorphism (RFLP)**. It is based on the presence or absence of a target for a restriction enzyme, usually due to a polymorphism at a single base pair (Figure 3.5).

RFLPs allowed the construction of a comprehensive map for the first time in 1987. This map had an enormous impact on human genetic research and was the basis for the cloning of many of the major monogenic disease loci. RFLP markers also provided an important tool for prenatal diagnosis, allowing a chromosome carrying a disease locus to be tagged and its segregation in a family to be followed.

3.3.2.2 Minisatellites

The problem with RFLP markers is that by definition there can be only two alleles at a locus, i.e. the restriction site is either present or absent. The maximum proportion of individuals in a population that are heterozygous is only 50%. Since genetic markers are only informative when they are heterozygous, on many occasions an RFLP marker will not provide useful information. In practice the situation is much worse, because usually one allele is less common and the frequency of heterozygotes is less than 50%. The proportion of meioses using a particular marker that will be informative is called its **polymorphism information content (PIC)**. The maximum PIC value of a biallelic marker is 0.375 (it is less than 0.5 because some

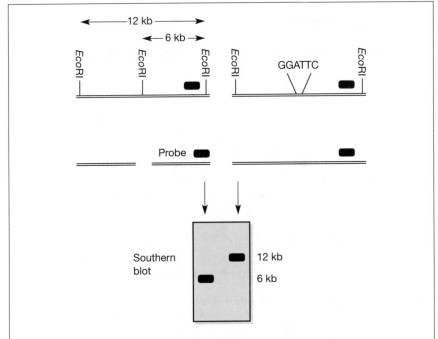

Figure 3.5 An RFLP marker. Parts of DNA molecules from two chromosomes differ from each other by a single base pair, which results in the absence of an *Eco*RI site in one of the chromosomes. Upon digestion with *Eco*RI, the chromosome without the extra *Eco*RI site produces a larger fragment than the other chromosome. The difference is recognised by Southern hybridisation using a probe (thick line) that hybridises within the region encompassed by the two flanking *Eco*RI sites present in both molecules.

meioses are not informative even if both parents are heterozygous). Minisatellite and microsatellite markers described below typically have much higher PIC values.

A more useful type of marker is one that is naturally multiallelic, so there is a much higher chance that an individual will be heterozygous. Minisatellite loci, also known as VNTRs (see Chapter 2), meet this criteria. The number of repeats in the locus is usually determined by Southern hybridisation. Figure 3.6 illustrates how the segregation of a VNTR may be followed in a family.

3.3.2.3 Microsatellites

Minisatellites have been used successfully in the construction of genetic maps. However, their use is limited by their distribution in the genome, as they tend to be clustered near telomeres. Furthermore, the Southern blot procedure needed to score them is both laborious and expensive. PCR can be used, but the large size of some minisatellites makes them difficult to amplify. A more useful type of marker is the microsatellite (see Chapter 2), which is more common and more evenly distributed than VNTRs. The

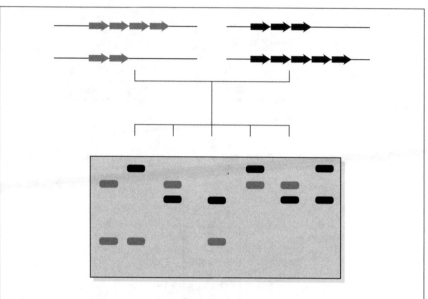

Figure 3.6 Segregation of a VNTR genetic marker in a family. The length of the VNTR depends on the number of repeats present. This can be recognised in a Southern blot using a sequence unique to the minisatellite as a probe. The genotypes of the two parents are shown in the left and right lanes respectively. Note that there are four alleles at this locus, segregating in a Mendelian fashion, with each child receiving one allele from each parent. This is indicated by colour coding.

microsatellite based on CA repeats has come to be the standard in the construction of genetic maps. The length of microsatellites is determined by PCR using primers designed from surrounding unique-sequence DNA (Figure 3.7). This makes them a special form of STS, which allows genetic maps based on microsatellite markers to be aligned with the different forms of physical map.

3.3.3 A comprehensive genetic map of the human genome

To be useful, a genetic map must be both comprehensive and detailed. Comprehensive means that it extends over the entire genome with no regions left unmapped. Detailed means that the distance between the markers is small, so that it is easy to detect linkage between the markers and biologically relevant loci, such as those which contribute to a familial disorder. Microsatellite loci have been used to construct a comprehensive map of the human genome, the resolution of which meets the goal set by the Human Genome Project. This map, constructed by the **Genethon** laboratory in Paris, is regarded as the definitive genetic map of the human genome.

The Genethon map is based entirely on microsatellite markers. The process of constructing the map began with identifying and sequencing

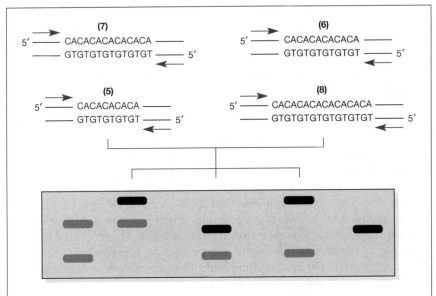

Figure 3.7 Segregation of a microsatellite locus in a family. The microsatellite locus is amplified using the PCR reaction with primers that anneal to the unique sequence either side of the CA repeats (shown by arrows). The size of the fragment produced depends on the number of CA repeats at each locus, indicated by the figure in parentheses above each DNA molecule. The products of the PCR reaction are separated by polyacrylamide gel electrophoresis, which is capable of resolving the difference in size between the different alleles. Note that there are four different alleles at the locus which show Mendelian segregation, each child receiving one allele from each parent (shown by colour coding).

5264 CA repeat loci. The genotype of these loci was then determined in members of large CEPH families. To map the autosomes, a total of 134 individuals in eight families were examined, allowing the outcomes of 186 meioses to be analysed. The mapping of the X chromosome required the examination of 20 families, 304 individuals and 291 meioses.

Once the genotypes of the markers had been obtained, they were analysed to determine first the order of the markers and then the genetic distance between them. This is done using specially developed computer programs. The 5264 markers analysed were found to define 2335 positions. The order of 2032 of these was determined with odds of 1000:1 against alternative orders. One of the most difficult and time-consuming steps in this process was checking for errors.

The length of the map in females is 4396 cM but only 2769 cM in males. The difference is due to an elevated rate of recombination in female meioses compared with male meioses. The sex-averaged length is 3699 cM. The length of the genetic maps for individual chromosomes generally reflects their physical size. The map of the largest, chromosome 1, is 292 cM; the smallest, chromosome 21, is 58 cM.

The average distance between markers is 1.6 cM. An important consideration is how evenly spaced these markers are. There are still some parts of the genome where there are not many markers, although these are a small minority of the total. For example, in 1% of the genome the distance between markers is over 10 cM. This actually consists of three positions where the interval between adjacent markers is 11 cM. This could be due to a large physical distance separating the markers, although it could also be due to enhanced rates of recombination in a particular region of the genome. In fact, the physical map showed that at least part of the distance is due to increased recombination. In contrast, there were some places where markers showed no recombination with each other but were physically separated by several megabases of DNA.

Such observations emphasise that measurements of genetic map distances assume that the rate of recombination between two markers depends only on the physical distance that separates them. This requires that the frequency of recombination is constant throughout the genome. In practice this is not true. There are recombination hotspots where recombination is more frequent and coldspots where it is less frequent.

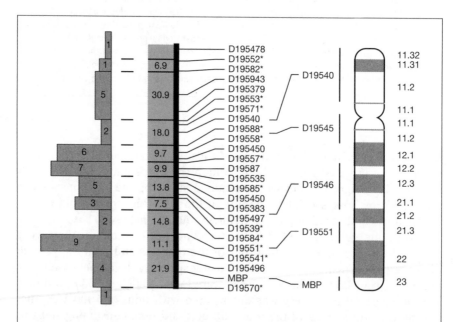

Figure 3.8 Genetic map of chromosome 18. The vertical histogram on the left shows the number of markers in each chromosome interval, shown in the centre (the figures show the size of the interval). The physical location of selected markers revealed by FISH is indicated on the cytogenetic map on the right. The map can be visited at the web site of the Cooperative Human Linkage Centre (http://www.chlc.org/HomePage.html). Clicking one of the intervals reveals more detail of the region.

The completion of a detailed and comprehensive genetic map is a milestone in human genetics. It provides an essential resource that will be used for the rapid identification of monogenic disease loci, allows the mapping of genes that make a contribution to multifactorial disease and provides a framework to check the authenticity of physical maps. Panels of markers evenly spaced through the genome may now be purchased from biotechnology companies and these are routinely used to hunt for genes of interest.

A sample of a similar genetic map constructed by the Cooperative Human Linkage Centre is shown in Figure 3.8, which shows an overall representation of chromosome 18 on a web page. Greater detail for any part of the map is obtained by selecting a particular chromosome interval.

3.4 Physical maps

Physical maps plot the actual location of DNA sequences in the genome. There are a number of different ways in which this can be achieved:

- **Clone maps** consist of libraries of overlapping clones where the relationship of each clone to the other clones in the library is defined.

- **Radiation hybrid (RH) maps** are based on the frequency with which markers are found on the same fragment of DNA after the genome has been fragmented by irradiation with X-rays.

- **Expressed sequence maps** plot the sites on the genome called ESTs, which are expressed in mRNA. Essentially this type of map is based on the physical location of genes. Remember that these constitute less than 5% of the total genome.

- **STS content maps** plot the location of a physical mapping marker (an STS).

- **Long-range restriction maps** are based on the position of the target sites of rare cutting enzymes determined by pulsed field gel electrophoresis.

- FISH locates the chromosomal position of cloned fragments of DNA.

In practice these overlap because clone maps, RH maps and EST maps all make use of STSs in their construction. Moreover, as we have seen, the latest form of genetic map is based on polymorphic STSs. This means that we can align all of these maps, so that it is possible to move from one format to another. This is illustrated in Figure 3.9, which should be referred to throughout the following sections. Originally the term 'physical map' was most often used to describe clone maps, but it is now more often used to describe a map based on the order of STS markers. This reflects the central importance of STSs in mapping. They are used to construct clone and RH maps and can be cross-referenced to expression maps based on ESTs and genetic maps based on microsatellites.

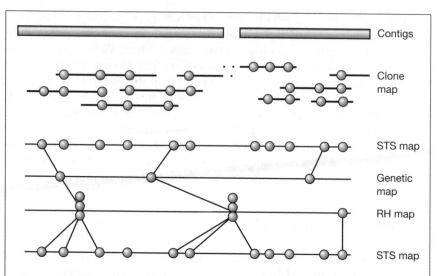

Figure 3.9 Different types of map can be aligned by STS content. Each STS is shown as a sphere; microsatellite-based STSs suitable for genetic mapping are shown in blue. The top part of the diagram shows the inserts of different clones aligned using STS content. The clones fall into two groups called contigs. Within each contig the overlap between clones may be traced through shared STSs, thus defining the spatial relationship of each clone to all the other clones in the contig. The extent of each contig is shown by the shaded bars. It is possible that the two contigs overlap (dotted line) but this has not been revealed by STS content. The overlap between the contigs allows the position of the STSs to be plotted to form the STS content map (shown twice, on the first and fourth lines). Genetic and RH maps are also based on STSs, allowing these maps to be aligned with the STS content map (shown by fine lines). Note that the distortion of map intervals in the genetic map is caused by variation in recombination frequencies. RH and genetic maps can link STSs together over long distances but the order of closely linked STSs may not be resolved. The genetic and RH maps serve as a framework for the clone maps. In this example, they place the two contigs next to each other in the absence of any physical overlap being detected between them.

3.4.1 Clone maps

The original concept of a clone map was to specify the relationship of the clones in a gene library by determining how they overlapped. A group of clones whose relationship is defined in such a manner is called a **contig** (Figure 3.9). Note that the overlap between clones allows the physical relationship between two non-overlapping clones to be specified because their proximity can be traced through the overlap of intervening clones.

A perfect clone map of the human genome would consist of 24 contigs, one for each of the 22 autosomes and each of the sex chromosomes. In practice this is difficult or impossible to achieve because there are always gaps where two contigs cannot be joined. This may be because some parts of the genome are unclonable so there are no clones in the library that cover the region, or because an overlap does exist but cannot be recognised by the methods employed to detect overlaps.

This problem may be solved by using other mapping formats, such as genetic maps or RH maps, to provide an overall framework for the position of the contigs and thus detect those adjacent to each other (Figure 3.9). In order to do this, it is necessary to locate the position of STS mapping markers within the clones so that they can be anchored to the framework provided by the other mapping formats.

There are three phases in the construction of a clone map:

1. construction of the genomic library;
2. detection of overlap between the clones to produce contigs;
3. closing the gap between contigs to produce a continuous map.

3.4.1.1 Constructing a genomic library

A gene library consists of a collection of clones that together contain all the sequences present in the donor genome. Figure 3.10 illustrates the process using a YAC vector. For these to be used to construct a clone map, it is essential that two conditions are fulfilled.

1. Every sequence from the donor genome is present so that the library is complete.
2. The **topology** (the geometrical arrangement) of sequences in the recombinant clones faithfully reflects that in the donor genome.

The completeness of the library is affected by two factors: the number of clones in the library and whether any sequences in the donor genome are unclonable in the host. The number of clones required is a function of the size of the insert in each clone related to the size of the total genome. Generally, the larger the insert that can be carried by the vector, the fewer clones are required for completeness. For this reason, YAC vectors have been used for the first clone maps of the human genome because they can carry large inserts and thus complete libraries contain far fewer clones (~10 000) than are necessary using other types of vector (see below).

Sequences from the donor genome may be unclonable for two main reasons. First, they may not be biologically neutral in the cloning host; for example, they may encode a protein whose biological activity slows cell growth. Second, recombination may take place between direct repeats that results in deletion of the intervening sequences (Figure 3.11). Such recombination can make it very difficult to clone highly repetitive DNA sequences arranged in tandem repeats.

The major reason that the topology may be disrupted is that fragments of DNA from different parts of the genome have ligated together to form chimeric clones. When analysed, such clones will falsely indicate that the two sequences are adjacent to each other in the genome.

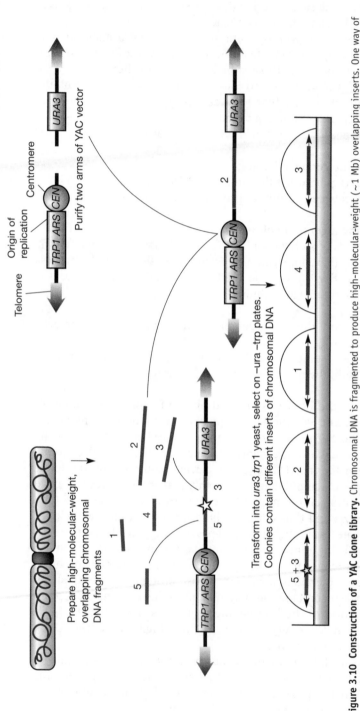

Figure 3.10 Construction of a YAC clone library. Chromosomal DNA is fragmented to produce high-molecular-weight (~1 Mb) overlapping inserts. One way of doing this is to partially digest with a restriction enzyme so that different copies of the same chromosome will be cleaved at different places. High-molecular-weight DNA is mechanically fragile and special precautions have to be taken to avoid shear; usually all manipulations are carried out with the DNA embedded in low-melting-point agarose, which acts as a mechanical support. The two arms of the YAC vector are prepared from a circular vector propagated in bacteria. The arms have cohesive ends that are compatible with the ends of the chromosomal inserts. Double selection on –ura –trp plates ensures that both arms of the artificial chromosome are present in the transformed yeast cells. A common problem with YAC vectors is that two DNA molecules can ligate together (shown by the star). If this happens, the resulting clone is chimeric. On analysis it will appear that two DNA sequences are adjacent to each other, when in reality they are located in completely different parts of the genome.

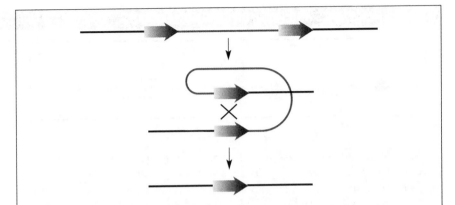

Figure 3.11 Recombination between direct repeats leads to deletions of DNA between them. Two direct repeats are shown by the arrows. DNA can loop back so that two sequences are aligned, allowing recombination to take place. As a result, the intervening DNA (shown in blue) and one copy of the direct repeat is lost.

These considerations determine the choice of vector for the library construction. Routine cloning plasmids, such as the pUC series, cannot be used for this purpose because the size of the inserts they can carry is far too small for the scale of the task. Vectors based on phage λ (maximum insert size 20 kb) and **cosmids** (maximum insert size 40 kb) have been used for many years to construct libraries. However, the maximum size of the insert they can carry is small in comparison with the size of the genome. This means that the number of clones necessary to have a reasonably complete genomic library is very large and ordering them is very difficult. Moreover, the clones are not always stable, and deletions and other rearrangements occur during propagation of the cloned DNA. Cosmids and λ vectors have been most useful for local genomic regions rather than whole genome libraries.

The next type of vector to be developed was the YAC (Box 3.3). YAC libraries have been made that contained very large inserts (up to 1 Mb). This meant that the whole genome could be encompassed in a few thousand colonies, allowing clone maps of the whole genome to be constructed. They have been the workhorse for the construction of the physical maps of the genome described below. However, there are several major drawbacks to YAC vectors, the most important of which is a very high rate of chimeric clones. **Bacterial artificial chromosome (BAC)** and **P1-derived artificial chromosome** (PAC) vectors (see Box 3.3) accept inserts in the 100–300-kb range and maintain the topology of inserts much more faithfully than YAC vectors. An ordered library of BAC clones was used to sequence the human genome by the clone-by-clone approach (see Section 4.3.1).

BOX 3.3: YACS, BACS AND PACS FOR GENOMIC LIBRARIES

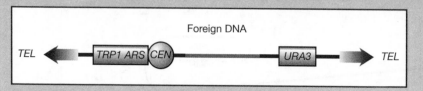

YACs contain all the elements found in a natural chromosome assembled into a vector that can propagate foreign DNA in a yeast cell. The *ARS* element is an origin of DNA replication, *CEN* is the centromere that ensures segregation to daughter cells and *TEL* is a telomere that protects the chromosome ends. *TRP1* and *URA3* encode enzymes required for the biosynthesis of tryptophan and uracil respectively and are used as selection markers (the host cell is *trp1⁻ ura3⁻*). These essential elements occupy only 6 kb in a circular vector that is maintained in *E. coli*. The two arms of the YAC are purified from this vector and ligated to high molecular weight DNA from the donor genome; the two selection markers ensure that each arm is present. Yeast chromosomes can be up to 1 Mb in size and so DNA in this size range can be carried in a YAC vector. This simplifies enormously the task of constructing and ordering a human genome library because so many fewer clones are required compared with a λ or cosmid library. However, they suffer from three major drawbacks.

1. Up to 60% of the clones are chimeric. It is not clear whether this occurs during the *in vitro* ligation process or *in vivo* by recombination between different vectors after transformation into the yeast cell. The transformation process involves the enzymic removal of the yeast cell wall. This is known to be a mutagenic process and it may stimulate mitotic recombination mechanisms. An alternative method of transformation that keeps the yeast cell intact results in a much lower level of chimeric clones.
2. Clones are unstable in the yeast host, with deletions occurring at a high frequency.
3. There is no easy way to purify the recombinant YAC vector from the remaining yeast chromosomes. It is necessary to fractionate the chromosomes by pulsed field gel electrophoresis and recover the chromosomes from blocks cut out of the agarose slab used for electrophoresis.

Phage P1 is a temperate bacteriophage. It can exist in one of two states in the host cell: in a lytic cycle where it multiplies and eventually lyses the cell, releasing progeny phage particles into the surrounding environment or in a dormant lysogenic state, where it is passively replicated along with the host chromosome. Lysogenic phages are induced to re-enter the lytic cycle by cues such as damage to the host DNA. It has been known for a long time that phage P1 mediates generalised transduction, i.e. it can mediate gene transfer from one *E. coli* cell to another. This formed the basis of one of the original methods used to map the

E. coli genome. What happens in generalised transduction is that fragments of host DNA are occasionally mistakenly incorporated into the phage head during a lytic cycle. This DNA is then injected into a new cell when the phage particle attaches to its surface. This ability to transfer non-phage DNA is the basis of PAC vectors. They can accept about 100 kb of insert DNA and maintain it in low copy number in the host cell as a replicating plasmid; when desired, the phage lytic system can be induced to increase copy number. The level of chimeric clones in libraries is less than 5%, which is much lower than YACs.

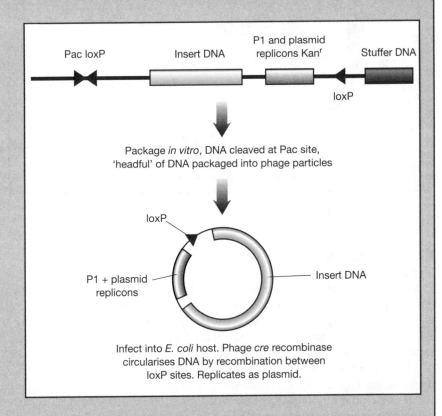

BACs are based on the F factor, the plasmid responsible for conjugation in *E. coli*. The *oriS* and *repE* elements mediate replication, and *parA* and *parB* maintain copy number at one or two per genome. *CM'* (chloramphenicol resistance) provides a means of selection. Insert DNA is cloned into the *BamHI* or *HindIII* sites and excised using the flanking *Not I* sites. The inserts can be transcribed *in vitro* to make RNA using the T7 and Sp6 promoters. BAC vectors can carry up to 300 bp DNA with a high level of stability and low rate of chimeric clones. One disadvantage of BAC vectors is that it is difficult to recover good yields of the recombinant DNA from the host because of the low copy number.

Continued ▶

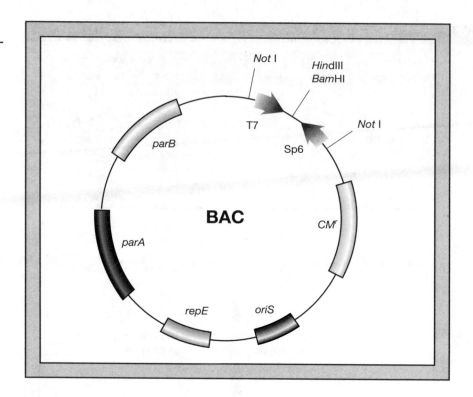

3.4.1.2 Ordering the clones

Once a complete gene library has been constructed, the clones must be ordered with respect to each other. This is done by detecting sequences present in two clones which indicate that they overlap. This can be done by a variety of methods, including STS content, hybridisation and fingerprinting.

STS content mapping seeks to identify clones that both contain the same STS site, indicating that they share an overlapping sequence. The use of STS content to order clones spanning part of human chromosome 21q is illustrated in Figure 3.12.

If two clones contain sequences in common, they will hybridise to each other. In principle one clone can be used as a probe to identify overlapping clones by using the DNA from one clone as a probe to hybridise to other clones. Hybridisation is more sensitive than fingerprinting or STS content and is often used to detect overlap between adjacent contigs that have been missed by these other methods. It thus provides a method of closing the gaps between adjacent contigs. A problem encountered with hybridisation is a high rate of false positive results. One reason for this is the presence of repetitive elements in the probe that hybridise to a large number of unrelated clones. One way to ensure that the probe carries only single sequence DNA is to use PCR to amplify the sequences between adjacent *Alu* elements (see Chapter 2) (Figure 3.13).

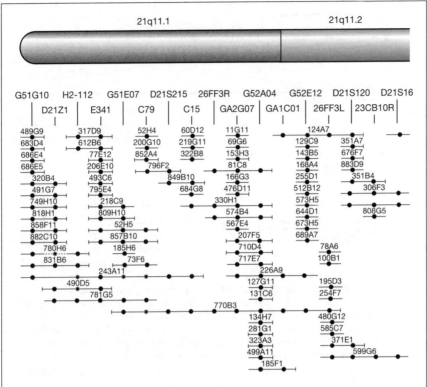

Figure 3.12 Part of a clone map of human chromosome 21q. The figure represents YAC clones, shown as lines, spanning the terminal G-band of the q arm (cytogenetic map is shown at the top). The clones are aligned by the overlap in the STS sites (shown as filled circles) in different clones. The names of the STSs are shown underneath the cytogenetic map in the order revealed by the STS content of the clones. Bars at the end of clone lines mean that the clone tested negative against adjacent clones. (From *Nature*, **357**, 540–1 (Chumakov et al. 1992), reprinted with permission from Nature, copyright 1992 with permission from Macmillan Magazines Limited.)

Fingerprinting methods aim to identify some property of a DNA sequence that allows it to be recognised where it occurs in two different clones. There are many ingenious ways in which this can be done. Figure 3.14 illustrates a simple method of how this might be achieved.

3.4.1.3 A clone map of the human genome

The first comprehensive clone map of the human genome was generated by international collaboration. The map consisted of YAC clones ordered into 225 separate contigs. The ordering was achieved by STS content, hybridisation and fingerprinting and was estimated to cover about 75% of the total genome. Only 786 independent loci were fully mapped so the resolution is much lower than the target of the Human Genome Project.

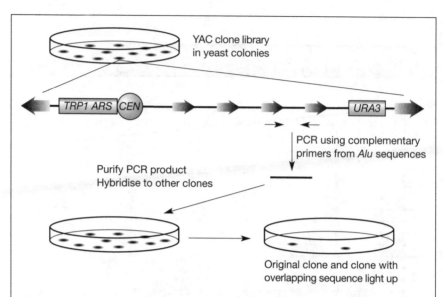

Figure 3.13 Detecting overlap between YAC colonies by colony hybridisation. Each clone contains a large insert of human DNA that has *Alu* repetitive DNA sequences along its length (large blue arrows). Complementary PCR primers (small black arrows) are designed from the *Alu* sequence and are used to amplify unique-sequence DNA located between the *Alu* sequences. The amplified products are used as a hybridisation probe to identify other clones in the library that contain the same sequence and therefore overlap with the first clone.

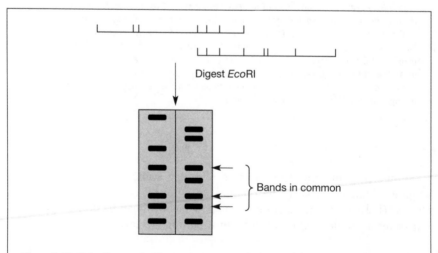

Figure 3.14 Detecting overlap between two overlapping clones by fingerprinting. The clones are digested with a restriction enzyme and the products run on a gel. Clones that overlap will display common bands. It is important to understand that there is no attempt to order the sites within the clones – overlap is indicated by the pattern of bands. Clearly, many clones will share bands by chance. The number of apparently common bands expected by chance can be calculated and overlap is only declared when the number of common bands is significantly greater. In practice, more complex protocols are used to derive arbitrary patterns that characterise clones.

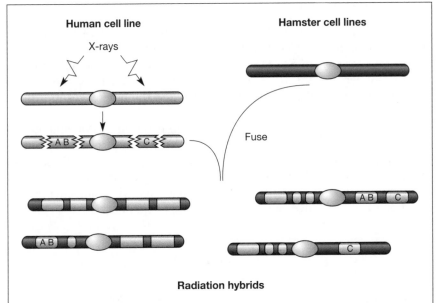

Figure 3.15 Construction of RH maps. Each chromosome represents a complete set of chromosomes in a cell line. A human cell line is irradiated with X-rays, which fragments the chromosomes. The irradiated cells are fused to a hamster cell line. The fragments of human chromosomes become incorporated into the hamster chromosomes. Markers adjacent to each other in the human genome (AB) show the same pattern of retention in the hybrid lines, i.e. they are co-retained. Marker C is distant from A and B and shows a different co-retention pattern compared with either A or B.

3.4.2 Radiation hybrid maps

In RH mapping a human cell line is irradiated with X-rays before fusion to hamster cells to make lines of somatic cell hybrids (Figure 3.15). The radiation fragments the human chromosomes so that they cannot be independently maintained in the hybrid unless they become incorporated into the hamster chromosomes. Each hybrid line retains about 20% of the donor human genome in fragments of about 10 Mb. A panel of lines is examined for co-retention of different STS markers. If two markers are co-retained more frequently than would be expected by chance, then this is evidence that they are linked. The distance between markers is estimated from the frequency of co-retention. The map units are Centirays, which are analogous to centimorgans but depend on radiation dose.

The value of RH maps is that they provide an independent method of mapping STS sites. Because it does not rely on the generation of clones, it is not subject to the same systematic errors such as chimeras and deletions that affect clone maps. Moreover, RH maps can detect linkage between STSs over a much longer distance than can be detected by the overlap of clones. They therefore provide a framework to link clone contigs together.

3.4.3 Expression maps

ESTs are a special form of STS that are generated from cDNA libraries, not genomic sequences. ESTs therefore identify coding regions and may be regarded as gene-based STSs (Figure 3.16). ESTs provide a wealth of information concerning the total number of genes encoded by the human genome, the nature of the proteins encoded and the differences in expression between different tissues.

The human genome is thought to contain between 30 000 and 35 000 genes (see Chapter 2). So far 16 000 of these have been placed on the integrated physical and genetic map screening panels of YAC clones and radiation hybrids for the presence of ESTs. This represents the first step in a programme to map the majority of human genes that will complete the construction of an integrated genetic–physical–expression map.

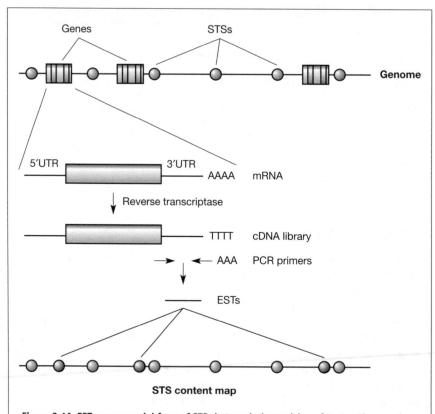

Figure 3.16 ESTs are a special form of STS that mark the position of genes. The top of the diagram shows the genome with isolated genes, each containing introns, shown as blue boxes. The STS content map is superimposed. A cDNA library is constructed from the mRNA derived from the genes. ESTs are prepared from the 3′ untranslated regions (3′ UTR). The 3′ UTR is used because it does not generally contain introns and tends to be unique, even in genes that have related coding sequences. These form STSs which mark the position of genes in the STS content map (filled blue circles).

Restriction maps have long been an important form of physical map for DNA sequences of up to a few tens of kilobases. The landmarks mapped are the sequences at which different enzymes cut the DNA molecule. Conventional methods of producing restriction maps cannot be used to generate longer maps, which might extend over hundreds of kilobases or even megabases, for two reasons:

1. Commonly used enzymes, such as *Eco*RI and *Bam*HI, cut at 6-bp target sequences. If random sequence DNA is digested with such an enzyme, the size distribution of the fragments produced would have a mode of 4096 bp (4^6). It would be completely impracticable to attempt to order the hundreds of fragments that would be produced if DNA in the megabase size range was digested with such an enzyme.
2. If a way could be found to produce larger fragments, conventional agarose gel electrophoresis (AGE) could not be used to characterise them as it becomes non-linear at large fragment sizes. As the size is increased, a point is reached where additional increase in size does not result in a decrease in migration rate. This is because the DNA molecules align along the electric field and migrate end-on through the agarose matrix.

The first of these problems was solved by the discovery of so-called **rare cutting enzymes**. Some examples of these are shown in Table 3.1. These cut less frequently in human DNA than normal enzymes for the following reasons:

- Some enzymes, such as *Not* I, have an 8-bp recognition target. In random sequence DNA this will only occur every 65 kb.

- They contain the dinucleotide CpG. This has two consequences. First, the frequency of CpG is depressed in human DNA (see section 2.2.1.2). Second, CpG is the target for DNA methylation, which prevents these restriction enzymes from cutting the DNA.

Table 3.1 Some examples of enzymes that cut human DNA rarely.

Enzyme	Target sequence
Mlu I	ACGCGT
Not I	GCGGCCGC
Nru I	TCGCGA
Sal I	GTCGAC
Sfi I	GGCCNNNNNGGCC
Xho I	CTCGAG

As a consequence, the restriction enzyme *Not* I cuts human DNA on average every 500 kb, and its target sites thus provide landmarks that are conveniently spaced for the analysis of human chromosomal regions. Other enzymes, such as *Nru* I and *Sfi* I, also produce fragments that are large enough that the target sites are useful landmarks for chromosomal maps. Note that enzymes such as *Xho* I and *Sal* I are comparatively rare cutters with human chromosomal DNA, even though they are commonly used in the construction of conventional restriction maps.

The problem of separating large DNA molecules was solved by a modification of AGE called **pulsed field gel electrophoresis (PFGE)**. In PFGE the direction of the electric field is periodically changed. In order to make progress in the new field, the DNA molecule must reorientate itself. The time taken for this reorientation is proportional to molecular weight, even with DNA molecules several megabases in size. The exact conditions of electrophoresis can be elaborate and have to be fine tuned for the particular size range employed.

Long-range restriction maps have proved very important during the course of positional cloning. They provide a framework with which to locate the position of clones derived from techniques such as chromosome walking. An example of a long-range restriction map constructed during the hunt for the cystic fibrosis gene is shown in Box 5.2. Another important practical consideration is that the sites of rare cutting enzymes will tend to be clustered in CpG islands that mark the beginning of genes. This allows the islands to be identified in long stretches of DNA.

3.5 An integrated physical and genetic map of the human genome

STSs can be used as a common marker to align genetic, RH- and YAC-based clone maps of the human genome. Genetic maps can detect linkage between markers over distances of 30 Mb, while RH mapping can detect linkage over distances of 10 Mb. As a consequence, comprehensive genetic and RH maps of the genome require only a few thousand markers. However, the resolution of fine structure mapping is low. In contrast, YAC-based clone maps detect linkage over the 1-Mb base range. Thus a large number of markers are required to provide adequate coverage of the genome, but the fine-structure resolution is high.

As we have seen, clone-based maps suffer from two particular problems: there are always gaps between the contigs, and false linkage between markers may be detected as a result of chimeric clones. RH and genetic maps may be used to provide a framework for clone-based maps, allowing adjacent clone contigs to be positioned with respect to each other and trapping errors that result from chimeric clones. An integrated map has been constructed that aligns the Genethon genetic map with an RH map and an STS-content map based on YAC clones (Figure 3.17).

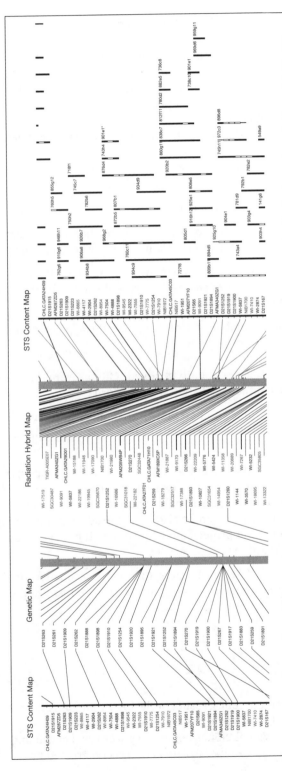

Figure 3.17 A small portion of the integrated map of chromosome 14. The figure follows the same principles as the idealised map shown in Figure 3.9. The STS content map is shown twice, flanking the genetic and RH maps. The lines connect markers found in more than one map. The bars on the right are the overlapping YAC clones which contain the STSs mapped. (From http://www.genome.wi.mit.edu/, Dr T. Hudson.)

3.6 Genome mapping is big science

A traditional view of the way that biological research is carried out would be of a small team of scientists working in a single laboratory. The scale of the effort required to produce whole genome maps has demanded a totally different method of working that is perhaps more familiar to the physical sciences such as particle physics. As an example, consider the following facts concerning the production of the integrated physical and genetic map (see Further reading):

- It involved 15 million separate PCR reactions.

- There were 51 authors to the paper.

- It required the construction of a robotic production line to carry out the PCR tests that was commissioned from a specialist engineering company.

- A full description of the results would have required 900 printed pages. For this reason the detailed results are published electronically and accessed via the world wide web (see Table 4.2 on p. 108).

These facts illustrate a number of aspects of how this type of work is carried out:

- Large numbers of personnel are involved. Often several laboratories or even institutions formally collaborate in consortia.

- Automation is essential to allow the large number of tests to be carried out. One value of PCR-based testing is that it readily lends itself to automation.

- Results are published electronically. This allows continuous publication, so that other researchers can exploit the results without the long delays of conventional means of publication. A list of web sites is given in Box 4.1 on p. 104. These can be readily accessed to monitor progress of the different mapping projects and to explore the maps in detail. The reader is encouraged to visit these sites.

- PCR-based tests allow researchers to use the sequence of the PCR primers to synthesise their own primers locally and so use the maps in their own particular research.

3.7 Summary

- Maps:
  - ○ Locate affected genes responsible for monogenic disorders and those contributing to the risk of complex diseases.
  - ○ Ordered individual clones before they were sequenced and acted as scaffolds to link together isolated contigs constructed by sequencing.

  - Provide a high-resolution description of a region identified by genetic or association mapping, allowing candidate genes to be selected for further study.
- Genetic maps are based on recombination at meiosis. They often provide the only connection between biological or medical reality and the underlying genome.
- A comprehensive genetic map of the human genome is based on polymorphic microsatellite markers.
- Because human families are generally small, it is necessary to combine data from many families to achieve statistical significance when testing for linkage between a mapping marker and disease or phenotypic trait. The most efficient way of doing this is LOD score analysis.
- LOD score analysis consists of a series of reiterative calculations. In each calculation, a particular value of recombination fraction (θ) between marker and disease is assumed. The log ratio of odds (LOD score) of the observed outcome in a family if there is linkage, compared with the odds of the outcome it there is no linkage, is then calculated. If the ratio exceeds 1000:1 at a particular value of θ, then linkage can be declared. The maximum LOD score shows the most likely value of θ.
- The human genome is composed of haplotype blocks; within each block there is limited diversity. A map of haplotype blocks in the human genome based on single nucleotide polymorphisms has been generated and is an essential tool in the search for alleles which increase the risk of complex disease.
- Physical maps locate features to a particular part of the DNA molecules that make up the genome. There are different types of physical map:
  - Clone maps use the overlap between clones in a genomic library to construct groups of clones called contigs, where the spatial relationship of each clone is defined with respect to the other clones in the contig. Comprehensive clone maps are constructed using yeast artificial chromosomes (YACs) as vectors because of the large insert size they can carry.
  - Radiation hybrid (RH) maps plot the position of STS markers by observing how often they are retained together on the same fragment when the genome is fragmented by X-rays and fused to hamster cells to make panels of radiation hybrid lines.
  - Expression maps plot the position of genes using a special form of STS called an expressed sequence tag (EST).
- These physical maps were an essential tool in sequencing and annotating the human genome. Now the human genome has been sequenced and annotated, a comprehensive description of every region is available in databases such as Ensembl. When a mutation causing a single gene disorder has been mapped by linkage analysis, the detailed physical map of the region can be used to identify candidate genes.

Further reading

LOD score analysis
RISCH, R. (1992) Genetic linkage: interpreting LOD scores. *Science*, **255**, 803–804.

The genetic map
DIB, C., FAURE, S., FIZAMES, C. *et al.* (1996) A comprehensive genetic map of the human genome. *Nature*, **380**, 152–154.

Physical maps
BOUFFARD, G.G., IDOL, J.R., BRADEN, V.V. *et al.* (1997) A physical map of human chromosome 7: an integrated YAC contig map with an average STS spacing of 79 kb. *Genome Research*, **7**, 673–692.

CHUMAKOV, I., RIGAULT, P., GUILLOU, S. *et al.* (1992) Continuum of overlapping clones spanning the entire human Y chromosome 21q. *Nature*, **359**, 380–387.

COHEN, D., CHUMAKOV, I. and WEISSENBACH, J. (1993) A first generation map of the human genome. *Nature*, **366**, 698–701.

FOOTE, S., VOLLRATH, D., HILTON, A. and PAGE, D.C. (1992) The human Y chromosome: overlapping clones spanning the euchromatic region. *Science*, **258**, 60–66.

The integrated map
HUDSON, T.J., STEIN, L.D., GERETY, S.S. *et al.* (1995) An STS-based map of the human genome. *Science*, **270**, 1945–1955.

ESTs and expression maps
ADAMS, M.D., KERLAVAGE, A.R., FLEISCHMANN, R.D. *et al.* (1995) Initial assessment of human gene diversity and expression patterns based upon 83 million nucleotides of cDNA sequence. *Nature,* **377** (Suppl.), 3–17.

SCHULER, G.D., BOGUSKI, M.S., STEWART, E.A. *et al.* (1996) A gene map of the human genome. *Science,* **274,** 540–545.

Sequence tagged sites
ROBERTS, L. (1989) New game plan for genome mapping. *Science,* **245,** 1438–1440.

Radiation hybrid mapping
LEACH, R.J. and O'CONNELL, P. (1997) Mapping of mammalian genomes with radiation (Goss and Harris) hybrids. *Advances in Genetics,* **33,** 63–101.

Commentaries on genomic maps
BOGUSKI, M.S. and SCHULER, G.D. (1995) ESTablishing a human transcript map. *Nature Genetics,* **10,** 369–370.

COX, D.R. and MYERS, R.M. (1996) A map to the future. *Nature Genetics,* **12,** 117–118.

JORDAN, E. and COLLINS, F.S. (1996) A march of the genetic maps. *Nature,* **380,** 111–112.

LITTLE, P. (1995) Navigational progress. *Nature,* **377,** 286–287.

SCHMITT, K. and GOODFELLOW, P.N. (1994) Predicting the future. *Nature Genetics,* **7,** 219–220.

Long-range restriction mapping
BARLOW, D.P. and LEHRACH, H. (1987) Genetics by gel electrophoresis: the impact of pulsed field gel electrophoresis on mammalian genetics. *Trends in Genetics,* **3,** 167–171.

Web sites
Visit the web sites listed in Table 4.2 on p. 108.

CHAPTER FOUR

The sequence of the human genome

Key topics

- Technology for sequencing the human genome
- The IHGSC clone-by-clone strategy
- The Celera shotgun sequencing strategy
- Draft sequence of the human genome
- Finishing the sequence
- Annotating the sequence
- Databases and browsers
- Sequencing other genomes
- Analysis of the structure and evolution of the human genome
- Identifying genes
- The total number of genes in the human genome
- The properties and evolution of the encoded proteins
- Oligonucleotide arrays
  - ○ Expression profiling
  - ○ Determining the boundaries of genes
  - ○ Genotyping SNPs
  - ○ Detecting and mapping copy number and structural variants
- The ENCODE project
- The cancer genome
- Next generation sequencing technologies

4.1 Introduction

The construction of comprehensive genetic and physical maps, elucidation of the entire sequence of the human genome, and detailed annotation of the genes and other elements that make up the genome underpin the research in all the different fields that together make up human molecular genetics. They have made the characterisation of genes that are defective in single-gene disorders a routine process – a very different situation from that in the 1980s and early 1990s when genes affected in diseases such as cystic fibrosis, Duchenne muscular dystrophy and Huntington's disease were first cloned. Furthermore, for the first time it has become possible to reliably identify alleles that contribute to the risk of complex disease.

When first proposed in 1988, the idea of sequencing the human genome was widely considered an overambitious and impractical task. The difficulties seemed immense. At that time, DNA sequencing was usually performed using radioactive isotopes and the sequence had to be read off the resulting autoradiograph manually. A skilled sequencer could generate reads about 300 bp long and needed four lanes to generate a single DNA sequence. Automated sequencing based on fluorescently labelled nucleotides was in its infancy and was still unreliable. The human genome contains 3.2×10^9 bp. With technology then existing, this implies a minimum of about 30 million gel reads for one-fold coverage; that is, on average every nucleotide will be sequenced once. In fact, the problem was very much worse than this for two reasons. First, sequencing was based on randomly derived clones. To be sure that every part of the human genome is represented in the clone collection, the laws of probability make it necessary to amass a total number of clones greater than the number of clones necessary to provide the one-fold coverage. Second, to be sure that the sequence is accurate, each base should be sequenced on multiple occasions. Some sceptics humorously proposed that sequencing the human genome should be a substitute for hard labour for long-term prisoners!

When enough sequence had been obtained, fitting it together to reassemble the entire sequence of each chromosome was also going to be logistically complex. In the first place, experience of creating clone libraries shows that, for various reasons, there will always be some fragments that will not be present (see Chapter 3). So fitting the overall sequence together was like doing a jigsaw with pieces missing. Second, a large part of the human genome consists of sequences that are repeated elsewhere (see Chapter 3). Thus, assembling a complete sequence from the fragments is very difficult because there are apparently many possible locations for each repetitive sequence.

Despite these difficulties, a large international consortium of laboratories called the Human Genome Sequencing Consortium (IHGSC) together formulated a plan to sequence the genome and formally initiated the Human Genome Project in 1990. The aim was to sequence the euchromatic portion of the genome consisting of 2.85×10^9 bps, which amounts to about 92% of the total genome. The remaining heterochromatic regions were not targeted because they consist of complex repeats refractory to normal sequencing technologies. They consist of the alpha repeat arrays around the centromeres, a large region on Yq, secondary constrictions near to the centromere on 1q, 9q and 16q, and repeats arranged in complex structures flanking the arrays encoding ribosomal RNA genes on 13p, 14p, 15p, 21p and 22p.

The project was planned in three phases. Phase 1 (1990–1997) was to construct detailed genetic and physical maps of the human genome so that an ordered collection of clones could be efficiently sequenced and assembled into a continuous final sequence for each chromosome. Alongside the mapping, new sequencing technologies would be developed and refined. Also in the first phase, the genome of simpler model organisms would be sequenced to test the new technologies and develop the best overall strategies for sequencing complex genomes. In the second phase (1997–2001), the sequencing would be carried out to produce a **draft sequence**, which would contain gaps and errors but would nevertheless serve as the basis for many new lines of research. The third phase (2001–2003) was to **finish** and annotate the sequence. Finishing the sequence involves closing the gaps and correcting the errors. The aim was to produce a sequence that would have less than one error in 10^4 base pairs. (In fact, the finished sequence exceeded this target as it is estimated to have an error rate of less than 1 in 10^5 bps.). This finished sequence would be the permanent foundation for virtually all research into human biology and biomedicine. Annotating sequence means identifying all the genes and other elements that together make up the genome.

The plan was accelerated by a private biotechnology company called Celera Genomics, which announced in 1998 that it was also going to sequence the human genome by sequencing random clones, called shotgun clones, considering that the careful assembly of ordered clones before sequencing was unnecessary. There was great concern that the human genome sequence could be patented before it was even in the public domain. IHGSC and Celera Genomics raced to produce the draft sequence. The race ended in an official draw when the completion of this draft sequence was announced by both groups in June 2000. The first printed accounts of the sequencing process and analysis of the draft sequence were published simultaneously in February 2001. A paper describing the sequence in what was considered its final finished form was published in October 2004.

We considered the process of producing genetic, physical and expression maps in Chapter 3. In this chapter we will consider in detail the two different strategies adopted by IHGSC and Celera genomics to produce the draft sequence. The IHGSC clone-by-clone strategy facilitated the finishing phase of the project and we shall see what was involved in this phase of the Human Genome Project. However, it should be pointed out that the shotgun approach has proved to be very efficient in rapidly generating draft sequences of a wide range of vertebrate genomes. The draft sequence provided enough information for a broad overview of the structure of the human genome and the different unique and repetitive elements of which it consists. Of course, the major focus of interest is to identify the genes and determine their roles. The draft sequence again provided an overview of the different classes of genes in terms of their function. However, detailed characterisation of the human gene complement required the finished sequence and careful annotation. We shall see two important conclusions about the nature of human genes. First, there are a lot fewer of them than anyone expected beforehand. The final count is stabilising at about 22 000 – only four times as many genes as it takes to specify a yeast cell! Moreover, only about 1.2% of the total DNA sequence encodes protein. Second, our conceptual view of a gene is rapidly changing. Many, or even most, genes do not produce a single unique RNA transcript. Instead, the different exons that can be present in a single mRNA molecule are often differentially spliced so that different combinations constitute a diverse pool of possible mRNA molecules. We shall look to the future as a third generation of sequencing technologies lowers the cost and effort of sequencing. This has raised the prospect of sequencing a variety of different individual genomes, perhaps as many as a thousand, to catalogue human variation in detail. Finally, we shall consider a project called ENCODE, which aims to produce an encyclopedia of the functional elements in the human genome. The first results of this project are challenging our overall perceptions of genome function.

4.2 Basic technology for sequencing the human genome

4.2.1 Sequencing technology

Both groups sequencing the human genome used technology that was based on the dideoxy method developed by Fred Sanger and his colleagues in Cambridge, UK, for which he received the Nobel Prize for Chemistry in 1980 (in addition to the Nobel Prize he received in 1959 for developing a method to sequence proteins). The original protocol is described in Figure 4.1. This is now normally referred to as 'manual sequencing'. It is a technically demanding procedure that requires the use of radioisotopes and autoradiography, and the experimenters must read and record the results of the autoradiograph themselves.

For the amount of sequencing that is needed for the whole human genome there was a need for a procedure that did not require unstable isotopes and the attendant autoradiography, and in which the sequence data were collected automatically, not manually. This led to the development of the fluorescence-based technologies described in Figure 4.2. These still made use of the basic concept of the Sanger method, but they replaced labelled isotopes with four fluorescent dyes that labelled each of the four dideoxy analogues a different colour. The product of these reactions could be run on a single lane of a polyacrylamide electrophoresis gel. At the bottom of the gel a laser excites the fluorophores incorporated into the DNA molecules. The order of coloured peaks passing a detector is recorded. A computer program then automatically converts the order of peaks passing the detector into a digital file of the DNA sequence. In the Human Genome Project a program called PHRED was commonly used; this is able to resolve possible artefacts, for example if the peaks are not regularly spaced or if two peaks have been merged into one. PHRED also assigns a numerical score

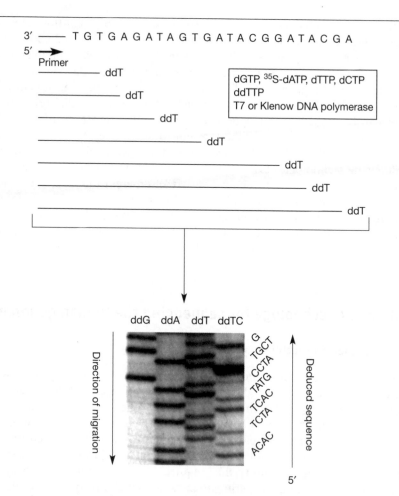

Figure 4.1 The Sanger dideoxy sequencing protocol. The DNA to be sequenced is either denatured or recovered in a single-stranded form from an M13-based vector. This is annealed to a primer that is designed to anneal to a fixed site in the cloning vector just outside the cloned fragment. The primer/template partial duplex is incubated in four parallel reactions, which each contain T7 DNA polymerase and a mixture of the four deoxynucleotides. In addition, each of the parallel reactions also contains a different 2′3′ dideoxynucleotide analogue that can be incorporated into a nascent chain, but cannot serve as a substrate for the next step in the polymerisation because there is no free 3′OH group. They are therefore said to be chain-terminating. In the example shown, dideoxy-TTP (ddT) is included in the reaction mix. At each A residue in the template either dTTP or ddT can be incorporated into the new chain. If dTTP is incorporated, the polymerisation proceeds until the next A in the residue in the template. If ddT is incorporated, the polymerisation of that particular chain is terminated. The size of the fragment upon chain determination reports the number of nucleotides from the primer to the position where ddT caused chain termination. The product is made radioactive by including ^{35}S-dATP in the reaction mix. The products of the four parallel reactions are fractionated by polyacrylamide gel electrophoresis and the molecules visualised by autoradiography. The sequence can be deduced by the order of bands in the four different lanes of the resulting sequencing ladder. On a good day about 400 bp of sequence can be read from a single gel.

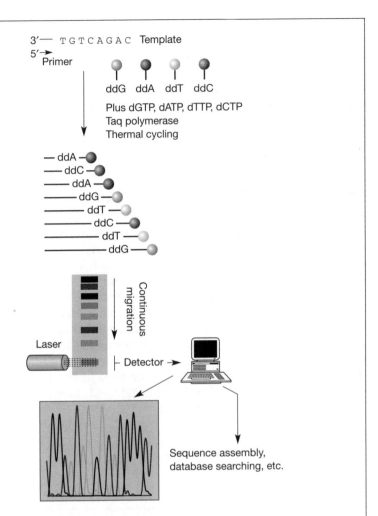

Figure 4.2 Automated sequencing using fluorescent dye terminators. The Sanger sequencing protocol described in Figure 4.1 is carried out with a different coloured fluorophore conjugated to each dideoxy analogue. A thermostable DNA polymerase is used that allows thermal cycling similar to that used in PCR. As a result, the products are amplified in a linear fashion to generate enough product for the fluorophore to be detectable. The products are fractionated on a polyacrylamide gel. As they pass a detector, a laser excites the fluorophore and a detector records the colour of each successive peak. The order of peaks therefore reports the DNA sequence, which is recorded and converted into a digital nucleotide sequence by an attached computer using a program such as PHRED.

of the sequence quality. This score is based on a logarithmic scale. Thus a score of 30 indicates the likelihood that a base has been called incorrectly to be less than 1 in 1000. The sequence produced by PHRED is ready for further analysis by programs that assemble the longer sequences from multiple gel readings. A commonly used program for this purpose is called PHRAP.

The speed at which the products of a sequencing reaction can be separated by electrophoresis is proportional to the applied voltage. This is physically limited in slab gels by the rate at which the resultant heat can be removed. A further time-saving development was the replacement of slab gels with capillary gels with a 150-μm internal diameter. These can be cooled much more efficiently, and so the applied voltage can be increased to the extent that they can be run up to 14 times faster. They are also much more sensitive, producing a readable signal with a much smaller amount of the sequencing reaction.

4.2.2 Sequencing factories

The sequencing reaction itself was not the only aspect of the process to become automated. Library production, template preparation, setting up sequencing reactions, loading and running gels, data collection, quality assessment, storage and processing were all carried out on robotic production lines. Celera were able to process 175 000 gel reads per day, operating 24 hours a day, 7 days a week. At the height of the IHGSC effort, the rate of sequencing was one-fold coverage of the genome every 6 weeks, or 1000 bp per second. The Celera operation was confined to one location, while the IHGSC consisted of a number of centres spread around the world, each of which operated as sequencing factories. Altogether, 21 centres contributed to the final sequence; however, 86% of the draft sequence was contributed by just six of these, listed in Table 4.1, together with the contribution they made to the final sequence.

Table 4.1 Top six sequencing centres and their contribution to the draft and finished sequence.

Sequencing centre	Draft (kb)	Finished (kb)
Whitehead Institute, Center for Genome Research	1 196 888	46 560
The Sanger Centre	970 789	284 353
Washington University Genome Sequencing Center	765 898	175 279
US DOE Joint Genome Sequencing Center	377 998	78 486
Baylor College of Medicine, Human Genome Sequencing Center	345 125	53 418

4.2.3 Shotgun sequencing

One aspect of sequencing that has been stubbornly difficult to improve is the length of the sequence that can be reliably read off each gel. Despite the improvements in throughput, the length of the sequence read off each gel is limited to about 800 bp. In order to sequence a larger piece of DNA it must be broken down into a random collection of small fragments that are cloned into sequencing vectors (Figure 4.3). Each sequencing reaction

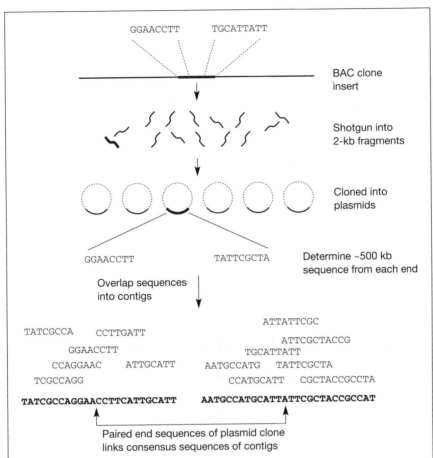

GGAACCTT   TGCATTATT

BAC clone
insert

Shotgun into
2-kb fragments

Cloned into
plasmids

GGAACCTT   TATTCGCTA   Determine ~500 kb
                                               sequence from each end

Overlap sequences
into contigs

                                    ATTATTCGC
TATCGCCA   CCTTGATT   ATTCGCTACCG
    GGAACCTT   TGCATTATT
  CCAGGAAC   ATTGCATT   AATGCCATG   TATTCGCTA
  TCGCCAGG   CCATGCATT   CGCTACCGCCTA

TATCGCCAGGAACCTTCATTGCATT   AATGCCATGCATTATTCGCTACCGCCAT

Paired end sequences of plasmid clone
links consensus sequences of contigs

Figure 4.3 Shotgun sequencing. The insert in each BAC clone is fragmented and the pieces cloned into a sequencing vector. The method of generating the fragments and their size varied in the different sequencing centres. One common way is to sonicate DNA, which generates fragments whose size is proportional to the length and intensity of the sonication. The fragments so generated have blunt ends; they can be joined to the cloning vector by blunt end ligation. Because the breaks are introduced randomly, the breaks will occur in different places in each insert molecule; the inserts in the clones will overlap. One gel's worth of sequence is obtained from each end of each clone using primers that anneal either side of the cloning site. Sequences that overlap are arranged into contigs, from which a consensus sequence is obtained (shown in bold). Gaps between contigs are bridged by paired end sequences that must be linked by virtue of being part of the same insert. The diagram follows the sequencing of one of the shotgun fragments (shown in blue), from its location in the original BAC clone, as an isolated fragment after sonication, as a sequence-ready clone and finally the location of the two end-paired sequences in different contigs. Only one strand of the DNA molecule is shown throughout the diagram. Often overlap will be detected between a sequence and its reverse complement that was sequenced in the opposite orientation in another clone. The computer software automatically reformats the gel readings to have the same orientation.

requires a primer of known sequence to initiate the reaction. Synthetic primers are used that anneal to sequences either side of the cloning site, so that about 800 bp from either end of the fragment are determined. The sequence of the starting fragment can then be assembled by looking for overlaps between the gel readings. A longer sequence that has been reconstructed in this way is called a **contig**. The contig may not extend the full length of the starting fragment because a section may be missing, so usually some form of external **framework** or **scaffold** is needed to locate the contigs with respect to each other. One very important piece of information is that the sequences at the two ends of a single clone are necessarily linked. So if the two end sequences from one clone fall into different contigs, the contigs must be adjacent to each other and the size of the gap can be estimated from the total size of the insert in the clone (Figure 4.3). The process of breaking down a larger sequence into random fragments, sequencing the smaller fragments and piecing together the sequence of the original DNA is called **shotgun sequencing**. Both Celera and the IHGSC used shotgun sequencing but differed in the size of the DNA that was sequenced by this method. Celera shotgun sequenced the whole genome; IHGSC applied this method to the 100–200-kb inserts of ordered BAC clones.

4.3 The IHGSC clone-by-clone sequencing strategy

4.3.1 Making a BAC clone map of the genome

The IHGSC project started by making genomic libraries in BAC or PAC vectors from blood and sperm samples that were collected from male volunteers. Because all identifying tags were removed and only about one-tenth of samples collected were actually used, the identity of the donors is unknown to both investigators and the donors themselves. So the published sequence comes from a mixture of different individuals whose identity is unknown. Male DNA was used because the female genome is missing the Y chromosome sequence.

The fragments for the genomic libraries were generated by partial digestion with *Hind*III so that the inserts would be overlapping. The inserts were relatively large (100–200 kb) so that the number spanning each chromosome would be manageable. BAC and PAC vectors were used because they can carry reasonably large inserts without compromising their integrity; that is, there is a low level of deletions, rearrangements and chimeric clones (see Chapter 3).

To order the clones, they were completely digested with *Hind*III and the products separated using agarose gel electrophoresis. The pattern of bands from each clone constitutes its fingerprint. Clones that overlap can be identified because their fingerprints will overlap (see Chapter 3). By overlapping the clones, they can be arranged into contigs called fingerprint clone contigs. Like all contigs, there will eventually be gaps because key clones will be missing or overlap between contigs exists but has not been detected. The long-range physical and genetic genomic maps described in

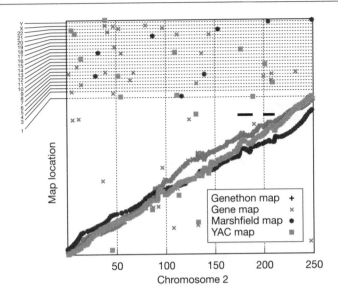

Figure 4.4 Alignment between BAC clones and other genomic maps. As the BAC clones are sequenced, STS markers are revealed by **ePCR**. The position of these in the genetic (Genethon and Marshfield), gene (radiation hybrid) and STS-content (YAC) maps anchors the sequence to chromosomal position and acts to trap errors in the BAC clone map. In this example of BAC clones mapping to chromosome 2, there are two apparent discrepancies between the sequence and all three of the other types of map (shown by black bar). Inverting the segment of the BAC clone map restored consistency between the sequence and the other types of map. A small number of markers from the other maps were found to have been misassigned to chromosome 2, their true chromosomal location revealed by sequencing (shown at the top of the figure). (Redrawn with permission from IHGSC (2001) *Nature*, **409**, 860–921.)

Chapter 3 were used to position the fingerprint clone contigs along the chromosomes. As illustrated in Figure 4.4, this allowed errors in the clone map to be identified. Clones were also hybridised to whole chromosomes using FISH so that the map was also integrated with the cytogenetic map (Figure 4.5).

The finished clone map contains 372 000 BAC clones in 965 contigs. As well as providing the foundation of the sequencing project, the clone map provides a permanent resource that can be used to retrieve any specified region of the genome. It is tightly integrated with the genetic, STS, expression and cytogenetic maps. Thus, once a gene, disease or phenotype has been mapped by linkage to a mapping marker, the DNA surrounding the marker is physically available for study, along with its sequence and knowledge as to which parts are expressed as mRNA. The clones also provide a physical source of DNA to act as a reference sequence when searching for mutations and SNPs.

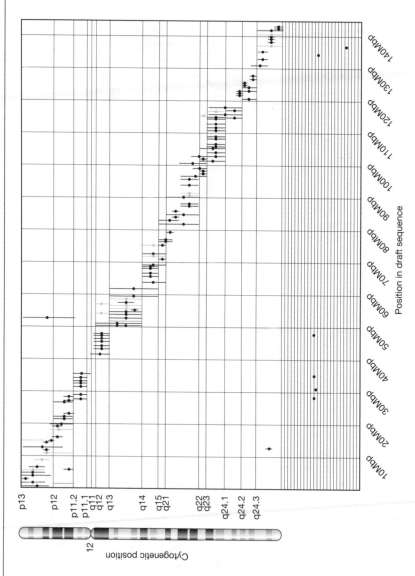

Figure 4.5 Correlation of BAC clone map with cytogenetic map. The cytogenetic map of chromosome 12 is shown on the left. The graph shows the range of bands to which each clone hybridises as determined by FISH. The position of each clone in the sequence is shown on the x-axis. A small number of clones hybridise to other chromosomes, revealing errors in the original chromosome assignment. (Redrawn with permission from the BAC Consortium (2001) *Nature*, **409**, 953–958.)

From each BAC fingerprint clone contig, clones representing the minimum overlapping set were selected for sequencing. One advantage of this scheme is that it is suitable for collaboration between different sequencing centres, because each can receive different BAC clones to sequence. The insert in these clones was then shotgun sequenced (Figure 4.3). Overall, in the draft phase the amount of sequence obtained was such that on average each nucleotide was sequenced four times, i.e. four-fold coverage. The gel reads were filtered for vector sequences and other extraneous DNA from the bacterial host or other non-human sources.

Sequences for each chromosome were then assembled in stages from the primary gel reads using the PHRAP computer program (Figure 4.6). First the gel reads were first assembled into sequence contigs for each BAC clone. These were then merged with those from adjacent BAC clones to

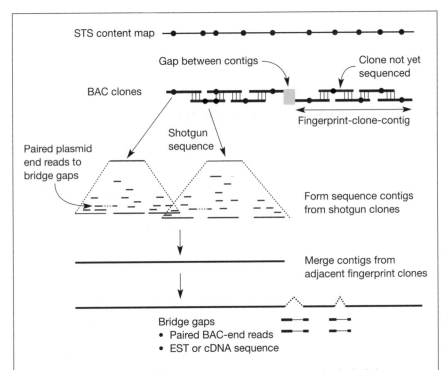

Figure 4.6 IHGSC clone-by-clone sequencing strategy. STS markers in the BAC clones, shown by solid circles, are aligned with markers in the STS content map. Merged contigs are located on a scaffold. Missing sequences, caused by gaps between fingerprint clone contigs or BAC clones not yet sequenced, are bridged by paired BAC-end reads or sequence from EST or cDNA sequence already in databases. Note that cDNA and mRNA are derived from mRNA sequences from which introns have been removed by splicing, so their sequences will not be continuous in nuclear DNA and may be widely separated. They are thus a useful resource to link and order sequences that originate in different contigs.

form longer contigs. One of the guiding principles of the IHGSC was that, as the primary gel readings were assembled into contigs, this information was released on the internet within 24 hours. The idea was to prevent patenting of DNA sequence information by placing it in the public domain as soon as possible. Ironically, this allowed Celera to make use of the information for its own privately funded sequencing of the genome (see below). In principle, the merged sequence contigs should extend the full length of the chromosome, but in practice there were gaps in the draft sequence. However, as shown in Figure 4.6, the contigs could be placed on a scaffold using information such as paired plasmid-end reads and data already in databases from cDNA and EST sequencing.

Overall, it was estimated that 88% of the human genome had been sequenced in the draft phase. The advantage of the clone-by-clone approach is that it is relatively easy to obtain most of the missing information because many of the clones that require extra sequencing are already available. Some of the sequence will be in the gap between BAC fingerprint clone contigs and this may be more difficult to retrieve, perhaps requiring the use of a different vector. The accuracy of the sequence is an important criterion. PHRED scores indicated that in the draft sequence 91% of the sequence was judged to have an error rate of less than 1 in 10 000, and 95% was estimated to have an error rate of less than 1 in 1000.

4.4 The Celera shotgun sequencing

Like the IHGSC, Celera constructed three different types of genomic library from anonymous volunteers. The three libraries differed in the size of their inserts, averaging 2 kb, 10 kb and 50 kb, respectively. Instead of ordering the clones, Celera moved directly to shotgun sequencing using gel reads from the end of each clone to amass a total sequence amounting to five-fold coverage. This was augmented by IHGSC gel reads, which they downloaded from the internet, so that in total the amount of sequence analysed amounted to eight-fold coverage.

4.4.1 Assembling the shotgun sequence for the whole genome

In order to assemble the shotgun sequence for the whole genome in one operation, a major problem must be solved – about 45% of the genome consists of repetitive elements (see Table 4.2). For example, the *Alu* family (see Chapter 2) consists of 1.5×10^6 copies of a 300 bp sequence and occupies 13% of the genome (Table 4.2). Each copy will differ slightly, but recognisably conforms to a consensus. Deriving a longer sequence by many overlapping smaller sequences will be a hopeless undertaking, because each sequence with an *Alu* element will overlap with over a million other sequences – only one of which will be the correct match.

The strategy Celera adopted to solve this problem is shown in Figure 4.7. There were two key elements in their strategy:

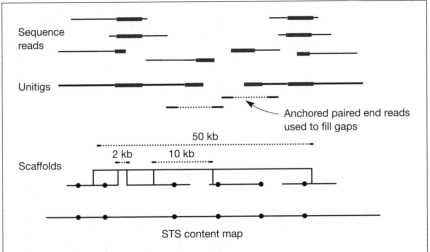

Figure 4.7 Celera whole genome shotgun sequencing strategy. Repeated sequence (blue bars) in the initial reads from the shotgun clones are masked. Unitigs are built up from overlapping the remaining unique sequence (black lines). Repeated sequence can be restored to the unitigs when its location has been unambiguously determined through overlap of the surrounding unique sequence. Unitigs ending in repeat sequence cannot be merged with the adjacent unitig because overlap of repeated sequences is disallowed. However, the unitigs can be placed on a scaffold using paired end reads from clones from the 2-, 10- and 50-kb libraries and located on the chromosomes through STS content maps. Once located on a scaffold, paired end reads can be used to fill the gap where one end is unambiguously located in a unitig. Multiple gel readings placed in the same gap were then assembled into contigs.

- Computer software was used to mask repeated sequences in the raw sequence reads. Only the remaining unique sequence was used to construct the contigs. Once the contigs were constructed, the repeated sequences were restored because their proper location had been determined.

- Three different libraries with inserts of 2 kb, 10 kb and 50 kb were constructed. The paired-end reads from these libraries were used to construct a scaffold that linked the contigs together.

Celera also independently assembled a sequence of the human genome using positional information from the BAC clone map constructed by IHGSC. Sequences from BAC clones, made publicly available by IHGSC, were matched against Celera's own gel reads. This allowed Celera's gel reads to be placed in bins defined by the BAC clones. The whole genome sequence was then assembled using Celera's sequence and the gel reads from the IHGSC downloaded from the internet. The initial construction of contigs only allowed matches between gel readings that came from the same BAC clone, thus simplifying the problem of repetitive sequences. When compared with the sequence derived from the whole genome shotgun approach it gave a slightly more complete sequence with fewer gaps. However, the important point is that it showed it was possible to use a whole genome shotgun approach to sequence large genomes with large fractions of repetitive DNA. Because it is faster and cheaper than the clone-by-clone approach, it can be used to determine the sequence of other large genomes (such as the mouse genome) that will be very important for bio-medical research and economically important animals and plants.

4.5 Finishing the sequence

The draft sequence of the human genome only covered 90% of the genome and contained 150 000 gaps. Moreover, 10% of the finished sequenced failed to meet the quality target of less than 1 error in 10^4 base pairs. So the gaps had to be closed and the quality made as near perfect as possible throughout the whole sequence. This was not an easy task. Many of the gaps and poorly sequenced regions were enriched in problematic sequences. The draft sequence not only revealed the presence of many repeat families, as expected, but also showed that there were many segmental duplications. Many of these duplicated sequences were near-perfect matches to each other, so distinguishing between them relied on a small number of base pair differences, placing a high premium on the accuracy of the sequencing. The task was further complicated by the fact that the sequence was being pieced together from the genomes of different individuals, which show natural sequence variation. Thus, to decide that two near-identical sequences were segmental duplications on the basis of a few base differences, it was necessary to be sure that the apparent differences were not sequencing errors or natural polymorphisms arising because the sequences were derived from different individuals.

The first stage in finishing was to construct a finished physical map consisting of a continuous series of overlapping large insert clones for each chromosome arm. This required filling the gaps in the BAC clone map and verifying the overlaps between each pair of overlapping clones. Closing the gaps was mainly done by walking outwards from the last clone in contig using the terminal sequence of the clone to identify an overlapping clone. The process is repeated until the walk comes to a dead end or a clone is identified that connects to a neighbouring contig. Overlaps between clones were recognised either by hybridisation using the terminal sequence to probe the BAC clone library, or by computation using the terminal sequence to search a database of end sequences from the BAC clone library. Occasionally, it was possible to parachute into the gap if a particular sequence was known to be located in the gap such as an STS, an EST or cDNA, or a region of homology to the mouse genome. This sequence could then be used to probe the BAC library by hybridisation. The second stage of the sequencing process was to sequence each large insert clone. This was done by generating a shotgun library of small inserts in plasmid vectors. The ends of each insert were then sequenced and the reads assembled into an initial sequence for the whole of the large insert. The initial sequence often had gaps and regions judged to be of low quality. Intensive work with each clone was then required to rectify these defects and produce a continuous high-quality sequence.

The finished sequence amounted to 2.85×10^9 bp. The number of gaps in the sequence was reduced 400-fold to just 341, of which 308 are in euchromatic DNA. A parameter which indicates the completeness of the genome is the N50, defined as the length x, such that at least 50% of nucleotides lie in a continous sequence greater than x. The N50 for the finished sequence is 38 Mb, which is 1000 times greater than the length of a typical human gene, and in three-quarters of cases is longer than half the chromosome arms in which it resides. For comparison, the N50 for the draft sequence is 81 kb. The missing sequence amounts to just 1% of the genome. The remaining gaps could not be closed, despite searching libraries with 30-fold coverage of the genome and libraries constructed with a variety of vectors. These sequences are simply difficult to clone and it was judged that it would be unproductive to continue using large scale sequencing strategies; instead, the remaining gaps will need to be targeted by specialist research. Quality control tests were carried out to verify the overlaps between the large insert clones, ensuring that every region was located in its proper place in the genome. The accuracy of the sequencing itself was quality controlled by extensive re-sequencing. It was estimated that the error rate was less than 1 in 10^5 bp, exceeding the original target by a factor of ten. The effort in overcoming the many difficulties involved in closing the gaps and sequencing what was often recalcitrant DNA, the meticulous care in which the finished sequence was assembled and the extensive quality control checks required roughly an equal amount of time and resources as it took to produce the draft sequence.

4.6 Single nucleotide polymorphisms

Both the Celera and IHGSC sequences were derived from a panel of different individuals. As the consensus sequences were assembled, differences between the nucleotide sequences of different individuals became apparent. If a difference is common in a population it is defined as a polymorphism. Thus, during the sequencing of the human genome **single nucleotide polymorphisms** (**SNPs**) were identified. Because they were identified in the course of sequencing their exact location is automatically known, so a map is generated of the common differences between individuals. SNPs have proved to be a key resource in mapping the alleles that contribute to the risk of common complex diseases. We shall consider them in much more detail in Chapter 6.

4.7 Annotating the sequence

The raw sequence data must be **annotated** to be useful. This involves determining and recording the genomic segments spanned by mRNA molecules, the boundaries of exons and introns, regions transcribed but not translated, the location of physical features such as chromosome bands, repetitive elements, CpG islands, mapping features such as STSs, genetic markers and SNPs and homologies to genes already studied in model organisms or duplicated copies of genes already studied. Analysis and annotation of the genome is carried out by teams of based at Bioinformatics institutes, such as EBI, NCBI and UCSC (Box 4.1). The annotation of the human genome is carried out alongside annotation of the sequenced genomes of other vertebrates, model organisms and prokaryotes, and there is much interaction between them. For example, a human gene may be recognised because exons in a genomic segment align with a cDNA sequence isolated and sequenced in mice. These teams construct genome browsers and databases. The browsers allow the sequence and the annotations to be accessed via a graphical interface on the world wide web. The searchable databases contain stored information about the elements in the genome such as RNA transcripts and encoded proteins. Much of the annotation can be carried out automatically by specially developed computer programs. However, manual review is necessary to ensure the annotation is of high quality and free from errors. When a database or genome sequence has been manually reviewed, it is said to have been curated. Box 4.1 describes these browsers and databases, and provides URLs to access them. Figure 4.8 shows an example from the Ensembl site showing the region around the *BRCA1* locus on chromosome 17q21.31. The web page from which this was taken has facilities to zoom in or out, navigate in either direction and view the DNA sequence itself.

BOX 4.1: GENOME DATABASES AND BROWSERS

Bioinformatics institutes

EBI (European Bioinformatics Institute) is part of the European Molecular Biology Laboratory (EMBL). It is based near Cambridge, UK, where it shares the Wellcome Trust Genome Campus with the Wellcome Trust Sanger Institute, which specialises in genome sequencing, which included a substantial part of the human genome.

NCBI (National Centre for Biotechnology Information) is a division of the National Library of Medicine at the NIH institute in Bethesda.

UCSC Genome Bioinformatics Group at University of California at Santa Cruz.

Browsers and databases

Ensembl Browser and gene database of humans and other eukaryote genomes constructed by the Wellcome Trust Sanger Centre and EBI. Gene database is generated by automatic annotation. **www.ensembl.org/**

Vega High-quality manual annotation of the human and other vertebrate genomes based on finished genome sequence. Generated by the Havana group based at the Wellcome Trust Sanger Institute. **http://vega.sanger.ac.uk/index.html**

EMBL A non-curated database which receives and stores submissions of DNA sequences from the primary researchers. May contain duplicates.

TrEMBL Non-curated translations of sequences submitted to EMBL.

SwissProt Manually curated protein database containing a single entry for each protein and in some cases additional information about protein structure and function.

Uniprot (Universal Protein Resource) Combined database of TrEMBL and SwissProt entries.

NCBI browser and genome maps Access to a variety of genetic, physical and expression genome maps as well as an integrated browser. **www.ncbi.nlm.nih.gov/genome/guide/human**

Entrez A cross database search engine that allows a genome element to be tracked with a single identifier through the different databases at NCBI.

GenBank A non-curated database at NCBI of nucleotide, mRNA and protein sequences. Entries are submitted by primary researchers who carried out the sequencing. Like EMBL, GenBank includes all entries submitted, so it may contain multiple entries for the same locus. Genbank exchanges data daily with EMBL.

Refseq (Reference sequences) A database that generates a single entry for each genomic DNA, mRNA and protein sequence from entries in GenBank. Refseq records are made automatically and then curated manually by NCBI staff who review GenBank records to produce the record based on best available information.

UCSC genome browser Based on the finished sequence of the human genome and browsers based on draft sequence and finished sequence of other vertebrates. **www.genome.ucsc.edu/**

Gene ontology (GO) Provides a consistent description of gene products in terms of biological processes, cellular components and molecular functions, using a controlled vocabulary. It allows a set of genes, which may be the complete set encoded by a genome, or a gene set detected in an experiment such as a microarray to be categorised into sets of related functions. **http://www.geneontology.org**

ENCODE Access to ENCODE data. **http://genome.ucsc.edu/ENCODE/** (see Section 4.10)

HapMap Homepage of the HapMap project (see Chapter 6). **www.hapmap.org/**

Mouse Knockout project (KOMP) A project coordinated by NIH to construct a collection of mouse embryonic stem (ES) cell lines containing a null mutation in every gene in the mouse genome. **www.komp.org** and **www.nih.gov/science/models/mouse/knockout/komp.html**

The Cancer Genome Project See Section 4.13. **www.sanger.ac.uk/genetics/CGP**

Cancer Genome Atlas **www.cancergenome.nih.gov**

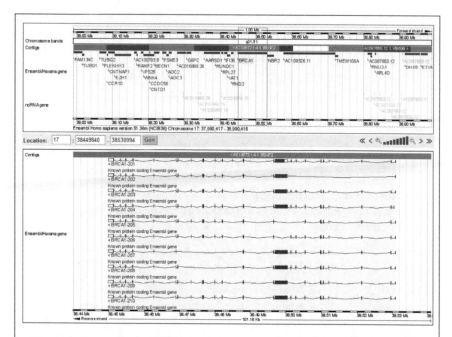

Figure 4.8 The entry for human *BRCA1* in the Ensembl website (www.Ensembl.org). The figure shows a screen grab from a page showing the detailed structure of the *BRCA1* gene. The browser presents an overview of the region, showing the position of adjacent genes and a detailed view of the gene. Both views can be zoomed in or out as desired. The browser allows the page to be configured to show a large number of genomic features such as regions transcribed, repetitive elements, SNPs, genetic and physical markers, regions of conservation compared to 21 eutherian mammals, GC content etc., clone contigs, etc. For clarity, only a subset of these features has been turned on in this example. Note the number of alternative transcripts and the small fraction of the gene forming exons (boxes) compared to the introns (wavy lines).

4.8 Sequencing other genomes

As well as the human genome, the genome sequences of a variety of organisms have also been determined. Model organisms such as *Saccharomyces cerevisiae* (1996), the bacterium *Escherichia coli* (1997), the nematode worm *Caenorhabditis elegans* (1998), the fruit fly *Drosophila melanogaster* (2000) and the mustard weed *Arabidopsis thaliana* (2000) were finished before the Human Genome Project and served as test beds to develop new sequencing strategies and technologies. Since the human genome was sequenced, these technologies have been applied to a wide variety of genomes from diverse organisms. The Ensembl site gives access to 39 eukaryote genomes, consisting of 23 mammalian genomes, including the chimpanzee, mouse and rat genomes, two marsupial and monotreme genomes, the chicken genome, the *Xenopus tropicalis* genome (African clawed toad) and five fish genomes. Together, these have proved an invaluable resource for a number of reasons:

- Comparison across the genomes identifies evolutionarily conserved sequences that are likely to have a functional significance. This has been an important resource in identifying coding regions which are expected to be conserved. Surprisingly, there are extensive conserved regions outside coding regions and adjacent regulatory sequences. This conservation in intergenic sequence is an indication that there is more to the genome than genes sparsely scattered among non-functional DNA – a conclusion that is reinforced by the results of the ENCODE project described below.

- The mouse is a tractable experimental organism with extensively developed technology for genetic manipulation that is widely used in biomedical research. The role of genes can be investigated by targeted deletions (gene knockouts), overexpression or ectopic expression and *in vitro* mutagenesis. This has already played a vital role in the investigation of the role of genes affected by disease-causing mutations. The availability of its genome sequence empowers these technologies. For example, projects are under way to construct a library of mice gene knockout strains to investigate the function of every gene.

- Comparison of genomes from different taxonomic categories illuminates evolution. Comparing the fly and worm genomes to vertebrate genomes traces the evolutionary innovations that led to vertebrates. The zebrafish is used for the genetic analysis of developmental processes. The availability of the human and zebrafish genomes will provide instant access to the human counterparts of zebrafish genes as their roles emerge.

- The chimpanzee is our nearest evolutionary relative. Comparison of human and chimpanzee genomes allows us to identify those genetic changes that evolved to give our species its particular special properties. Moreover, it allows us to identify the ancestral allele at polymorphic sites because the ancestral is likely allele to be the allele with the same base as the corresponding base in the chimpanzee genome. This has proved very important in reconstructing recent human evolution using the Y-chromosome genealogy as described in Chapter 10.

- The puffer fish genome is unusually small and is much more compact than the human genome, with a much higher proportion of coding DNA. It is therefore easier to identify these coding regions. As they are likely to have a similar complement of genes as humans, this has aided the identification of human coding regions.

4.9 Analysis of the human genome sequence

With the determination of its sequence, the human genome becomes a finite entity that can be completely and accurately described. In principle, we now know the location and sequence of every gene and repetitive element. We can precisely enumerate the vital statistics of the genome: its size, the fraction that codes for proteins, the fraction that constitutes repetitive elements, the location of CpG islands, the correlation of physical and genetic maps and so on. Before the sequence was determined, all of these features could only be studied indirectly, albeit with increasing accuracy and confidence. The reader should refer to Chapter 2 for an introduction to the main features of the genome. This section summarises some of the main conclusions that emerge from an analysis of the sequence of the human genome. Most of the analysis is derived from the analysis of the draft sequence which provided a general overview of the genome that has not substantially changed as more detailed information was published on a chromosome-by-chromosome basis. An important exception to this statement is the identification of genes, which has been radically changed by the availability of the complete sequence and its careful manual annotation. Table 4.2 provides a summary of the vital statistics of the human genome.

Table 4.2 Vital statistics of the human genome. Sources shown in right column. Ensembl data was accessed July 2008 (stable archive link http://jul2008.archive. ensembl.org/Homo_sapiens/index.html). Note the data for number of transcripts and the proportion of genome transcribed is likely to be considerably revised as a result of the ENCODE project. The repetitive elements *Alu*, LINE1, MaLR and MER1-Charlie are the most common in each respective class.

Total genome size	3258 Mb	Ensembl
Chromosomes		IHGSC draft
Largest (chromosome number)	279 Mb (1)	IHGSC draft
Smallest (chromosome number)	45 Mb (21)	IHGSC draft
X	163 Mb	IHGSC draft
Y	51 Mb	IHGSC draft
Fraction CpG	41%	IHGSC draft
Number of CpG islands	28 890	IHGSC draft
Proportion of genome that encodes proteins	1.2%	Ensembl
Proportion of genome transcribed	33%	IHGSC draft
Known protein-encoding genes	20 067	Ensembl
Novel protein-encoding genes	1461	Ensembl
Estimated total number of protein-encoding genes	21 528	Calculated from Ensembl
Total number of exons	294 420	Ensembl
Pseudogenes	6282	Ensembl
RNA genes	4810	Ensembl
Gene transcripts	61 318	Ensembl

Table 4.2 *(cont'd)* 109

SNPs	13 099	Ensembl
Transcripts per gene	2.84	Calculated from Ensembl
Mean of size of genes by genomic extent (median)	27 kb (14 kb)	IHGSC draft
Largest gene by genomic extent (name)	2.4 Mb (DMD)	IHGSC draft
Mean gene transcript size (median)	1340 bp (1100 bp)	IHGSC draft
Largest gene transcript (name)	80.8 kb (Titin)	IHGSC draft
Mean proportion of each gene that is coding	5%	IHGSC draft
Proportion of genome that consists of repetitive elements	44.8%	IHGSC draft
SINEs (number in 1000s)	13.4% (1558)	IHGSC draft
Alu elements	10.6% (1090)	IHGSC draft
LINEs	20.42% (868)	IHGSC draft
LINE1	16.89% (516)	IHGSC draft
LTRs	8.29% (443)	IHGSC draft
MaLR	3.65% (240)	IHGSC draft
DNA transposons	2.84% (294)	IHGSC draft
MER1-Charlie	1.39% (182)	IHGSC draft
Recombination per Mb (cM/Mb)		Celera
Male	0.88%	Celera
Female	1.55%	Celera
Sex-averaged	1.22%	Celera

4.9.1 Overall sequence architecture

4.9.1.1 GC content

The human genome has an average GC content of 41%. However, this is not uniformly distributed throughout the genome. Figure 4.9 shows that, when examined in 50-kb windows, the GC content of each window can vary from 30% to over 60%. This is 15-fold greater than the variation that would be expected by chance in random sequence DNA. Using probes taken from regions of high and low GC content, FISH shows that G-bands of chromosomes correspond to regions of low GC content. GC content is strongly correlated with gene density, with GC-rich areas containing more genes than AT-rich regions (Figure 4.9).

4.9.1.2 CpG islands

The dinucleotide CpG (i.e. C followed by G on the same strand) occurs less often than would be expected by chance in the human genome. As described in Chapter 2, the cytosine base in CpG is the target for DNA methylation, which inhibits gene transcription. Scattered throughout the genome are islands of sequence where the occurrence of CpG is not depressed from that expected and which are unmethylated. These **CpG islands** tend to occur at the start of genes, spanning the first exon and the

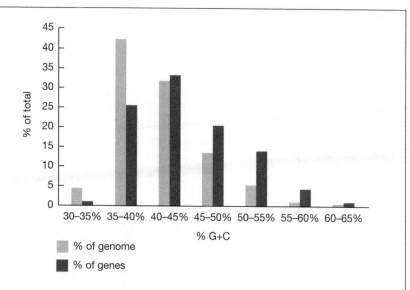

Figure 4.9 Distribution of GC content in the genome and the relationship between genes and GC content. Genome GC content was determined in 50-kb windows and divided into GC intervals of 5%. The histogram shows the fraction of total DNA (blue bars) and percentage of genes in each interval. There are proportionately more genes in the intervals above 50% GC and proportionately less in the intervals with less than 40% GC. (Reprinted with permission from Venter, J.C. *et al.* (2001) *Science*, **291**, 1304–1351, © 2001 American Association for the Advancement of Science.)

promoter sequences immediately upstream. The genome sequence alone does not give information about methylation but is informative about the number and distribution of CpG dinucleotides. There are about 28 000 CpG islands across the genome; the exact number is sensitive to the percentage GC used to define an island. This figure is close to the predicted number of genes. Most of the predicted islands are between 500 bp and 1500 bp long and there is a strong tendency to overlap the first exon and promoter sequence of known or predicted genes.

4.9.1.3 Repetitive DNA

About 45% of the human sequence is derived from one of four types of transposable element, described in Chapter 2. This figure is likely to be an underestimate, because many repetitive elements will have diverged so much that they are no longer recognisable. Furthermore, many of the gaps in the draft sequence will be repetitive DNA. The distribution of repetitive DNA among these types is shown in Table 4.2. The most common families are the *Alu* family, occupying 11% of the genome, and the LINE1 family, occupying 17%. Repeated sequences are interspersed continually throughout the genome, as illustrated in Figure 4.8.

Because the sequence of every copy in a family is known, it is possible to construct a phylogenetic tree and deduce the sequence of the original

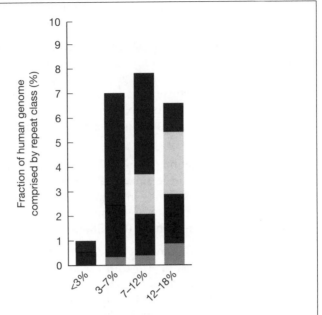

Figure 4.10 Sequence variation in repeated sequences. The percentage variation from the consensus sequence for *Alu* (blue), LINE1 (white), LTR elements (black) and DNA transposons (grey) is plotted in 3% intervals. Each interval represents about 25 million years of evolution. Few elements are in the <3%, indicating that most are older than 25 million years. (Redrawn with permission from IHGSC (2001) *Nature*, **409**, 860–921.)

progenitor of each family. So for each individual element the number of nucleotide changes from the inferred progenitor can be determined, forming a molecular clock that allows the birth and subsequent fate of these elements to be determined. The clock can be calibrated by measuring the average difference in nucleotide sequence between humans and Old World monkeys, which are known from the fossil record to have diverged 25 million years ago. Figure 4.10 shows the range of substitutions found in each of the four types of transposable element. Each interval of 3% substitution represents about 25 million years of evolution. This analysis not only provides the age of the elements, but also allows the periods during which the transposable elements were active to be calculated. In other words, we can view the transposable elements as if they were independent organisms and chart the rise and fall of each type as if in a fossil record. A number of striking conclusions emerge from this study:

- In the lineage leading to humans, transposition of all elements has been virtually absent in the past 25 million years. Nearly all the repetitive elements in the genome are no longer living, but are fossils. After transposition, they accumulated mutations that prevented further transposition events from occurring. Most repetitive elements remaining in the genome are ancient and pre-date the eutherian radiation (mammals that are not marsupials).

- There were two peaks of DNA transposon activity, one before and one after the eutherian radiation. These involved two families of DNA transposons called Charlie and Tigger, respectively. There has been very little activity of DNA transposons in the human lineage in last 50 million years. Only two classes of DNA transposons, MER75 and MER85, have been active in this period. Together they occupy less than 0.00041% of the genome.

- Only one transposition, an event involving an LTR retrotransposon, is known to have occurred since the divergence from chimpanzees 6 million years ago. This results in a single LTR element, found in the HLA locus, that is not common to all humans. LTR transposons were apparently most active more than 100 million years ago.

- The *Alu* and LINE1 families are 80 and 150 million years old respectively. The last burst of *Alu* transposition was about 40 million years ago. Before that the most common elements were LINE2 and a SINE element called MIR. MIR transposition used the LINE2 reverse transcriptase and so became inactive alongside LINE2.

- Mechanisms for removing inactive elements are very inefficient. The half-life of a transposable element in the mammalian genome is 800 million years. This is in contrast to other types of animals. For example, the half-life of a repeated sequence in *Drosophila* is only 12 million years.

- Transposable elements are still active in the mouse genome, and so most repetitive elements in the mouse are younger. This correlates with the observation that 10% of mouse mutations, but only 1 in 600 human mutations, are caused by repetitive elements. It is not clear why there should be such a difference in the evolution of hominids and rodents. It could be to do with the size of the breeding group, which is much smaller in hominids. This may have resulted in bottlenecks that would have reduced diversity (see Chapter 10).

The distribution of repetitive elements is not even across the genome. Some regions are very rich; others are almost totally devoid of such elements. For example, a 525-kb region on the X chromosome, Xp11, consists of 89% repetitive DNA. A striking contrast is found in the four Hox regions that each contain a cluster of homeobox genes that play a crucial role in organising development. These four regions are thought to be the result of ancient duplications and are homologous to each other. All four regions contain less than 2% repetitive elements. The regions contain complex *cis*-acting regulatory elements. It is possible that insertion of transposable elements would disrupt the complex regulation shown by these genes. If so, it will be interesting to investigate the function of other areas of the genome that show a low density of repeat DNA. Another factor affecting the distribution of *Alu* elements is GC content. Young *Alu* elements are more elements show a progressively increasing bias towards GC-rich DNA. Apparently, *Alu* elements preferentially target AT-rich DNA, but some form of selection operates to retain them in GC-rich DNA. The nature of this selective pressure is unclear, but its existence suggests that *Alu* elements are not selectively neutral.

The spread of repetitive elements throughout the genome is generally thought to have occurred not because they are phenotypically beneficial, but because they are essentially selfish elements that evolved only through selection within the genome that favoured their transposition. Nevertheless, repetitive elements have played a role in the evolution of genome function. The reverse transcriptase of LINE elements can act on mRNA to produce a DNA copy, which can insert into the genome to produce a copy of the original gene, called a **retrotransposon** (see Chapter 2). Because the mRNA has been spliced after transcription, the retrotransposon will lack introns. It will also have a poly-A encoded in the DNA from the poly-A tail of the mRNA. If it loses function as a result of the accumulation of inactivating mutation, it is known as a **processed pseudogene**. Retrotransposons can be identified because they will be single exon genes that show sequence homology to other multi-exon genes in the genome. Sequence analysis identified nearly 300 such examples, of which 97 are already known to be expressed and to have some function. In this set, genes involved in translation (ribosomal proteins and elongation factors) and nuclear regulation (non-histone chromatin proteins) are over-represented. Because of the mechanism by which they arose, retrotransposons do not carry upstream regulatory sequences from the gene that was copied. Thus at the new site where they insert they may acquire different regulatory regions, potentially changing the pattern of expression.

Repetitive elements may contribute to genome function in other ways. A total of 47 human genes are homologous to the proteins encoded by DNA transposons. There are also examples where repetitive elements are present in regulatory regions, suggesting that they have been co-opted to increase the repertoire of the patterns of gene expression.

4.9.1.4 Correlation of physical and genetic distance

Genetic maps are based on the proposition that the amount of crossing over between two polymorphic markers at meiosis is proportional to the distance separating them. The units of genetic maps are centimorgans (cM), corresponding to a recombinant fraction of 0.01. The location of polymorphic microsatellites used to construct genetic maps can be determined in the genomic sequence. This allows the correlation between physical and genetic distance to be examined. Across the whole genome a general correlation between genetic and physical distance is evident (Figure 4.4). It was already known that more recombination occurs in female than in male meioses. Averaging between sexes, 1 cM on the genetic map corresponds to 1.2 Mb of sequence. However, the actual correspondence varies at both small and large scales.

- At the local level, when recombination is measured in windows of 3 Mb, there is an almost five-fold variation in the rate of recombination from the hottest to the coldest regions. This greatly exceeds the two-fold difference between males and females.

- Two types of variation are seen at the chromosomal level:
  - Recombination is higher near telomeres and lower near centromeres, especially in males.
  - Proportionately more recombination is seen on short chromosomal arms compared to long arms. This may be due to the requirement for at least one recombination per chromosome arm, which is necessary for disjunction of homologous chromosomes at meiosis.

4.9.1.5 Chromosome evolution

Comparing the sequence of a segment with the rest of the genome makes it possible to identify examples of DNA duplication. However, segments that have duplicated recently will have almost identical sequences. This results in two problems:

1. During the assembly, sequences that are located in different parts of the genome may be combined into the same contig. This is much more of a problem for the whole genome shotgun approach, where there is no independent positional information. In fact it was one of the most difficult problems to solve, and required special programs that assessed whether the coverage of a contig was greater than would be expected, given the average coverage of the whole genome.
2. If the assembly of the final sequence was imperfect and gel reads that actually come from the same chromosomal location were assembled into different contigs, then the final sequence will appear to have a duplicated segment. The solution to this is to reject apparent duplications where the match is over 99.5% in the finished sequence (98% in unfinished sequence), suggesting that they are really the same sequence that has been misassembled. The penalty is that some genuine recent duplications will be missed.

This analysis showed that about 5% of the genome is duplicated in blocks of between 10 kb and 50 kb. In addition, the regions around the centromeres (**pericentric**) are very rich in interchromosomal duplications, giving rise to the suggestion that they may be involved in some repair mechanism for chromosome breakage.

An important practical reason for unravelling chromosomal changes during evolution is that the order of genes in chromosomal segments is likely to be conserved between humans and experimental animals such as the mouse. Genes whose chromosomal positions are conserved in this way are said to be **syntenic**. Experience with yeast and other organisms shows that the local order and orientation of genes can change, so the criteria for synteny is the presence of multiple homologous genes in the same chromosomal segment in both organisms or in duplicated blocks within the same organism. Knowledge of synteny between humans and model organisms will aid biomedical research in different ways. For example, intensive genetic analysis in mice may identify a genomic region involved in susceptibility to a mouse model of a complex disease. Synteny will allow the same region to be studied in humans to see if there is any evidence of involvement of a human counterpart.

Another reason to study chromosome evolution is to test a theory of vertebrate evolution, which holds that there were two rounds of whole genome duplication around 500 million years ago, at the time that jawed fishes appeared. This theory is based on the existence of the quadruplicated Hox gene clusters mentioned above. However, an alternative explanation is that they arose through localised interchromosomal duplications. Clearly this is an interesting question that may be resolved by study of the genome. A precedent for whole genome duplication is the yeast genome, where sequence analysis has revealed that about 100 million years ago it duplicated and then lost about 85% of the duplicated genes, leaving the remainder as duplicated blocks distributed around the genome.

Comparing DNA sequence allows only recent chromosomal events to be analysed. More ancient chromosomal changes cannot be followed because most non-coding DNA, which occupies >98% of the genome (see below), is not subject to sequence conservation. One way to approach the study of long-term chromosome and genome evolution is to look at the distribution of genes rather than compare DNA sequences. In this way, larger blocks of duplicated chromosomal evolution emerge. For example, the Celera analysis revealed the following:

- A total of 1077 blocks of duplicated sequence were identified in the human genome, of which 781 contain five or more genes.

- Three very large blocks of duplication were detected, two of which involved chromosome 2.
  - Between 2p and chromosome 14: the duplication represents 70% of the sequence from the latter.
  - Between 2p and chromosome 12: this duplication includes two of the Hox gene clusters referred to above.
  - Between chromosomes 18 and 20 (Figure 4.11): this duplication includes 64 genes. After discounting a clear insertion between 'Krup rel' and 'collagen rel' on chromosome 18, the duplication covers about half of each chromosome. Only a fraction of each duplicated segment was found in the other chromosome. For example, the 64 genes occurring on both chromosomes 18 and 20 are found in a block of 217 genes on chromosome 18 and 322 genes on chromosome 20. The chances of this occurring by chance are very low. The most likely explanation is that a large segment was originally duplicated, then genes were lost from both segments, resulting in some genes still being present on both chromosomes, while other genes

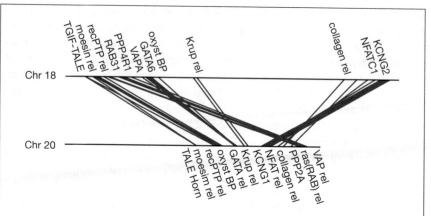

Figure 4.11 Duplicated blocks on chromosomes 18 and 20. The map compares a block of 36 Mb on chromosome 18 and 28 Mb on chromosome 20. Twelve of the 64 duplicated genes are shown. A large insert is present between Krup rel and collagen rel on chromosome 18. Apart from the insert it can be seen that there is a copy of each gene present on each chromosome. The order is not exactly conserved, but rearrangement of local gene order after duplication is commonly observed in chromosome evolution. (Reprinted with permission from Venter, J.C. *et al.* (2001) *Science*, **291**, 1304–1351. © 2001 American Association for the Advancement of Science.)

are present only on one or the other chromosome. Regions in the mouse genome can be identified that are syntenic to the duplicated blocks on chromosomes 18 and 20. Gene order and sequence conservation is much more similar in the pairs of syntenic mouse/human partners than in the pairs of duplicated blocks on the human chromosome. It follows that the duplication preceded the divergence of the lineages that led to rodents and primates. Although this picture is consistent with ancient whole genome duplications, it does not really distinguish between the competing hypotheses of piecemeal duplications of chromosomal segments. As more vertebrate genomes are sequenced, the evolution of chromosomes should become clearer.

Interspecies comparisons allow comprehensive maps of synteny to be constructed. Comparison with the mouse genome has led to the identification of 183 syntenic segments, including some very large ones. For example, the whole human chromosomes 17 and 20 are syntenic to segments on mouse chromosomes 13 and 2, respectively. Human and rodent lineages are thought to have diverged 100 million years ago, so there has been a combined total of 200 million years of evolution since then, giving an approximate overall rate of 1.0 rearrangement per 1 million years of evolution. There is some evidence that the rodent genome shows an unusually high rate of rearrangement. Comparison of the human and zebrafish diverged 400 million years ago, so the rate of rearrangement is about 0.5 rearrangements per 1 million years of evolution, i.e. half as high as in the rodent lineage. Although the genome sequences of other mammals are not yet available, much information can be gained from comparison of the cytogenetic, physical and genetic maps. Some startling conclusions emerge – for example, the organisation of the cat genome can be transformed into the human genome organisation by as few as 13 translocations. These and other similar observations suggest that the rate of rearrangements among some mammals may be as low as 0.2 rearrangements per 1 million years of evolution.

4.9.2 Genes

The main interest of the human genome is clearly focused on genes which generate our phenotype and whose malfunction leads to disease. We can divide genes into protein-encoding and non-coding or RNA genes. For the purposes of this discussion, a protein-encoding gene is defined as a genomic segment spanned by an mRNA molecule in which a group of exons have been spliced together. There are two complications to this definition. First, primary transcripts often originate from the same genomic region but are spliced to generate mRNA molecules containing different combinations of exons. Genomic segments giving rising to primary transcripts that are differentially spliced count as a single gene. Second, the definition ignores the regulatory regions that are necessary for the gene to be expressed, which cannot be defined without extensive further experimentation (discussed further below in Section 4.11 discussing the ENCODE project). Restricting the definition of a protein-encoding gene to the genomic region spanned by an mRNA transcript provides a secure and workable definition. Protein-encoding genes need to be distinguished from pseudogenes, which will have a similar sequence to protein-encoding genes but are not functional either because they are not expressed or because they containing inactivating nonsense and frameshift mutations. Processed pseudogenes arising by retrotransposition can be recognised because they encode sequences similar to a protein-encoding gene but lack introns.

RNA genes are sequences that are transcribed and contribute to the phenotype, but are not translated. Classically, these are RNA molecules with a known function such as rRNA, tRNA, 5s RNA, small nuclear RNAs (snRNAs) that are part of the splicing machinery and microRNAs that perform a regulatory role (see Chapter 7). Recent studies have shown that the genome-wide pattern of transcription is much more extensive than expected and the role of many of the resulting transcripts is presently unclear. We shall discuss these findings in more detail in the section on the ENCODE project. This section considers how protein-encoding regions were identified, what they reveal about the parts list required to build the human organism and, by comparison to the genomes of model organisms, what they reveal about evolution.

4.9.2.1 Identifying coding regions

Because coding sequences are interrupted by introns it is not possible to recognise genes simply by searching for long open reading frames. Indeed, in many genes exons are small islands of coding sequences embedded in much larger stretches of intronic DNA. An example of this is shown in Figure 4.8 for the *BRCA*1 gene. A typical gene will extend for 30–40 kb of the genome and contain a mean of 10.4 exons whose mean size is 147 bp. Some genes are much more extreme. The largest gene in terms of chromosomal coverage is the dystrophin gene covering 2.4 Mb on the X chromosome. The largest in terms of coding sequence is the titin gene, whose mRNA is over 80 kb long, spliced together from 178 exons. Overall, only 1.2% of the genome is coding sequence. The annotator draws gene models or structures, which identify the exons, and defines their boundaries, and specifies which exons are combined together in mRNA transcripts to form genes.

Three lines of evidence are used to construct gene structures:

1. Computer programs can be used that recognise exons on the basis of features such as length, open reading frames, codon bias, splice sites (GT at the 5′ splice site and AG at the 3′ site). Such programs are said to work *ab initio*, because they use only sequence and do not consider other types of experimental data. An example of such a program is GenScan. Such programs are reasonably successful in identifying exons, but less successful in predicting how the exons are combined into complete **gene structures**. The results of these *ab initio* programs are augmented by **BLAST** searches that identify homologies to known proteins or protein domains in protein databases such as Uniprot. (BLAST refers to a family of programs that search databases for similarity to a query sequence which may be either protein or DNA).

2. Coding sequences are expected to be evolutionarily conserved, whereas non-coding sequences should be less conserved. Thus the human sequence can be searched for segments that show conservation with other animals. The mouse and puffer fish (*Tetraodon nigroviridis*) genomes are used for this. The latter has a very compact genome, with exons constituting a much higher fraction of its genome. For example, the Exofish program identifies regions of conserved sequence between the puffer fish and human genomes.

3. The exons of a gene will match the sequence of ESTs and cDNAs from human and other mammals in databases such as Refseq. EST and cDNA clones may have been derived from contaminating genomic DNA or hnRNA that does not appear in spliced mRNA, especially from sequences containing a poly-A tract that will hybridise to the poly-T primer used to prime reverse transcription. To eliminate these clones, evidence of splicing is required. That is, the EST or cDNA is composed of sequences that are non-contiguous in genomic DNA. It is important to be sure that the full length cDNA has been recovered. The true 3′ end can be identified by polyadenylation – a sequence of at least four A residues present on the cDNA 3′ terminus that are not present in the

genomic DNA; alternatively, the presence of an AATAAA polyadenylation signal within 50 nucleotides of the 3' end. The start of a coding
region can be identified by a **Kozak** consensus translation initiation
sequence immediately adjacent to the first ATG.

All of these criteria can be used together to identify genes, but there is no
'one size fits all' approach that can be used. Of course, many genes have
already been identified by the piecemeal research that has been carried out
over the years by a wide variety of different laboratories interested in different aspects of human biology and medicine. Records of the nucleotide
sequence of known genes and the amino acid sequence of the encoded proteins are present in databases such as Refseq and Uniprot/Swissprot and
can be readily mapped onto the genome sequence. Automated computer
programs can be used to suggest candidate genes, as in the set of genes present in the Ensembl genome browser but, in the end, each candidate gene
needs expert human examination such as has been carried out for vertebrate genes in the Vega database (see Box 4.1). A match to a full length
human cDNA annotated in one of the databases is the best criteria. Such a
gene is said to be a known gene, supported by experimental criteria. Other
genes identified by a match to a mammal cDNA transcript are categorised
as novel or predicted genes. A match to an experimentally detected cDNA
not only ensures that the putative gene is a real biological entity that is
expressed, but also reveals gene structure; that is, which exons are grouped
together to form the gene. Full annotation of the human genome has occu-
pied complete large teams of annotators for several years. It is now nearly
complete, but it will probably be subject to modification and improvement
for some time yet.

Because of the importance of identifying full length cDNAs, the
Mammalian Gene Collection (MGC) was established. This is a multi-
institute initiative that aims to identify, clone and sequence full length
human, mouse and rat cDNAs. To do this, MGC has established over 629
cDNA libraries derived from human cell lines and mouse and rat tissues.
The diversity of the different libraries and special experimental technologies
to target rare cDNAs and long cDNAs should minimise the chance that genes
escape detection because they are not abundant, or are expressed in only a
limited number of tissues or for only a brief time during development. For
predicted genes, where there is evidence that a gene exists but the clone is
not present in the MGC collection, the missing cDNA clones are being
targeted for isolation by PCR using primers designed from the predicted
sequence. The cDNA clones in the MGC collection are freely distributed to
research laboratories so, in addition to identifying genes and resolving gene
structure, the cDNA clones are a powerful resource for further gene function studies.

4.9.2.2 How many genes?
It is clear from the above that this apparently simple question is not easy to
answer definitively. The initial estimate from the draft sequence was about

30 000 protein-encoding genes, which was a major surprise as previous estimates before genome sequencing were in the region of 100 000. As the sequence was finished, it was subjected to intensive manual curation using the criteria outlined above to identify protein-encoding genes. As this process continued, the estimated number of protein-encoding genes reduced even further. According to statistics listed on the Ensembl web site (July 2008), the database contains 21 528 protein-encoding genes, consisting of 20 067 known genes and 1461 novel genes. In addition, the Ensembl database contains 6287 pseudogenes, 4810 RNA genes and 2931 RNA pseudogenes. (Ensembl genes are automatically annotated; manual annotation by the Vega group lists 19 356 protein-encoding genes.) The exact figures may change with ongoing annotation, but the estimates of the gene number have now stabilised and true figures will not be very different from those listed above. The change in the estimated number of genes from the draft sequence highlights the improvement made to the accuracy of the nucleotide sequence and annotation by the finishing process. When the current gene models were mapped back to the draft sequence, 58% had errors, of which 39% had missing exons or incorrectly ordered exons. The inflated estimates in the draft sequence resulted from **fragmentation**; that is, where two separate genes were predicted but were eventually judged to be the same gene.

What is the significance of the estimated number of human genes? The yeast genome contains 6200 protein-encoding genes, the fly genome 13 000 and the worm genome 19 000. Thus it takes roughly 1.1, 1.6 and 3.5 times the number of genes to make a human compared to a worm, fly and yeast respectively. Although an anthropomorphic judgement, it is generally accepted that a human is more far complex than these organisms. A very important point to note is that 61 000 different transcripts have been identified. So on average, there are approximately three different transcripts for every gene. (The first results of the ENCODE project described below suggest that there are 5.4 transcripts per gene.) This implies that, while the total number of genes is much smaller than we might have expected, the exons that make up the genes are combined in different combinations by alternative splicing, greatly increasing the diversity of proteins encoded by the 21 528 genes. Understanding how the encoded proteins combine to produce the final complex human form defines human research for the foreseeable future. We can make a start by examining the properties of these proteins and comparing them to the proteins encoded by these model organisms.

4.9.3 The human proteome

Just as the total aggregate of genetic material in an organism forms the genome, so the total sum of proteins encoded is known as the **proteome**. Much can be learnt about the likely functions of the proteome by comparing the predicted sequences of the proteins with those already studied, whether of human origin or from model organisms. The molecular function of a protein can be discerned if its sequence or parts of its sequence are similar to proteins whose function is known. On this basis, 58% of the proteins in the human proteome can be assigned a likely molecular function, a fraction that is similar for the yeast, fly and worm proteomes as well. Figure 4.12 summarises these functions.

The human proteome can be compared to the proteome of other organisms whose genomes have been sequenced. About 74% of human proteins are recognisably similar to proteins from other organisms. Conversely, 61% of the fly proteome, 43% of the worm proteome and 46% of the yeast proteome show sequence similarity to proteins in the human proteome.

Similarity in the sequence of a protein in two different organisms arises because the encoding gene can be traced by descent to the common ancestor of the two organisms. Such proteins are said to be **orthologs**. An organism may have two or more genes that encode similar proteins because of gene duplication. Such proteins are said to be **paralogs**. The distinction between orthologs and paralogs is important because orthologs are likely to carry out similar functions in the two different organisms, whereas paralogs can diverge to acquire different functions. Often, pairs of proteins showing sequence similarity are referred to as **homologs**. This term also implies descent from a common ancestor, but it is normally used in a way that does not attempt to distinguish between orthologs and paralogs. If the paralogs had already arisen in the common ancestor of two organisms, then the proteomes of both may contain the descendants of the paralogs. It can thus be difficult to distinguish between orthologs and paralogs. If the two proteomes each only contain one copy of a pair of homologous proteins, then they are likely to be orthologs playing a similar role in each organism. Orthologs may also be recognised if a pair of proteins from the two different organisms are more similar to each other than to other paralogs in their respective proteomes.

When the yeast, fly, worm and human proteomes were compared, the IHGSC identified a set of 1308 proteins where there is at least one ortholog in each species. This set included cases where paralogs were present in one or more of the proteomes. When these were excluded, 564 proteins contained only one ortholog and no paralogs. This represents an evolutionarily conserved set of core proteins that are necessary for the operation of a eukaryotic cell. A function could be assigned to 305 of these proteins. They mediate housekeeping functions in the cell such as metabolism, DNA replication, intracellular signalling, protein folding, membrane transport and the cytoskeleton. If the comparison is limited to multicellular animals – worm, fly and humans – a set of 1195 proteins can be identified where there is only one ortholog and no paralogs in each. The extra proteins are those that are required to organise cells into a multicellular organism, and include proteins required for the complex intercellular signalling necessary for metazoans such as *src*-like tyrosine kinases.

A similar exercise by Celera identified a set of 1523 proteins that are each present in the human, fly and worm genomes and can be confidently defined as orthologs (this figure is higher than the one given above because

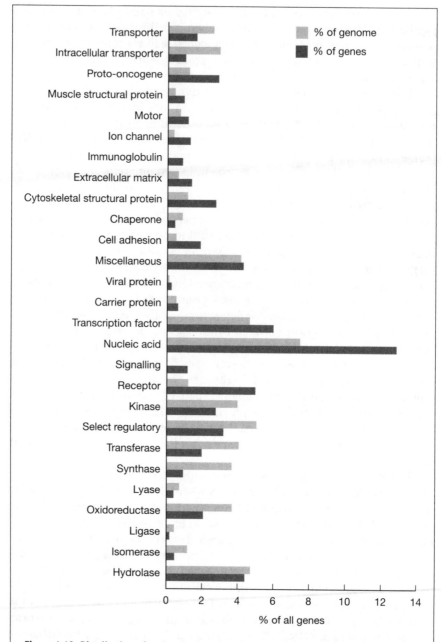

Figure 4.12 Distribution of molecular function in known and predicted human genes and putative metazoan orthologs. Pale bars: the likely function of either known human genes or predicted genes where supporting evidence for the prediction was strong. Select regulatory refers to: (i) proteins involved in signal transduction such as heterotrimeric G-proteins; (ii) cell cycle regulators; (iii) proteins that modify kinases, phosphatases and G-proteins. Dark bars: function of orthologs that are present in worm, fly and human genomes. (Data re-plotted from Venter, J.C. *et al.* (2001) *Science*, **291**, 1304–1351. Reprinted with permission from AAAS.)

it includes cases when paralogs are also present). The distribution of the functions of these conserved proteins is shown in Figure 4.12. When compared to the composition of the human proteome it can be seen that the following classes are over-represented in the conserved set:

- Enzymes for nucleic synthesis and metabolism.

- Enzymes involved in intermediary metabolism (transferases, oxidoreductases, ligases, lyases and isomerases).

- Special regulatory molecules such as cell cycle regulators, heterotrimeric G-proteins and protein kinases.

- Protein transport and chaperones.

Analyses such as these do not identify all the conserved functions, because others will be mediated by members of protein families where the distinction between paralogs and orthologs can not be clearly defined. Nevertheless, they offer us a glimpse of those functions conserved in evolution from the common ancestor of eukaryotes.

So far we have considered the functions of the human proteome that are conserved in the evolution of metazoans and eukaryotes in general. It is of equal interest to consider the nature of the innovations that marked the evolution of vertebrates, mammals and ultimately humans. About 7% of the human proteins and protein domains are found only in vertebrates. As might be expected, these are mainly made up of proteins involved in immune defence and the nervous system. So there has been relatively little innovation in the evolution of proteins with totally novel activities.

What has occurred is a great increase in the complexity of protein structure. Proteins have a modular structure in which domains with different activities are combined in different combinations to give a protein its final function. The human proteome contains more types of protein architecture, i.e. the linear order of domains in a protein, than the proteomes of the other eukaryotes. For example, a domain called the trypsin-like serine protease domain occurs with 18 other domain types in the human proteome to form proteins found in the complement fixation system, blood coagulation and fibrinolytic (clot-dissolving) enzymes. This domain occurs with only eight other domains in the fly proteome, five in worm and one in yeast.

As well as an increase in protein complexity, there has been a selective expansion in protein families and domains associated with vertebrate specialisations, particularly with respect to immune function, intercellular signalling, olfactory receptors, haemostasis, apoptosis, neural function, splicing and translation. Some examples are given in Table 4.3. Expansion of protein families is not limited to human evolution; selective expansion can be seen in the proteomes of the worm, fly and yeast genomes. Often there is evidence of independent expansion of the same family in the evolution of different groups of eukaryotes.

Table 4.3 Examples of expansions in protein families or domains in humans compared with other organisms. (Data from Venter, J.C. *et al.* (2001) *Science*, **291**, 1304–1351.)

Function	Example	Number of proteins			
		Human	Fly	Worm	Yeast
Intercellular signalling	Fibroblast growth factor	24	1	1	0
Immune defence	Immunoglobulin domain	765	140	64	0
Haemostasis	Matrix metalloprotease	19	2	7	0
Apoptosis	Calpain	22	4	11	1
Neural function	Voltage-gated K$^+$ alpha	33	5	11	0
Transcription factor	Zinc finger containing	607	232	78	28
	Homeotic	168	104	74	4

Therefore, although there are only 1.1–1.6 times more genes in the human genome compared to the fly and worm genomes, the biological complexity that can be programmed by this genome is greatly increased by novel combinations of protein domains and selective expansion of protein families. Moreover, about 40% of human genes show alternative splicing, which gives these genes the capability of producing a number of different proteins. Finally, complexity can arise from patterns of gene regulation both at the transcriptional and post-transcriptional level, chemical modifications of proteins after translation and regulation of protein stability. The expansion of proteins involved in transcription, splicing and translation will all contribute to an increase in the overall capacity for regulating gene expression.

4.10 Genomics

The determination of the human genome sequence will have a profound effect on the future direction of both fundamental and biomedical research. Knowing the complete sequence of the human genome, and the genome of other representative organisms, has led to the emergence of a new science called **genomics**. Genomics allows the properties of the genome to be considered holistically instead of one gene at a time. We have already seen how we can define the human proteome, which represents a complete list of proteins present in the human organism. The human proteome can be compared to that of other eukaryotes to define the minimum protein set needed to build a eukaryotic cell or a metazoan animal and to investigate the innovations that arose in the evolution of vertebrates, mammals and humans. As well as the proteome, we can define the human **transcriptome**, i.e. the sum total of all the RNA transcripts. We can investigate the range of transcripts produced in each tissue and how that varies during development or in response to environmental changes. We can also investigate the range of differently spliced transcripts to define the different proteins that can be expressed by each gene. The function of each gene is investigated by **functional genomics**. Although many types of investigation are precluded using humans, 74% of human genes have a homolog in a model organism, where sophisticated molecular genetics can be used to investigate their function. Another way to investigate function is **structural genomics**, which aims to elucidate the three-dimensional structure of each protein. Proteins with no discernible sequence similarity may nevertheless have a similar structure and therefore function. So structure determination may help reveal the function of the 40% of genes whose sequence gives no clue as to their function. Determination of structure is carried out by X-ray crystallography, or for small proteins (<20 kDa) by NMR. Both methods require milligram quantities of purified protein. Expression of human genes in heterologous hosts such as bacteria, yeast or insect tissue will provide recombinant proteins for these studies. It is hoped that it may not be necessary to determine the structure of every protein. As the database of determined structures expands, algorithms to convert primary sequence into secondary and tertiary structure may become efficient enough to allow reliable predictions of protein structure to be made from the primary sequence. Although long sought after by protein chemists, such algorithms are presently, at best, imperfect.

4.10.1 Oligonucleotide arrays

The range of technologies that involve hybridising DNA or RNA samples to a large number of oligonucleotides arrayed on a solid support is collectively known as oligonucleotide microarray technology. There are a large number of different technology platforms that fall into this category, but the general principle is the same in each case. Oligonucleotide microarrays contain a large number of oligonucleotides arrayed in a known order. Each oligonucleotide is known as a cell or probe. The array is hybridised to a fluorescently labelled DNA or RNA sample, and automated laser confocal microscopy is then used to determine in each cell of the array whether hybridisation has occurred and to record the result.

Most microarrays are one of two types: those where each cell is formed by spotting a very small amount of oligonucleotide solution on a glass slide using similar technology as used in inkjet printers (known as spotted or printed arrays), and those where the oligonucleotides are synthesised on the array. One example of such technology is that used to fabricate the popular Affymetrix GeneChip® arrays. Here photolithography techniques used in the construction of computer chips in conjunction with solid-phase oligonucleotide synthesis can be used to build very precise regular arrays of oligonucleotides in a very small surface area (Figure 4.13). More than 10^6 different oligonucleotides can be arrayed on a 1.28 cm × 1.28 cm glass surface in regular square cells as small as 5 μm. For this reason, microarrays are sometimes known as DNA chips.

The fabrication of spotted arrays is generally less reproducible than for arrays where the oligonucleotides are synthesised on the array. In order to

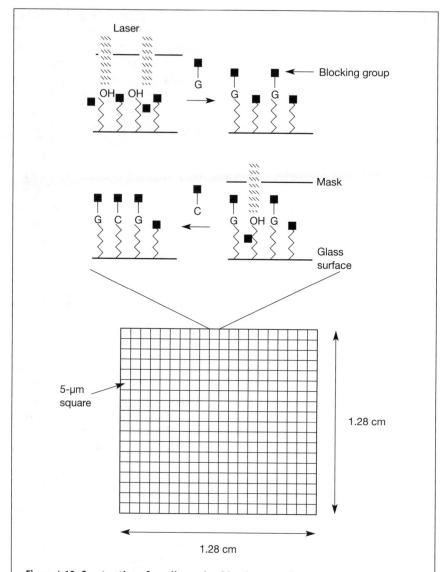

Figure 4.13 Construction of an oligonucleotide microarray (DNA chip). Solid-state oligonucleotide synthesis is carried out in cycles. In each cycle a light is used to remove a photolabile blocking group from a linker molecule attached to the glass substrate. A mask is used to direct the light to the selected cells and a protected nucleotide is coupled to the unprotected linker. The mask is then moved to initiate the next cycle where a different oligonucleotide will be coupled.

account for this, different coloured fluorophores are used to label control and experimental samples, the samples are competitively hybridised to the array, and the relative hybridisation of control and experimental samples is measured – thus array-to-array variation is less important. Hence spotted arrays are also referred to as two-colour arrays. Because arrays where the oligonucleotides are synthesised on the array are more reproducible, there is no need to have an internal control with a different colour. In these so-called one-colour arrays, only one sample is hybridised to the array, so the array measures the absolute hybridisation to each probe.

Spotted arrays are generally printed by the individuals carrying out the experiment, or in a centralised facility in the same institute. They are cheap and allow a high degree of flexibility and customisation as a design can be changed simply by acquiring new oligonucleotides. Other types of arrays require very specialised equipment and are fabricated by specialised companies to a small number of standard designs. Creating a new design is a time-consuming and expensive process.

Microarrays allow a very large number of hybridisations to be carried out in parallel. In the following sections we shall see how this technique can be applied to investigate patterns of gene expression in different tissues, determine which parts of the genome are expressed, allow large-scale population surveys of sequence variation in SNPs, and discover regions which show copy number variation.

4.10.1.1 Expression profiling

Oligonucleotide arrays can be used to investigate the pattern of gene expression. Such expression profiling is probably the most common use of microarray technology. Arrays with oligonucleotides representing each gene (or even each exon) are hybridised to fluorescently labelled cDNA prepared from different tissues. If hybridisation is stronger for one sample, the gene is said to be differentially expressed in that sample. This sort of expression allows comparison of which genes are switched on in two tissues, in diseased versus normal tissues or at different times in development. Figure 4.14 illustrates the use of a two-colour array to compare gene expression profiles in two different tissues. A typical example of a one-colour array measured the expression of known and predicted human genes in 79 different tissues. Knowledge of which tissue a gene is expressed in helps to suggest the function of the gene. In addition, genes that are expressed together are likely to have a similar function. Figure 4.15 shows an example of coordinated expression in brain tissue of a group of three genes in adjacent locations on chromosome 13. This observation suggests that there may be a previously unrecognised **locus control region (LCR)** that promotes the expression of multiple genes in a single chromosomal region. Moreover, it strongly suggests that products of the three genes are involved in the development and function of the brain.

4.10.1.2 Identifying transcribed regions of the genome

Microarrays can also be used to provide a map of the genomic regions that are transcribed. A tiling array is synthesised consisting of a series of probes that are evenly spaced across the genome or a part of the genome. RNA is isolated from cells and converted to cDNA using reverse transcriptase. The use of random primers provides a non-biased representation of all the

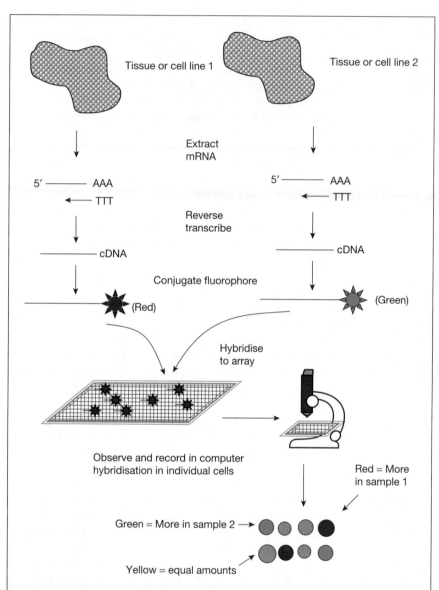

Figure 4.14 Hybridisation of fluorescently labelled samples to a two-colour array. cDNAs are prepared by RT-PCR from mRNA from two different samples. They are conjugated to two different fluorophores that are excited at particular wavelengths by lasers and emit or fluoresce at a different wavelength, resulting in a characteristic colour. Commonly used fluorophores are cy3 (fluoresces red) and cy5 (fluoresces green). Because of the limitations of colour availability, blue is used in this diagram as a representation of red fluorescence and grey as a representation of green flourescence. The labelled sample cDNAs are competitively hybridised to the array and then examined with a laser scanning confocal microscope. This illuminates the array with laser light filtered to the excitation wavelength of the fluorophore. The emitted light in each cell is then recorded and can be quantified. Where there is more of a particular sequence in sample 1, the corresponding spot on the array will appear red; if there is more of sample 2, the spot will appear green. If there are equal amounts in the two samples, then the sport will appear yellow. Genomic DNA amplified by PCR can also be used as a sample in a similar way.

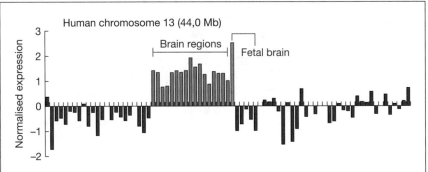

Figure 4.15 Coordinated expression in brain tissue of a group of adjacent genes on chromosome 13. The average expression (*y*-axis) of three adjacent genes on chromosome 13 was determined in 79 different tissues (*x*-axis). The genes are coordinately upregulated in brain tissues and a sample from fetal brain. (Redrawn from Su, A.I. *et al.* (2004) *Proceedings of the National Academy of Sciences (USA)*, **101**, 6062–6067.)

RNA sequences present, whereas the use of oligo-T primers enriches for poly-A+ mRNA. The term cDNA is normally used to refer to DNA generated from mRNA in this manner but, strictly speaking, cDNA or copy DNA refers to a DNA molecule that has been generated using any RNA molecule as a template. cDNA molecules that originate from mRNA detect protein-encoding sequences in the genome and, as discussed above, are the best criteria for detecting protein-encoding genes and defining gene structure. However, such samples will not detect genomic regions that are transcribed but not included in mRNA molecules, either because the sequence is removed by splicing from the primary transcript, or because the transcript has some other function. To recognise all transcripts in an unbiased fashion, cDNA originating from total RNA must be hybridised to the tiling array.

The resolution of the information is determined by the distance between the probes. When the probes are close or even overlapping, then the boundaries of the expressed sequence can be determined with a high degree of accuracy. High-resolution tiling arrays hybridised to cDNA probes can scan a sequence to detect the start and stop of exons (Figure 4.16). A large space between probes allows a larger section of the genome to be covered. The complete human genome can now be covered in a set of seven tiling arrays with 10-bp gaps between probes. Hybridisation of cRNA to these arrays reports which parts of the whole genome are transcribed. Tiling arrays have revealed that a far higher portion of the genome is transcribed at a very low level than had been previously thought (see Section 4.11).

4.10.1.3 Detecting sequence variation

The reference sequence of the human genome was derived from a panel of different individuals. As the consensus sequence was assembled, differences between the nucleotide sequences of different individuals became apparent. If a difference is common in a population, it is defined as a polymorphism.

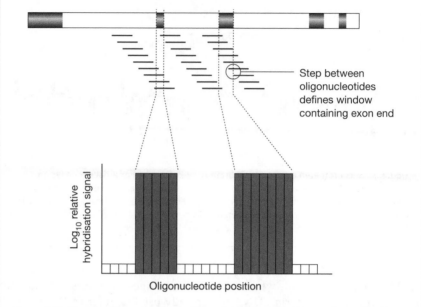

Figure 4.16 **Use of a tiling array to identify and define exons.** 60-mer oligonucleotides are synthesised so that the start of each oligonucleotide is staggered by 10 bp, forming a tiling array that spans a genomic region. Labelled cDNA is hybridised to the array. Only oligonucleotides that partially or completely overlap an exon (shown in blue) will hybridise. The beginning and end of the exons will be defined to within the 10-bp step within each nucleotide. The exact borders of the exon will be evident from the location of the GT/AG rule.

Thus, during the sequencing of the human genome single nucleotide polymorphisms (SNPs) were identified. As we shall see in Chapter 6, a major proportion of population variation in susceptibility to common diseases is due to genetic variation, of which SNPs are expected to form a large part. There was thus great interest in identifying and mapping SNPs to see which were associated with these common diseases. After the draft genome sequence was completed, sections of genomes were re-sequenced in different individuals drawn from a variety of different populations to systematically identify SNPs, and a database of common SNPs was established. Once the SNPs have been identified, there is a need for technology that can efficiently and cheaply genotype SNPs in a large number of experimental subjects at a large number of loci to see which are associated with common diseases. Oligonucleotide arrays are ideally suited to this purpose.

A number of commercial companies have developed platforms for the genotyping up to 1 000 000 SNPs on a single chip. One approach is based on hybridisation to oligonucleotide arrays after multiplex PCR has been used to amplify the whole genome using primer pairs and conditions that amplify long tracks of DNA (Figure 4.17). The amplified DNA is labelled and hybridised to arrays which contain cells complementary to the alleles

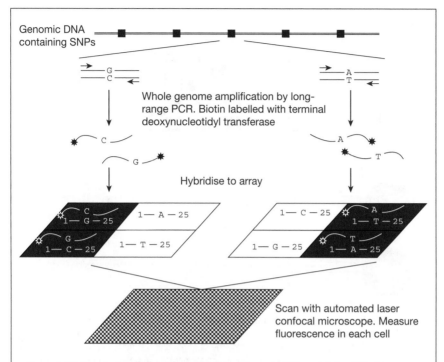

Figure 4.17 Genotyping SNPs with a microarray. Genomic DNA containing SNPs is amplified by PCR using biotinylated nucleotides. The diagram shows a set of four cells on the microarray chip that contain 25-residue oligonucleotides that anneal to each of the two DNA strands of the two alleles of an SNP. Hybridising probes are visualised by the use of streptavidin-Cy3, which binds tightly to the biotinylated nucleotides and fluoresces when excited with lasers of the appropriate wavelength. If the source genome is homozygous for one or other of the alleles, then only two cells fluoresce as appropriate. If the genome is heterozygous, all four cells fluoresce, but with only 50% of the intensity of the homozygous alleles. In practice, the arrangement of the cells is more complex, with controls providing the background fluorescence from the hybridisation of probes with a single mismatch to the genomic DNA. Details of the microarray design are described in Hinds, D.A. *et al.* (2004) *American Journal of Human Genetics*, **74**, 317–325.

found at each SNP site. An alternative approach is the bead chip technology developed by a biotechnology company called Illumina, in which each cell on a chip contains beads to which oligonucleotides corresponding to a particular SNP have been attached. Genomic DNA spanning SNP sites is amplified as before and hybridised to the beads (Figure 4.18). If a match is found, the genomic DNA hybridises to the oligonucleotide, which can then serve as a primer for DNA synthesis which incorporates fluorescent nucleotides. These technologies formed the basis of the HapMap project that mapped the distribution of SNP alleles across the genome in different human populations, and formed the basis for population surveys to identify alleles which are associated with complex disease. We shall discuss how this was done in detail in Chapter 6.

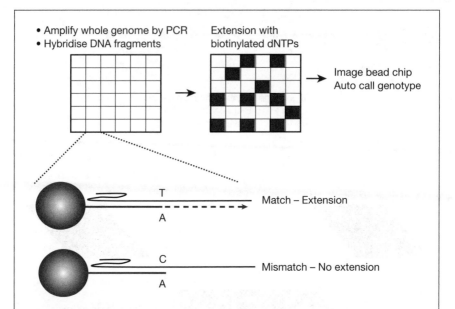

Figure 4.18 The Illumina bead chip for genotyping SNPs. Whole genomic DNA is amplified by PCR and hybridised to the bead chip. This consists of an array of cells, each of which contains beads attached to oligonucleotides. The oligonucleotides match sequence preceding the SNP, with the last residue being complementary to one or other of the bases in the SNP. Genomic DNA is amplified by PCR and hybridised to the chip. The reagents for DNA synthesis, containing biotinylated nucleotides, are added to the chip. If the genomic DNA is a match to the SNP, the oligonucleotide acts as a primer for DNA synthesis which incorporates the biotinylated nucleotides, providing the basis for their detection by fluorescence as described in Figure 4.17.

4.10.1.4 Copy number and structural variants

As well as sequence variation at the level of single nucleotides, variation in the human genome can occur on a larger scale consisting of deletions and insertions ranging from kilobases to megabases in size. Such variation results in a change in copy number, so this type of change is called a **copy number variant (CNV)**.

The completion of the human genome sequence provides the means to systematically map copy number variation in human populations. Two general approaches have been taken (Figure 4.19). In one, called comparative genome hybridisation (CGH), total genomic DNA isolated from individuals from diverse populations was hybridised either to a set of BAC clones representing a whole genome tile path, or to a 500-kb oligonucleotide array that had been prepared to investigate SNP haplotypes. In each case the intensity of the hybridisation signal from each clone or oligonucleotide was compared to the signal from the hybridisation of a reference genome. Changes in copy number in the test genome compared to the reference genome show up as a deviation from a ratio of 1.0; heterozygous autosomal

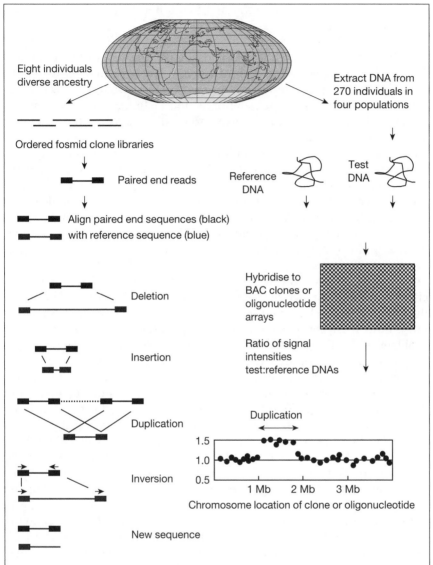

Figure 4.19 Identifying copy number and structural variants. Two approaches are illustrated, both comparing multiple genomes from diverse populations. On the left, fosmid clone libraries were prepared from eight individuals, the ends of each clone were sequenced (black boxes), and the paired end reads were used to order the clones by overlapping the ends. The paired end reads of each clone were then aligned with the reference sequence (blue). The size of the inserts in the fosmid libraries is tightly controlled, so the distance between the paired end reads in the reference sequence can be accurately predicted. Anomalies are revealed by discrepancies between the predicted and observed locations as illustrated. On the right, cell lines were established from a geographically diverse set of test subjects. After checking the karyotype was normal, DNA was extracted and labelled with a fluorescent dye (cy3 or cy5) and reference DNA labelled with a different colour from the test DNA (all hybridisations were repeated with the dyes swapped round). The labelled test and reference DNAs were then competitively hybridised either to an array of BAC clones consisting of the minimum tiling path covering the whole genome or to an oligonucleotide array designed to identify with SNPs across the whole genome. Deviations from a ratio of 1.0 between test and reference DNA indicate a copy number change.

duplications will result in a ratio of 1.5 and heterozygous deletions will result in a ratio of 0.5.

The second approach, called end sequence pair (ESP) mapping, is based on fosmid clone libraries. Fosmids are cloning vectors that use the F-factor origin of replication, which results in a low copy number within the *E. coli* host. This promotes a high degree of insert stability, preserving the sequence topology from the donor genome. Importantly, the size of the insert in the fosmid clone is fixed, so all the inserts are the same size. The clones are sequenced at either end to produce paired end reads which allow the clones to be ordered and organised into a minimum tiling path. The paired end reads are then aligned with the reference human genomic sequence. Duplications, deletions, inversions and more complex structural changes in the test DNA are revealed as anomalies in the alignment (Figure 4.19). The advantage of this approach is that it gives more detailed information about structural changes than just copy number. For example, an inversion will not affect the copy number and so would not be revealed by CGH analysis.

The results from the BAC array CGH suggested that the regions affected by copy number changes were much larger than those detected by the ESP approach. Hybridisation to the BAC arrays gives a higher signal-to-noise ratio than hybridisation to oligonucleotide arrays. However, this problem with oligonucleotides can be overcome by constructing high-density arrays containing over 10^6 probes so that multiple oligonucleotides cover each region, which allows the consensus result from multiple oligonucleotide probes to be used to decide whether the copy number of the region is departing from two for autosomes and one for the X chromosomes in men. When CGH was carried out with these high-density arrays it was found that the size of affected regions was much smaller than those detected with the BAC clones, and these high-density arrays gave similar results to the ESP approach. The high-density array was used to examine the genomes of 270 individuals used in the HapMap project (see Chapter 6). It was estimated that less than 5% of the genome is affected by CNVs larger than 100 kb and that 90% of the CNVs were relatively common (population frequency >1%). It will be important to see if these CNVs are associated with a predisposition to common diseases. This issue will be discussed in Chapter 6.

4.11 The ENCODE Project

We saw above that only 1.2% of the human genome encodes proteins. What does the rest do? Does some or all of it have a function, or is it junk, as one scientist memorably called it? We would certainly expect some of it to have a function in promoting and controlling the expression of the protein-coding sequences. We know that there are many regions that are transcribed but not translated. Some of these, the RNA genes, we know have a function. Perhaps our knowledge of the types and functions of

RNA-coding genes is incomplete – it is easier to spot protein-coding regions than RNA genes that have a novel function. We would also expect parts of the genome to have a structural role in aiding chromosome replication, stability and orderly segregation to daughter cells in mitosis and meiosis. To address these issues, an international consortium has initiated a project to construct an ENCyclopaedia Of DNA Elements (ENCODE). As its title suggests, the aim of the ENCODE project is to construct a catalogue of all the structural and functional elements of the genome. To do this, the ENCODE consortiums collaborates on a variety of complementary strategies that will identify and map the following:

- All the transcribed regions of the genome, including transcripts that do not appear in cDNA derived from mRNA.

- All regulatory sequences, such as enhancers, silencers, promoter binding sites, etc.

- Regions showing evolutionary conservation assayed by:
  - Sequence conservation in a variety of non-human vertebrates, including sequencing of ENCODE targets in 10 species in addition to completed sequencing projects or those in progress.
  - Sequence variation within human populations by re-sequencing the ENCODE targets in 48 individuals from the HapMap project.

- Origins of replication.

- Sites of DNA methylation.

- DNase-1-sensitive regions, which indicate where chromatin has been opened up to allow transcription.

- Sites of histone modification.

The project is planned to be carried out in three phases. In the first pilot phase, already completed, 44 regions totalling 30 Mb, amounting to 1% of the total genome, were examined. About half the target regions were selected because they had already been intensively studied (such as the globin genes and the CFTR region). The other half of the target regions were chosen to provide a random selection of different representative types of genomic regions. Investigations into which regions of the genome were transcribed into RNA were designed to identify transcripts arising from non-coding regions as well as coding regions, represented in cDNA libraries. To this end tiling microarrays were used that span the whole genome, not just protein-encoding regions. These were hybridised to cDNA as well as to total RNA. In addition, a special form of PCR called RACE was used to detect the 5′ end of primary transcripts to independently verify the results of the tiling array experiments. This illustrates a general principle of the project that, as far as possible, a variety of different experimental approaches will be adopted to address each of the targets listed above. This will allow the efficiency of the different methods to be evaluated.

Furthermore, consistency between experimentally independent methods gives confidence in the reliability of the data generated. The second and third phases of the project, called the development and production phases respectively, will expand the project to the rest of the genome in progressively greater detail, using the experience gained from the pilot stage.

The results of the pilot stage show that the genome is more complex than a simple idea of isolated regions of protein-coding sequence, which so far we have tended to equate with 'genes' surrounded by large tracts of non-functional sequence. A few of the many interesting conclusions are listed below:

- The majority of the genome is transcribed.

- Transcripts of coding regions can be initiated a considerable distance a way from the coding region. Sometimes these are even initiated upstream of the start of an adjacent gene (Figure 4.20). Such transcripts can initiate

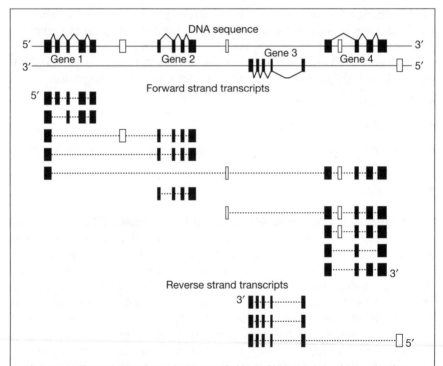

Figure 4.20 Transcriptional complexity revealed by ENCODE. *Top* Genomic map showing the structure of four genes; exons are represented by black boxes. *Below* Transcripts detected by ENCODE; spliced regions are shown as dotted lines. In addition to the expected transcripts which just span the exons, novel transcripts have been detected which span unexpectedly extensive regions of the genome and can start upstream of adjacent genes. These novel transcripts are known as TARs (transcriptionally active regions) and are represented in the figure by open boxes. (Reproduced from Gerstein, M.B. *et al.* (2007) What is a gene, post-ENCODE. History and updated definition. *Genome Research,* **17**; 669–681.)

at previously mapped enhancer sites that were thought to be located away from the transcription start site.

137

Next-generation sequencing technologies

- On average there are about 5.4 different transcripts per coding region – this is a higher figure than previously estimated.

- Sixty-three per cent of transcripts do not encode proteins, 41% of all transcripts span intronic regions and 23% span intergenic regions.

- Transcription start sites are associated with regulatory features such as transcription factor binding sites and sites of chromatin remodelling. However, some regulatory sites are not associated with a transcript.

- About 5% of the genome shows sequence conservation. About 60% of the conserved sequence is associated with regions identified as functional – for example, they align with protein-coding regions or annotated regulatory sequences.

- However, many apparently functional regions are not conserved.

These observations challenge us to reconsider our view of genome function and raise many interesting questions. Why is so much of the genome transcribed if it does not encode protein – is this transcription background noise or does it have a function? The fact that most transcripts start at sites that have the hallmarks of regulatory regions suggest that it is not background noise. What is the role of the 23% of transcripts that span intergenic regions? Interestingly, some SNPs associated with complex disease have been mapped to intergenic regions (see Chapter 6), so perhaps these RNAs have an important functional role. Why do some transcripts map so far from the start of the coding sequence? Is our definition of the gene, based on the mRNA transcript, an oversimplification?

4.12 Next-generation sequencing technologies

For the past 30 years, the vast majority of DNA sequencing, including that for the Human Genome Project, was performed using various modifications of the Sanger method, first introduced in 1975. Sequencing chemistries based on the latest version of this technology allow the generation of pieces of sequence (or **reads**) of up to 800 bp in one reaction. The latest sequencing instruments allow up to 384 such reactions to be carried out in one **run**, and for several such runs to be carried out one after the other without any intervention from the scientist. Despite the vast improvement this represents from the older iterations of the technology, producing large amounts of sequence is still expensive and time-consuming, with the latest human genome to be sequenced in this way, that of the genome entrepreneur Craig Venter, reportedly costing nearly $10 million and taking years to complete. However, a new generation of sequencing technologies are vastly reducing the cost and time per base of sequencing. The principle behind some of these technologies varies (see Box 4.2), but they share two important ideas.

BOX 4.2: NEXT-GENERATION SEQUENCING TECHNOLOGIES

Roche-454 sequencing

The genomic DNA to be sequenced is fragmented, linkers are ligated to each end of the resulting fragments and the fragments are attached to a bead through a biotin molecule. Treatment with NaOH generates a library of single-stranded molecules with primers in a fixed orientation. These single-stranded molecules are diluted so that

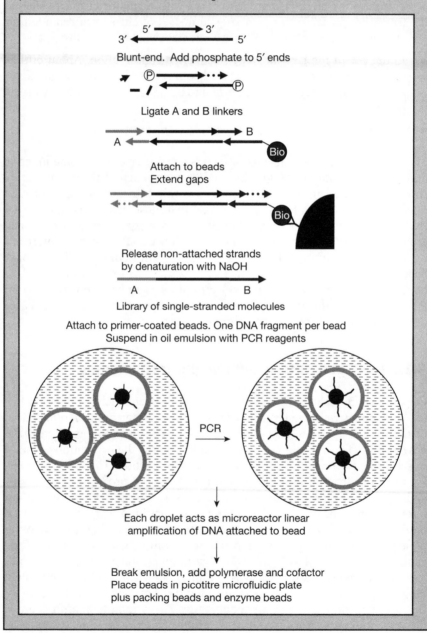

when attached to primer-coated beads there is only one molecule per bead. The beads are suspended in an oil emulsion so that each bead is surrounded by an aqueous layer containing PCR reagents. Each droplet acts as a very small bioreactor, allowing the single-stranded molecules to be linearly amplified, and the products attach to the bead. In this way the bead becomes coated with copies of the original template. The beads are then suspended in a microfluidics plate and, in turn, the four deoxynucleotides are flowed over the plate. As well as the beads containing the amplified template, each well contains enzyme beads containing DNA polymerase and an enzyme mixture that generates a flash of light if a deoxynucleotide is incorporated. If there is a run of bases in the template (i.e. a homopolymer), then more than one base will be incorporated when the complementary nucleotide is flowed over the well. This is recognised by the amplitude of the light signal.

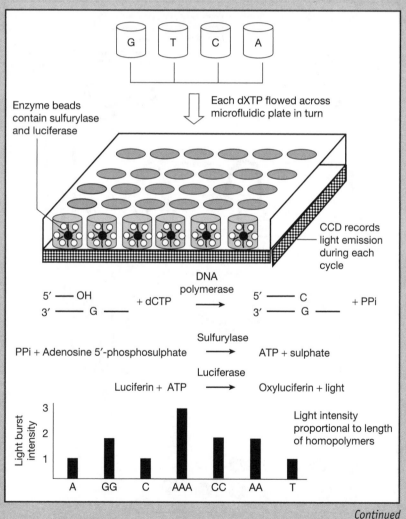

Continued ▶

Illumina reversible terminators

The process again starts with fragmented DNA to which adapters are ligated. These are attached to the inside channels of a flow cell to which primers have also been attached. The free end of the DNA fragment can form a bridge by annealing to a complementary primer, which then serves to prime PCR. Cycles of denaturation and synthesis produce a cluster of copies. The sequencing phase comes after the amplification phase. Primers and reagents for DNA synthesis and the four nucleotides are flowed over the cell. Each nucleotide is labelled with a different colour fluorophore and the 3'OH group is chemically blocked, so that only one nucleotide can be added in each cycle. Imaging of the cell reveals which nucleotide has been added in each cluster. The key to the process is that the 3'OH blocking is reversible, which allows the cycles to be repeated, building up the DNA sequence at each cluster.

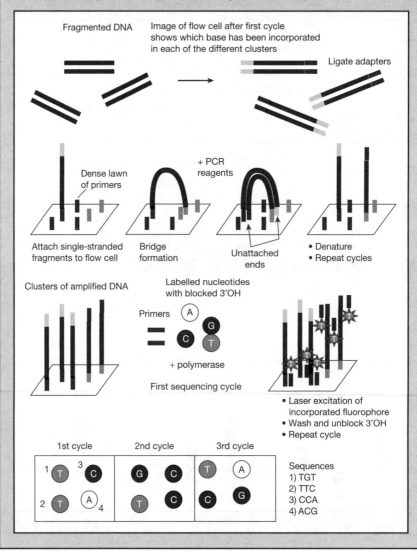

ABI ligation-based Solid™ system

The target DNA is fragmented, ligated to adaptors and amplified by emulsion PCR as in the Roche-454 sequencing. The beads are attached to a glass slide. Dinucleotides are flowed over the slide together with ligase and a universal primer that recognises the adaptors which were ligated to the fragments. The DNA fragments act as a template so that when a dinucleotide is complementary, ligase will ligate it to the universal primer. Each dinucleotide is labelled with a different colour fluorophore through an extension which is degenerate and will bind to any sequence on the target. The colour of the fluorophore indicates which dinucleotide has been incorporated. The dinucleotide is cleaved off and the cycle repeated up to seven times. As shown in the diagram, the process results in the dinucleotides which are read being separated by dinucleotides which have not been determined. To read these nucleotides, the whole process is repeated using a different primer that binds at N-1 compared to the first primer. The process is repeated until each nucleotide has been interrogated twice.

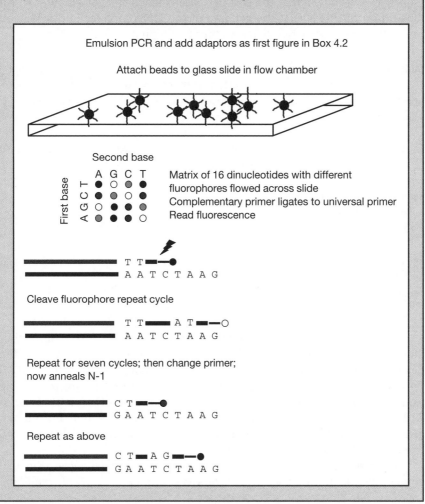

First, they are based on fragmenting the target DNA and amplifying individual fragment molecules to produce clusters consisting of copies of the fragment on a solid support or a bead. Second, instead of producing a small number of long reads in one run, a very large number of shorter reads are produced in each run. For instance, a single run of sequencing based on technology developed by Solexa (known as sequencing-by-synthesis) can produce millions of 36-bp reads (or millions of 36-bp read pairs using 'paired end' sequencing; see Box 4.2); using technology based on that developed by 454 Life-Sciences (known as pyrosequencing), 400 000 reads (or read pairs) of 200 bp each can be produced in one go. For this reason, these technologies are collectively known as massively parallel sequencing technology. Since the sequencing protocol for these technologies generally does not include a cloning stage, they eliminate the problems in traditional approaches associated with the clonability of sequences and the stability of clones. While a single run using this technology is more expensive than a single run on an instrument based on Sanger technology, such a vast amount of sequence is produced that the cost per base is much smaller (Table 4.4). An illustration of this is that the genome of the DNA pioneer James Watson was sequenced by 454 Life-Sciences for $2 million in just 2 months.

Table 4.4 Performance of different sequencing technologies. Figures show the amount of sequence generated in a single run and the cost per Mb of generating that sequence from three massively parallel sequencing technologies, as well as traditional Sanger sequencing.

Technology	Average read length (bp)	Average number of reads	Total amount of sequence	Cost per Mb
Sequencing-by-synthesis (Solexa/Illumina)	36	35 000 000	1.3 Gb	$6
Pyrosequencing (454/Roche)	200–300	400 000	100 Mb	$85
Sequencing by ligation (ABI SOLiD)	35	85 000 000	3 Gb	$6
Sanger method	800	384	300 kb	$600

Several other technologies are being developed that will further increase the speed and reduce the cost of sequencing. Competition in the development of sequencing technologies is being stimulated by the X PRIZE Foundation, which is offering $10 million to the first group to be able to demonstrate that they can sequence 100 human genomes in 10 days for $10 000 per genome. The final goal is to be able to sequence a human genome for $1000 in 1 day. At this cost level it would be possible for everyone to have their genome sequenced.

These new sequencing technologies also pose new challenges. Principal among these is the short lengths of the reads produced. A set of traditional

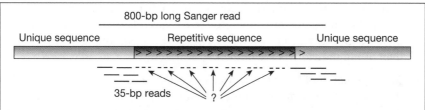

Figure 4.21 The length of a DNA sequence read determines the length of a repetitive DNA region that can be sequenced. The figure shows a set of short sequence reads aligned to a sequence that contains a repetitive region. The aligned reads have matches to some unique sequence. However, the sequence highlighted in blue could map anywhere within the repetitive region. Reads that are longer than the repeat, such as the 800-bp read, can match unique sequence on either end of the repetitive region, allowing the position in the genome of the repetitive sequence to be determined.

Sanger method sequence reads can be assembled into a longer sequence by looking for overlap between the sequences. Since the reads from these new sequencing technologies are so much shorter, the overlap between adjacent reads is shorter, making it difficult to decide unambiguously which reads belong together. The shorter the read, the greater the chance that the sequence will be present in more than one place, and so it will be difficult or impossible to position a read correctly. The length of a repetitive region that can be sequenced is directly related to the length of a read (Figure 4.21). Using paired end reads can help here as it can in using traditional sequencing techniques (see Figure 4.3). For this reason the use of these technologies for determining completely novel sequences is restricted to cases where the sequence is short, such as sequencing pathogen genomes, or determining the sequence of cDNAs. Many pathogen genomes have now been sequenced using these new technologies. When new strains of pathogenic bacteria or viruses arise, they can now be sequenced quickly and cheaply. A big application of the new techniques is determining a sequence that can be compared to a pre-existing or reference sequence (known as re-sequencing). Even in re-sequencing projects, a higher degree of coverage is required than in traditional approaches.

These new technologies are allowing a large range of new sequencing based applications. These include:

- **Re-sequencing of sequenced genomes:** the human genome sequence, as defined by the Human Genome Project, represents just one possible genome. Further, since the DNA came from a range of sources, it represents a patchwork of genomes rather than a genome that actually exists. Cheap and rapid sequencing allows the genomes of individuals to be determined. In turn, this provides comprehensive information on the places in which the genome varies between people, both in terms of SNPs and also structural and copy number variations. The 1000 genome project is a major new project aiming to determine the sequence of 1000 human individual genomes to a 2× coverage (that is, each base is sequenced twice on average). While this is insufficient data to be able to write down the complete sequence of each individual confidently, it should provide sufficient data to discover variants that are found in more than 1% of individuals whereas currently only variants present in more than 5% of individuals are known.

- **Re-sequencing of targeted sections of sequenced genomes:** determining the complete sequence of an individual is still a major undertaking; however, various techniques allow for sequencing to be targeted at particular regions, such as coding exons. In this way, the sequence of all exons in a large number of samples can be determined. These samples might be from individuals with a particular disease or phenotype, such as cancer.

- **Sequencing of genomes closely related to sequenced genomes:** although it is difficult to assemble complete long genomes from the short reads that are produced by new sequencing technologies, if a related genome is available, this can be used as a scaffold to arrange the short reads. For example, this is the case in the sequencing of the mitochondrial genome from fossilized Neanderthal samples; it is now envisaged that the whole Neanderthal genome may be sequenced (see Chapter 10).

- **Determining of expressed sequences (transcriptomics):** traditionally, in order to determine expressed sequences, RNA is isolated and turned into DNA and then cloned into plasmid vectors. These plasmid vectors are then sequenced. New sequencing technologies do not require the cloning stage. The RNA is converted into DNA and then directly sequenced. In this way a far larger number of sequences are determined and so the chance of sequencing a piece of RNA that is only found at very low levels within the sample is much greater. This exhaustive sequencing of all RNA or DNA in a sample is sometimes known as 'deep sequencing'.

- Almost any application of oligonucleotide array technology can be achieved using massively parallel sequencing, with the levels of each sequence being determined by the number of times it is sequenced rather than the level of hybridization to oligonucleotides arrayed on a chip. For example, digital expression profiling is a replacement for expression arrays; the relative abundance of each expressed sequence is determined by the number of times a particular sequence is seen. Another example is a technique known as ChIP-seq (chromatin immunoprecipitation sequencing), which is used to determine the DNA sequences with which DNA binding proteins are associated. In this technique, DNA is isolated and fragmented. The protein in question is immunoprecipitated using an antibody, and the sequence of associated DNA is determined using massively parallel sequencing. For example, ChIP-seq could be used to determine the DNA sequences recognised by a transcription factor. This would reveal all the places in the genome where the transcription factor bound and thus reveal which genes are regulated by that transcription factor.

The important point about these applications is that they do not require the design and manufacture of arrays for each different application.

Over a lifetime the cells of the body will accumulate mutations in their genomes. So long as these mutations are not in the germ cells, they will not be passed on to the next generation. Such mutations are known as **somatic mutations** and generally have little effect. However, if a somatic mutation allows the uncontrolled expansion of the carrying cells, the result is a tumour. Knowledge of the 'normal' human genome sequence allows us to compare the sequence of genes in cancer cells to look for important somatic mutations. Projects such as the Cancer Genome Project and the Cancer Genome Atlas (see Further reading for web links) aim to produce catalogues of the somatic mutations found in cancer cells by using a variety of methods, including re-sequencing exons to look for SNPs and using **comparative genome hybridization** (see Section 4.12) to look for amplified and deleted regions.

Such studies have found that cancer cells harbour a higher than expected number of mutations. One difficulty with interpreting such information is that mutations found in cancer cells can be either driver or passenger mutations. **Driver mutations** are those mutations that are causative – they drive the expansion of cells harbouring them. In contrast, **passenger mutations** are those mutations that were present in the cell which originally acquired the driver mutations and so are present in the population of cancer cells. This is made worse by the fact that cancer cells tend to harbour a larger number of mutations than normal cells, most of which will be totally irrelevant to the cancer phenotype. Distinguishing functionally important changes is difficult. One approach is to compare the number of **synonymous** changes (where the genetic change does not lead to a change in the amino acid sequence of the encoded protein) to the number of **non-synonymous** changes when DNA from a large number of different cancers is sequenced. Driver mutations are expected to be non-synonymous, so a large number of non-synonymous changes indicate the presence of a driver mutation. Using such an approach, the Cancer Genome Project sequenced 510 genes in tumour samples, finding 910 mutations, of which they predicted 66 were driver mutations and that roughly one-third of their samples would contain a driver mutation in one of the genes sequenced.

It may not be possible to identify unambiguously particular mutations that lead to cancer without performing the sort of **genome-wide association studies** that are performed using heritable mutations to define the causes of complex inherited disease. However, studies can implicate biochemical pathways. For example, when the Cancer Genome Atlas studied 143 samples from a particular type of cancer known as glioblastoma, they found that between 78% and 88% of samples contain mutations in each of three well known biochemical pathways (the RB tumour suppressor pathway, the p53 tumour suppressor pathway and the RTK oncogenic pathway). Of these samples, 74% contained changes in all three of these pathways.

New sequencing technologies (see Section 4.12) promise to rapidly increase the scope of such studies, allowing the rapid sequencing of many more genes in many more cancer samples.

4.14 Summary

- When comprehensive maps of the human genome had been constructed, the International Human Genome Sequencing Consortium (IHGSC) started the sequencing phase of the Human Genome Project. The IHGSC strategy was to use the maps to construct and order overlapping series of BAC clones that spanned each chromosome with the minimum tiling path. Once this was done, each BAC clone was sequenced by a shotgun strategy. At the same time, a private company (Celera Genomics) started sequencing the human genome using a fundamentally different strategy of shotgun sequencing the whole genome without first ordering large clones.

- IHGSC and Celera jointly announced the completion of the draft sequence in 2000.

- The draft sequence contained errors and gaps and covered only 90% of the total euchromatic sequence. Closing the gaps and correcting the errors produced the finished sequence, which was completed in 2004. Finishing the sequence took roughly an equal amount of time and resources as it took to produce the draft sequence. The finished sequence has an estimated error rate of less than 1 in 100 000 bases, which is an order of magnitude better than the original target. The complete sequence has only a small number of gaps in regions that have proved particularly difficult to clone and sequence.

- To be useful the sequence must be annotated, which means identifying all the different repetitive and unique elements that make up the genome. Genome browsers allow the sequence and the annotations to be accessed via a graphical interface on the world wide web. Searchable databases contain stored information about the elements in the genome such as RNA transcripts and encoded proteins.

- The complete sequence allows an overall view of genome structure and evolution. This allows:
  ○ The long-range variation of GC content to be determined. Genes are more dense in GC-rich regions.
  ○ The total number of CpG islands to be obtained.
  ○ A complete census of transposable repetitive elements. Most transposable repetitive elements are no longer active in the human genome.
  ○ Identification of duplicated blocks of chromosomes.
  ○ Comparison with the mouse and other vertebrate genomes to identify syntenic regions.

- Protein-encoding exons are difficult to recognise because they are short sequences embedded in much longer stretches of intronic DNA.

- A gene is defined as a group of exons that are spliced together to generate an mRNA molecule. The best way of characterizing gene structures is matching cDNAs generated from mRNA to the genome sequence. Because of the importance of cDNA in characterising genes, the Mammalian Genome Collection was formed as a collaboration between different institutes that aims to generate a comprehensive library of cloned and sequenced full-length human, mouse and rat cDNAs.

- The total number of genes in the human genome is estimated to be about 22 000. This is much lower than the estimates of 100 000 before the genome was sequenced, and lower than the estimates of about 30 000–35 000 from the draft sequence.

- Transcripts generated from genes often undergo alternative splicing events, so for each gene there are on average at least three alternative transcripts, and the actual figure may well be much higher. Each alternative transcript results in a protein with a different structure, greatly increasing the diversity of proteins encoded by the 22 000 genes.

- Conceptual translation allows the amino acid sequence of the encoded proteins to be determined. The total sum of encoded proteins is called the proteome and constitutes a parts list to make a human organism. Comparison with other proteins allows the function of about 60% of the proteins to be guessed.

- Comparison of the yeast, worm, fly and human proteomes allows the evolution of proteins to be studied and identifies a core group of proteins that are present in all eukaryotes and a second group that is common to all metazoans.

- Determination of the nucleotide sequence of entire genomes has established a new science of genomics, which analyses genomes holistically. Functional genomics studies the functions of genes, and structural genomics aims to elucidate the structure of the encoded proteins.

- DNA or RNA can be hybridised to microarrays, oligonucleotides or cloned DNA. This technology has a variety of uses:
  - Study of the expression of all genes in the genome in a variety of different tissues in one operation.
  - Definition of the extent of DNA sequences transcribed into mRNA, i.e. define the position and boundaries of exons.
  - Allows mass screening of sequence variants such as SNPs.
  - Allows characterisation in copy number and structural variants.

- The ENCODE project aims to produce an encyclopaedia of all the elements in the human genome. This includes characterising all the regions that are transcribed as well as elements such as regulatory regions, origins of replication, sites of DNA methylation and sites of histone modification. The results of the pilot study have been surprising: the human genome is much more extensively transcribed than previously envisaged, and transcripts of protein-encoding regions can be initiated a long way from the sequences that remain in the mRNA after splicing.

- Efforts to catalogue somatic mutations found in cancer, such as the Cancer Genome Project and the Cancer Genome Atlas, help to define the ways in which cancer cells differ from normal cells, reveal mechanisms by which cancer functions, and suggest new treatment approaches.

- Next-generation sequencing technologies are based on a variety of techniques that exploit massively parallel technologies. These technologies open up new vistas for future research:
  - They will allow shallow (~two-fold) re-sequencing of entire human genomes in large numbers of individuals and deep sequencing of targeted regions.
  - Deep sequencing (that is, exhaustive sequencing of all DNA and RNA molecules in a sample) will replace many of the applications of microarrays such as expression profiling.
  - They allow the fragmentary remains of DNA in Neanderthal fossils to be sequenced, which has been exploited to sequence the mitochondrial genome of Neanderthals and has opened up the prospect of sequencing the entire nuclear genome of Neanderthals.

Further reading

Sequencing and annotating the human genome

BAROSS, A., BUTTERFIELD, Y.S., COUGHUN, S.M. *et al.* (2004) Systematic recovery and analysis of full-ORF human cDNA clones. *Genome Research*, **14**, 2083–2092.

CURWEN, V., EYRAS, E., ANDREWS, T.D. *et al.* (2004) The Ensembl automatic gene annotation system. *Genome Research*, **14**, 942–950.

Describes the Ensembl automated process to annotate genomes and identify protein-encoding genes.

Human Genome Sequencing Consortium (2004) Finishing the euchromatic sequence of the human genome. *Nature*, **431**, 931–945.

The official publication describing the finished sequence of the human genome.

KITS P. Genome assembly and annotation process, Ch. 14. *The NCBI Handbook.*

http://www.ncbi.nlm.nih.gov/books/bv.fcgi?rid=handbook.chapter.ch14

A chapter in an online electronic book that describes the process of gene annotation in the NCBI databases.

Nature Genome Collection

http://www.nature.com/nature/supplements/collections/humangenome/

A web page that gives access to all the papers published by *Nature* concerned with sequencing the human genome. Provides access to papers describing the finished sequence and analysis of each individual chromosome.

The International Human Genome Consortium (2001) Initial sequencing and analysis of the human genome. *Nature*, **409**, 860–921.

VENTER, J.C., ADAMS, M.D., MYERS, E.W. *et al.* (2001) The sequence of the human genome. *Science*, **291**, 1304–1351.

These two papers describe the sequencing strategy and analysis of the draft sequence of the human genome. Other papers in the same issues of *Science* and *Nature* provide further commentary and analysis.

Structural variation

KIDD, J.M., COOPER, G.M., DONAHUE, W.F. *et al.* (2008) Mapping and sequencing of structural variation from eight human genomes. *Nature*, **453**, 56–64.

Describes mapping of CNVs using ESP mapping of fosmid clones.

MCCARROLL, S.A., KURUVILLA, F.G., MCCARROLL, S.A. *et al.* (2008) Integrated detection and population-genetic analysis of SNPs and copy number variation. *Nature Genetics*, **40**, 1166–1174.

Describes the construction of a high-density oligonucleotide array that can be used to genotype CNVs and SNPs.

REDON, R., ISHIKAWA, S., FITCH, K.R. *et al.* (2006) Global variation in copy number in the human genome. *Nature*, **444**, 444–454.

Describes CGH using BAC and oligonucleotide arrays.

ENCODE

GERSTEIN, M.B., BRUCE, C., ROZOWSKY, J.S. *et al.* (2007) What is a gene, post-ENCODE? History and updated definition. *Genome Research*, **17**, 669–681.

An interesting discussion of how the definition of a gene needs to be modified in light of the data emerging from the ENCODE project.

The ENCODE Project Consortium (2004) The ENCODE (ENCyclopedia Of DNA Elements) Project. *Science*, **306**, 636–640.

Mission statement of the ENCODE project.

The Encode Project Consortium (2007) Identification and analysis of functional elements in 1% of the human genome by the ENCODE pilot project. *Nature*, **447**, 799–816.

Detailed report of the pilot stage of the ENCODE project.

Next-generation sequencing technologies

MARGULIES, M., EGHOLM, M., ALTMAN, W.E. *et al.* (2005) Genome sequencing in microfabricated high-density picolitre reactors. *Nature*, **437**, 376–380.

Describes the Life-Sciences 454 pyrosequencing sequencing process.

http://www.nature.com/nrg/posters/sequencing/
Sequencing_technologies.pdf

A poster produced by *Nature Genetics* in collaboration with Applied Biosystems illustrating the different technologies available for next-generation DNA sequencing.

http://www.illumina.com/pages.ilmn?ID=203

A description of the Illumina sequencing by reversible terminators.

Construction and use of oligonucleotide arrays

DUGGAN, D.J., BITTNER, M., CHEN, Y., MELTZER, P. and TRENT, J.M. (1999) Expression profiling using cDNA microarrays. *Nature Genetics*, **21**, 10–14.

HACIA, J.G. (1999) Resequencing and mutational analysis using oligo-nucleotide microarrays. *Nature Genetics*, **21**, 42–47.

LIPSHUTZ, R.J., FODOR, S.P.A., GINGERAS, T.R. and LOCKHART, D.J. (1999) High density synthetic oligonucleotide arrays. *Nature Genetics*, **21**, 20–24.

These three papers review the basic technology for the construction and use of oligonucleotide arrays.

SU, A.I., WILTSHIRE, T., BATALOV, S. *et al.* (2004) A gene atlas of the mouse and human protein-encoding transcriptomes. *Proceedings of the National Academy of Sciences (USA)*, **101**, 6062–6067.

Describes the use of microarrays to analyse the expression of all human protein-encoding genes in 79 human tissues.

Cancer genome/atlas

COLLINS, F.S. and BARKER, A.D. (2008) Mapping the cancer genome. *Scientific American*, **18**, 22–29.

Describes the rationale of characterising the changes that occur in cancer cells and the way that strategy will be used to detect the changes.

The Cancer Genome Atlas Research Network (2008) Comprehensive genomic characterization defines human glioblastoma genes and core pathways. *Nature*, doi:10.1038/nature07385.

The first results of the cancer genome atlas analysed 206 glioblastomas, the most common type of adult brain cancer.

CHAPTER FIVE

Single-gene disorders

Key topics

- Positional cloning of disease genes
- Cystic fibrosis and the action of CFTR
- Duchenne muscular dystrophy and the dystrophin gene
- Trinucleotide repeat expansions
  - Fragile X syndrome
  - Huntington s disease
  - Mytotonic dystrophy
- Haemoglobinopathies
- Inherited predisposition to cancer
  - Retinoblastoma and the tumour suppressor hypothesis
  - Li—Fraumeni syndrome
  - Hereditary breast cancer: *BRCA1* and *BRCA2*
  - Colorectal cancer: FAP and HNPCC
  - Neurofibromatosis
  - Ataxia telangiectasia

5.1 Introduction

One of the benefits of the development of detailed maps of the human genome and its sequence is the ability to characterise genes involved in medical disorders. With all the resources of detailed genetic and physical maps and the availability of the complete nucleotide sequence, it is now relatively straightforward to identify and characterise genes mutated in single-gene disorders. This chapter is concerned with how genes mutated in single-gene disorders are mapped and cloned and, using some well-known monogenic disorders as case studies, we shall examine what this tells us about the normal function of the affected gene and the molecular pathology of the disorder in affected individuals. In the next chapter we shall consider the much more difficult task of identifying alleles that contribute to the risk of complex disorders.

Often, cloning a gene makes it possible for the first time to know about the nature of the protein whose malfunction causes a disease. The amino acid sequence of the protein encoded by the gene can be deduced from the nucleotide sequence of the gene. The amino acid sequence may show similarities to other proteins which have been studied either directly in humans or in model organisms such as bacteria, yeast, nematode worms, fruit flies and mice. This will provide important clues to its cellular function and illuminate the molecular pathology arising from its malfunction. Knowing about the molecular defect helps research to develop better and more rational therapies for the disease. The type of mutation in affected individuals may be determined, leading to diagnostic tests for the presence of the mutation in families at risk. The uses of this information in diagnostic tests are considered in Chapter 8. Finally, it may become possible to cure the disease by gene therapy, in which a functioning copy of the gene is introduced to complement the defective gene. Progress towards gene therapy is considered in Chapter 9.

5.2 Cloning disease genes

If the protein encoded by a gene has been characterised, a number of methods are available to identify the correct clone in a gene library. First, a partial amino acid sequence may be used to predict the nucleotide sequence of part of the gene. A synthetic oligonucleotide may then be synthesised to use as a hybridisation probe to recognise a clone with complementary DNA sequences. Alternatively, the protein can be used to raise an antibody. This can be used to screen a cDNA library constructed in an expression vector to ensure the encoded protein is expressed in the host. Such methods were used to clone important genes such as the α- and β-globin genes where the encoded globin proteins were abundant and well characterised, and the gene encoding human blood clotting factor XIII, where previous research had managed to purify some of the protein.

In most cases, however, the protein had not been characterised before cloning the gene which encodes it. So how can the gene be cloned in the first place? The only information that may be used to find the gene is to map its position in the genome. This is exploited in an approach known as positional cloning, which has been the method used to clone many genes. Box 5.1 illustrates a recent application of positional cloning to identify an autosomal dominant mutation in a gene called TRPV4, responsible for one form of the bone disease brachyolmia. The identification of the mutation in this gene uses all the modern genomic resources, and the relative ease of its identification should be contrasted with the difficulty of cloning of the gene mutated in cystic fibrosis described in Box 5.2, which was carried out before these genomic resources were available.

BOX 5.1: POSITIONAL CLONING OF THE GENE AFFECTED BY AN AUTOSOMAL DOMINANT MUTATION CAUSING BRACHYOLMIA

Brachyolmia is a bone disease manifested by dysplasia of the vertebrae of the lower spine, resulting in pronounced lateral spinal curvature. It is a heterogeneous disease with at least three types being documented. Types 1 and 2 are autosomal recessive disorders. Type 3 is an autosomal dominant disorder. The figure shows a pedigree where type 3 is segregating.

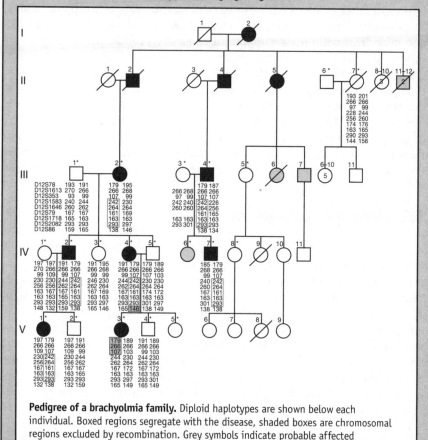

Pedigree of a brachyolmia family. Diploid haplotypes are shown below each individual. Boxed regions segregate with the disease, shaded boxes are chromosomal regions excluded by recombination. Grey symbols indicate probable affected individuals. (Reprinted by permission from Macmillan Publishers Ltd: From Rock, M.J. et al. (2008) *Nature Genetics*, **40**, 999—1003.)

A genome scan was carried out on members of this family using LOD score analysis specifying that the disease was autosomal dominant, fully penetrant with an allele frequency of 0.0001. The table shows markers co-inherited with the disease. Three markers, D12S1583, D12S1646 and D12S79, result in a maximum LOD score at $\theta = 0$, and with two of them the score is higher than 3.0, the threshold for significance.

Microsatellite markers shown in the pedigree identify an ancestral extended haplotype that co-segregates with the disorder and so marks the chromosome carrying the causative mutation (see figure). During the meiosis that generated

LOD score values between brachyolmia and mapping markers on chromosome 12 the chromosome that III2 transmitted to her affected daughter IV4, recombination occurred within the extended haplotype so that the region of chromosome 12 marked by allele 138 at D12S86 on the affected chromosome was replaced by the region from the unaffected homolog. Since IV was still affected, this region of the chromosome can be excluded (shown as a grey box). Similarly, a second recombination event during meiosis generated the chromosome that IV4 transmitted to her unaffected daughter, V3. The region that originated from the ancestral affected haplotype on this chromosome extended from D12S78 to D12S353. Since V3 is unaffected, this region can also be excluded. The mutation must therefore be located between D12S1583 and D12S2082.

LOD score at recombination fraction (θ):

Locus	Z_{max}	0.00	0.01	0.05	0.1	0.15	0.2	0.3	0.4	0.5
D12S78	1.45	−0.02	0.93	1.42	1.45	1.35	1.19	0.78	0.37	0.00
D12S1583	3.02	3.02	2.97	2.74	2.45	2.14	1.82	1.18	0.56	0.00
D12S1646	2.15	2.15	2.11	1.94	1.72	1.50	1.27	0.82	0.39	0.00
D12S79	3.04	3.04	2.99	2.76	2.46	2.16	1.84	1.19	0.57	0.00
D12S86	1.48	0.01	0.96	1.45	1.48	1.38	1.22	0.83	0.42	0.00

Genome annotation shows many genes in this region. A gene expression analysis was carried out on cartilage and non-cartilage tissues using 230 probes from this region. One probe, derived from *TRPV4*, was differentially expressed in cartilage. *TRPV4* was sequenced in affected members from the index family and another family where an apparently identical disease was segregating. In each family, a variant was found in *TRPV4* that segregated exclusively with the disease. In each family the variant resulted in an amino acid substitution at a site that was completely conserved in the *TRPV4* orthologs from the chimpanzee, Rhesus monkey, cow, rat, mouse, chicken, stickleback, zebrafish and clawed toad.

TRPV4 encodes a calcium channel expressed in a wide variety of different cell types. It opens to allow calcium ions to enter cells in response to stimuli such as hypotonic cell swelling, heat, 4α-phorbols and endogenous ligands such as arachidonic acid. Both mutations affect a transmembrane domain that forms part of the Ca^{++} ion pore. To investigate the effect of the mutations, the mutant genes were transfected into human embryonic kidney cells (HEK). Cells expressing the mutant proteins showed a much higher constitutive and agonist-induced activity, so the mutant proteins show the gain of function predicted from the autosomal dominant pattern of inheritance. Moreover, overexpression of mouse *TRPV4* in zebrafish caused a similar phenotype to human brachyolmia. The link between increased calcium channel activity and skeletal abnormalities is not yet understood, but there are several plausible hypotheses. For example, it is known that *TRPV4* activation induces Sox9 expression. Sox9 is a transcription factor with an essential role in chondrocyte differentiation and Sox9 deficiencies lead to skeletal malformations.

The first step in positional cloning is to identify linkage with genetic mapping markers. This is done by a genome scan where markers evenly spaced across the genome are tested to see whether they co-segregate with the disease in family pedigrees affected by the disease. The analysis uses LOD-based linkage analysis, as described in Chapter 3. Usually about 300 markers, spaced about 10 cM apart, are used in the first instance. When evidence of linkage is found, more closely spaced markers are used to narrow the limits of the genetic region identified. Box 5.1 shows the result of the genome scan in the case of brachyolmia, defining a region where several markers produced a statistically significant, maximum LOD score value at a recombination fraction (θ) of zero. Often an important clue is provided by recombination events which define the limits of the chromosome region carrying the mutation. In the case of brachyolmia, the chromosome carrying the mutation is marked by an extended haplotype that passed through most of the meioses in the pedigree without being disrupted by recombination. However, two recombination events allowed regions at either end of the haplotype to be excluded. Another event that can be helpful in identifying the locations of genes is a chromosomal mutation such as a translocation or deletion that can be defined by cytogenetic analysis. For example, when the end point of a translocation, which can be mapped, occurs within a gene it will disrupt its function. Such an event was critical in the mapping of the Duchenne muscular dystrophy gene to Xp21 (Figure 5.1).

Genetic mapping has a relatively low resolution, and the region defined will often contain many genes. The next task is to identify the one which is affected by the mutation. There is no single way of doing this. The availability of the detailed genome annotation at least means that all the genes in the region will be defined and, for many of these genes, their likely biological role may be known, either because a gene homolog has already been studied in a model organism, or because the amino acid sequence of the encoded protein provides a clue as to its biochemical activity. This information may be used to select candidate genes for more intensive study. For example, a gene mutated in HNPCC, a form of colon cancer, was identified because genetic mapping located it to a region on chromosome 2 that contained a homolog of a DNA mismatch repair gene studied in yeast and bacteria (see Section 5.7). Another useful clue may be provided by expression analysis using microarrays, which may identify genes that are preferentially expressed in tissues affected by the disease. In the case of brachyolmia, the genetic analysis defined an 11.3-Mb region on chromosome 12q24.1–12q24.2 which contained 230 genes (Box 5.1). Because the brachyolmia phenotype is restricted to the skeleton, microarray analysis was used to examine whether one of these genes was preferentially expressed in cartilage cells and in this way *TRPV4* was selected as a candidate.

Once a candidate gene has been identified, the next task is to show that a mutation in the gene co-segregates exclusively with the disease: that is, all affected individuals in a pedigree carry the allele, which is not present in

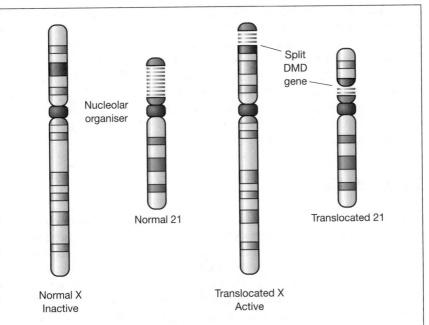

Figure 5.1 Localisation of the DMD gene to Xp21 using a translocation. A female patient with DMD was found to have a translocation between the X chromosome and chromosome 21. The breakpoint on chromosome 21 was within the 28S ribosomal locus. The breakpoint on the X chromosome was at the cytogenetic band Xp21. Although there was still a normal X chromosome, DMD resulted because of unfavourable Lyonisation, i.e. in most of her cells the normal X chromosome was inactivated. The reason for this is not clear, but it could be connected with some form of dosage compensation: if the translocated X chromosome was inactivated there would be two active copies of the X chromosome between Xp21 and the telomere of the p arm (one on the X chromosome and one on the translocated 21 chromosome). Probing a gene library made from this individual with a 28S rDNA probe yielded a junction clone called XJ. This clone contained sequences from the Xp21 region, which proved to be closely linked to DMD mutations.

any non-affected members of the population. The problem here is that many sequence variants will be harmless polymorphisms rather than harmful mutations. Deletions and nonsense or frameshift mutations will obviously have a profound effect on protein structure and are convincing evidence that a sequence variant is harmful. Non-synonymous changes (i.e. changes that alter the amino acid sequence) are more likely to be significant than synonymous changes or variants that occur in introns, although it is possible that the latter could still affect RNA expression, RNA splicing or mRNA stability. Another clue may be the likely effect on protein structure; for example, the introduction of a polar amino acid into a predicted transmembrane domain is likely to be disruptive. A useful indicator is if the mutation disrupts a sequence highly conserved in orthologs from other organisms. In the case of brachyolmia, two mutations in independent cases both changed amino acids that are completely conserved in nine vertebrate orthologs.

A final proof that is not always feasible is the effect of introducing the gene into tissue cells *in vitro*. In the case of an autosomal recessive mutation the test is whether the phenotype is rescued in a mutant cell line by the introduction of a wild type gene. This was demonstrated in the case of cystic fibrosis (see Box 5.2). In the case of a dominant mutation, the test is whether the introduction of the mutant allele into a wild type cell line causes a phenotype. This was demonstrated in brachyolmia, where the mutation was shown to cause gain of function in the calcium permeable ion channel encoded by *TRPV*4 (see Box 5.1).

5.3 Cystic fibrosis

Cystic fibrosis (CF) is the most common monogenic disorder affecting Northern Europeans. It affects approximately 1 in 2000 people and thus has a carrier frequency of 1 in 22. Its primary symptom is chronic bacterial infection and inflammation of the lungs. CF may also result in pancreatic exocrine insufficiency because the pancreatic duct is blocked, obstruction of the bowel in the newborn (**meconium ileus**), diabetes mellitus, liver cirrhosis and male infertility owing to congenital bilateral absence of the vas deferens (CBAVD). A diagnostic feature of CF is elevated salt content of sweat which provides a preliminary diagnostic feature in newborn babies.

The surface epithelium of the small airways of the lungs is bathed in an airway surface liquid (ASL) secreted by the epithelial cells that line the airways. ASL consists of two layers, the periciliary layer (PCL) and the mucus layer. The PCL extends from the surface of the epithelial cells as far as the tips of the cilia. It supports the cilia and acts to reduce friction with the mucus layer that is continually cleared from the lung by the action of the cilia (Figure 5.2). The mucus layer lies on top of the PCL and contains secreted mucins that trap foreign particles and bacteria inhaled into the lungs. In patients with CF the ASL is underhydrated and the sticky mucus that results cannot be cleared by the action of the cilia, so foreign particles such as bacteria cannot be removed. This leads to chronic bacterial infections in which *Pseudomonas aeruginosa* and *Staphylococcus aureus* predominate. Dehydration of epithelial cell surfaces lining other ducts in the body, such as the pancreatic duct and the vas deferens, results in blockages that cause the additional symptoms of CF.

5.3.1 *CFTR* gene and protein

The gene responsible for CF was cloned exclusively from its genetic map position without the aid of cytogenetic rearrangements or knowledge of the protein involved. Box 5.2 describes the process in detail as an example of the difficulties encountered in such an undertaking in the absence of modern genetic, physical and expression maps and an annotated genome sequence. The protein encoded by the gene is now known as the cystic

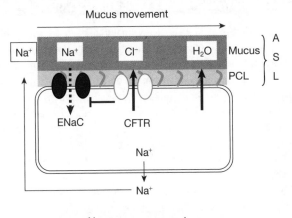

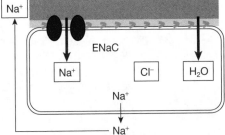

Figure 5.2 Hydration of lung airway epithelia. In a normal lung airway epithelial cell (top), Na$^+$ ions secreted by the basal lateral ATP-dependent Na$^+$/K$^+$ are absorbed by the ENaC Na$^+$ ion channel in the apical membrane. Inhibition of this pump by CFTR reduces the inward flow of Na$^+$ ions and Cl$^-$ ions flow outward through the CFTR channel to maintain ionic balance. The increased electrolyte content in the airway surface layer (ASL) results in the osmotic flow of water out of the cell to hydrate the ASL. The periciliary layer supports the cilia and reduces friction of the mucus layer as it is cleared by the beating action of the cilia. In the CF cell (bottom), the absence of the inhibitory action of CFTR on the ENaC increases the reabsorption of Na$^+$ ions and removes the channel for the outward flow of Cl$^-$ ions. As a result the electrolyte content of the ASL decreases and water flows inward rather than outward. The consequent lack of hydration of the ASL causes the PCL to collapse, impairing the action of the cilia and resulting in a thick mucus adhering to the epithelial cell surface.

fibrosis transmembrane conductance regulator (CFTR) (Figure 5.3). The *CFTR* gene is a large gene, occupying approximately 230 kb of chromosome 7, and consists of 27 exons ranging in size from 38 bp to 724 bp. Like other large genes, only a small fraction is actually coding sequence. It produces a 6.5-kb transcript that is found in those tissues affected by CF such as the lungs, pancreas, sweat glands, liver, nasal polyps, salivary glands and colon.

BOX 5.2: THE LONG WALK TO THE CYSTIC FIBROSIS GENE

The gene responsible for CF proved exceptionally difficult to clone. This was partly because of the lack of any cytogenetic rearrangements that could be used to map its location accurately and partly because of technical difficulties encountered during cloning. The first step was to map the gene to the chromosomal region 7q21. This was done by observing linkage between CF and two RFLP markers (MET and D7S8). An intensive search for new markers yielded two more, D7S122 and D7S340, that were more closely linked to the *CFTR* gene. Unfortunately, they were only 10 kb apart so effectively they marked the same point on the chromosome. Bidirectional chromosome walks were initiated from the landing places of chromosome jumps from these two markers. In total, this involved cloning 280 kb in 49 λ clones and making nine chromosome jumps. In the figure, each clone is shown by an arrow, indicating the direction of the step that was made with it. Double-headed arrows indicate clones that were used to walk in both directions. Each jump is shown as an arc. A long-range restriction map was also generated using rare cutting enzymes and pulsed field gel electrophoresis (see Section 3.4.4). The map showed that the *CFTR* gene (the name now given to the gene affected in CF) was about 250 kb in size and was entirely contained within a 380-kb *Sal* I fragment. Progress of the walk was monitored as sites were passed in this long-range restriction map. As well as these clones recovered by walking and jumping (shown in the left-hand part of the diagram), further clones were recovered by probing genomic libraries with cDNA clones (those clones on the right above the *CFTR* gene), so that in total 500 kb of DNA was cloned.

As the walk proceeded, most of the techniques described earlier were applied to identify the *CFTR* gene. Zoo blots proved to be effective and identified four regions whose sequence was conserved in other mammals. Any transcripts encoded by these regions were sought by using the cross-hybridising clones to probe Northern blots and cDNA libraries prepared from tissues affected by CF. Four clones cross-hybridised to other species, shown by the numbered vertical arrows in the figure. Region 1 could be eliminated on genetic grounds. Region 2 corresponded to a gene called *IRP* (INT-related protein) that was known not to be responsible for CF. However, because *IRP* was known to map to the D7S8 side of CF, this indicated the direction in which walking and jumping should continue. Although region 3 contained a CpG island that often marks the start of a gene (see Chapter 2), no transcripts could be detected.

Region 4 eventually turned out to be the 5′ end of the *CFTR* gene, but this only became apparent after a frustrating series of experiments. The region was identified by two clones called H1.6 and E4.3. The sequences of these clones were rich in undermethylated CpG dinucleotides, suggesting that they were CpG islands. However, they did not contain any ORFs and nor did they detect any transcripts in Northern blotting. Eventually, after screening seven different libraries, clone H1.6 hybridised to a single cDNA clone, called 10.1, in a cDNA library made from sweat glands. Clone 10.1 hybridised to a 6.5-kb transcript in a Northern blot and the signal was stronger in cells expected to express the *CFTR* gene. Sequencing showed that the 920-bp insert in clone 10.1 contained a long ORF. Only 113-bp of this

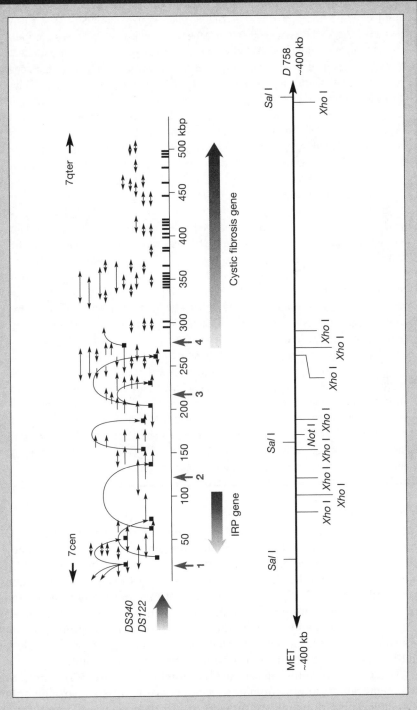

Continued ▶

sequence overlapped with H1.6. The small overlap explained the difficulty experienced in identifying a transcript. The *CFTR* gene had been found, but only just! Less perseverance or a slightly smaller insert in clone H1.6 and it would have been missed.

Obtaining a full-length cDNA clone again proved more difficult than expected. Clone 10.1 was used to re-probe cDNA libraries; 18 clones were recovered but none contained the full 6.5-kb gene identified in Northern blots. The number of clones recovered was less than expected, and they grew slowly and contained rearrangements. This suggested that clones containing the full-length gene were selected against in the formation or propagation of the cDNA libraries. The consensus sequence of the gene had to be pieced together from the partial sequences of the inserts of these 18 clones. A clone containing the full sequence was then assembled by subcloning from these fragments.

This gene was re-isolated from cDNA libraries prepared from sweat glands of affected and unaffected individuals. A common 3-bp deletion (ΔF508) was exclusively associated with the occurrence of the disease: 145/214 CF chromosomes carried this mutation (68%); in contrast 0/198 normal chromosomes had the mutation ($P < 10^{-57.5}$!). The protein encoded by this gene showed strong similarities to the protein superfamily known as the ABC (ATP-binding cassette) membrane transporters. These two observations strongly suggested that the gene responsible for cystic fibrosis had indeed been isolated. The final proof came later with the demonstration that, when transfected into epithelial cell cultures derived from CF patients, it corrected the chloride ion conductance defect.

The work was carried out as a collaboration between the groups of Lap-Chee Tsui and John Riordan in the Hospital for Sick Children, Toronto, and Francis Collins in Michigan. Lap-Chee Tsui s group identified the closely linked D7S122 and D7S340 markers and carried out the chromosome walking and physical mapping in collaboration with the group of Francis Collins, who had developed the chromosome jumping technique. John Riordan s group provided the cDNA libraries and identified the crucial cDNA clones. The outcome was victory in a race with many other groups around the world who were seeking the gene.

Although existing research had suggested that the primary defect in CF was chloride ion transport, prior to cloning the gene there was no direct knowledge of the protein affected. Conceptual translation of the nucleotide sequence showed that the CFTR protein consists of 1480 amino acid residues. It is similar to the ABC (ATP-binding cassette) superfamily of membrane transporters. These proteins are found in all living organisms, from bacteria to man. As their name implies, they are responsible for pumping small molecules in and out of cells. Members of this family can be responsible for multidrug resistance in human cells undergoing cancer chemotherapy.

The CTFR protein consists of five domains (Figure 5.3). There are two membrane spanning domains, each consisting of six transmembrane segments. These domains are thought to form a membrane pore through which the chloride ions are transported out of the cell. There are two

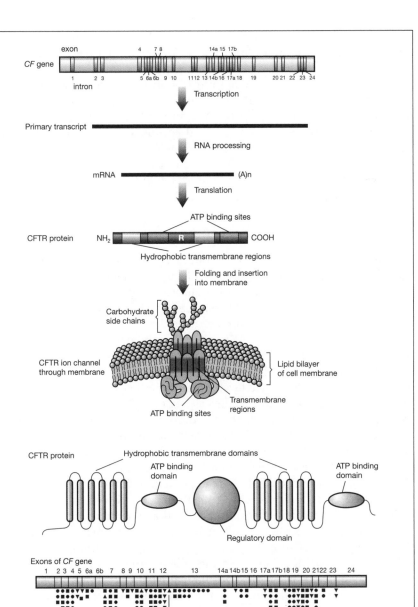

Figure 5.3 Structure of the *CFTR* gene, the transmembrane protein it encodes and the spectrum of mutations that affect its function. *Top* Artist s impression of the CFTR protein inserted into cell membrane. *Middle* Schematic representation of its structure. *Bottom* Spectrum of mutations that have been found to inactivate gene function. (From Collins, F.S. (1992) Cystic Fibrosis: Molecular biology and therapeutic implication, *Science*, **256**, 774–9. Reprinted with permission from AAAS.)

domains called nucleotide binding folds (NBF) which bind and hydrolyse ATP, essential for the transport process. A domain that has a high proportion of charged amino acid residues separates the two halves of the molecule. This domain is unique to the CFTR protein and has a regulatory function. It contains several sites which are phosphorylated by cAMP-dependent protein kinase (protein kinase A). This phosphorylation is essential for the protein to function and accounts for the prior observation that chloride ion transport is dependent on cAMP-dependent protein kinase. The carboxy terminal of CFTR, consisting of the amino acids threonine, arginine and leucine (TRL), interacts with the cytoskeleton and other proteins in the plasma membrane, including ion channels and receptors.

5.3.2 The molecular pathology of CF

The hydration of the ASL is regulated by the sodium ion absorption through the ENaC sodium ion channel located in the apical membrane (the side of the cell adjacent to the airway surface) (Figure 5.2). Sodium ions are secreted from epithelial cells through a Na^+/K^+ ATPase pump from the epithelial cell basolateral membrane (distant to the airway surface). When more water is required in the ASL, ENaC is inhibited, resulting in the accumulation of sodium ions in the ASL; ionic balance is maintained through the outward flow of chloride ions through the CFTR channel in the apical membrane. As the electrolyte level increases in the ASL, water flows out of the epithelial cells by osmosis. A key aspect of CFTR function is its capacity to repress ENaC activity, which is additional to its role as a chloride ion channel. Knocking out the CFTR gene in mice does not recapitulate the symptoms of CF. However, mice engineered to overexpress ENaC in lung airway epithelial cells suffer from a very similar disease to CF in humans, showing the importance of ENaC regulation in maintaining the ASL.

Because CFTR inhibits ENaC, its absence in patients with CF causes increased sodium ion reabsorption as well as a decrease in the outward flow of chloride ions. As a result, ASL electrolyte content is reduced and the consequent loss of osmotic water flow results in the ASL becoming dehydrated; the PCL collapses so the cilia are unable to function and the concentrated mucus layer sticks to the epithelial cell surface. Apart from a failure to clear, the thickening of the mucus has other detrimental consequences. It favours the formation of biofilms by *P. aeruginosa* bacteria where layers of bacterial cells are enclosed within a polysaccharide capsule that provides protection against host defences and antibiotic treatment. The thickened mucus also inhibits innate host antibacterial defences such as the action of neutrophils that patrol the mucus layer, and antibacterial peptides and proteins such as lactoferrin.

5.3.3 Mutations which affect the *CFTR* gene

Over 1000 different mutations of the *CFTR* gene have been catalogued (Figure 5.3). However, over two-thirds of CF mutant alleles are exactly the same mutation, known as ΔF508. As discussed in Chapter 9, there is evidence that this mutation is extremely ancient, possibly occurring in one of the very first anatomically modern humans that lived in Europe, 50 000 years ago, and passed on to succeeding generations from this founder. It has been suggested that carriers of a CF mutant allele enjoy some selective advantage. Two theories have been advanced for the basis of this advantage. First, mouse CF/cf– heterozygotes secrete less water from intestinal epithelial cells when treated with cholera toxin. This may allow carriers a better chance to survive the diarrhoea caused by cholera infection that is often fatal, especially in infants. Second, it has been shown that CFTR acts as a receptor for the uptake *Salmonella typhi* bacteria into epithelial cells and that heterozygous CF/cf– mouse epithelial cells take up fewer *S. typhi* cells than wild type CF/CF cells. Thus, CF carriers may have been more resistant to typhoid fever.

The ΔF508 mutation is a 3-bp deletion that results in the loss of a phenylalanine residue at position 508 in the polypeptide chain in one of the two NBFs described above. (Δ is a symbol commonly used to signify a deletion; F is the standard single-letter abbreviation for phenylalanine.) The CFTR protein undergoes a complex maturation and folding process before the active form is finally located in the cell membrane. ΔF508 polypeptides are defective in this process as they do not undergo an ATP-dependent folding step in the Golgi apparatus and are rapidly degraded. However, if the protein does fold, the mutation does not affect its activity. When cells carrying the ΔF508 mutation are cultured at subphysiological temperatures, folding takes place and the protein is active. This provides a target for therapeutic intervention – small drugs may be found that aid the folding of ΔF508 CFTR polypeptides *in vivo*. Other CF mutations do not affect protein folding or localisation of the protein to the plasma membrane, but result in inactivating protein activity.

5.4 Duchenne muscular dystrophy

DMD is characterised by progressive failure of muscle growth and wasting (dystrophy), leading to weakness, paralysis and respiratory difficulties. In addition, the heart, smooth muscle and central and peripheral nervous systems may be affected. It is a sex-linked disease affecting 1 in 3300 boys. Symptoms usually become apparent in boys by the age of 3; they become confined to a wheelchair in their teenage years and usually die by the age of 30. A milder form, called Becker muscular dystrophy (BMD), is caused by mutations that map to the same locus.

5.4.1 DMD gene

The gene was cloned by positional cloning in 1987. Cytogenetic rearrangements resulting in DMD always have one breakpoint in the region Xp21. One rearrangement resulted in a fusion of Xp21 with the 28S ribosomal

RNA locus on chromosome 21, which allowed a marker called XJ1.1, closely linked to DMD, to be isolated (see above). Another important strategy was the use of DNA from a boy identified as 'BB' who suffered from three X-linked disorders: DMD, chronic granulomatous disease and retinitis pigmentosa. Because the genes responsible for these disorders are closely linked, it seemed likely that BB's X chromosome carried a deletion that affected at least part of all three genes. DNA from the region spanned by this deletion was isolated from an XXXXY cell line using a technique called subtractive hybridisation, which preferentially enriched for the region missing in BB's DNA. One of these clones, called pERT87, was shown to be closely linked to the site of DMD mutations. Clones such as pERT87 and XJ1.1 were used to map the genomic locus, including the construction of a long-range restriction map. pERT was also used to probe cDNA libraries, resulting in the isolation of a number of cDNA clones that together spanned the 14-kb mRNA of the locus. These cDNA clones were then used to make a detailed map of the genomic locus, including the positions of exons.

The DMD gene is notable for its huge size. It is composed of 79 exons that, together with promoter regions, are scattered across 2.4 Mb, about 2% of the whole X chromosome! Transcription of this region is complex, with different cell types using at least seven distinct promoters between them. A full-length 14-kb mRNA is produced from different promoters in muscle cells, cortical cells and Purkinje cells. A shorter transcript is produced by internal promoters in Schwann cells and glial cells, encoding a protein that is correspondingly smaller.

5.4.2 Dystrophin

The protein encoded by the DMD gene, called **dystrophin**, is composed of 3685 amino acid residues. Its role is shown in Figure 5.4.

Dystrophin is a long rod-like molecule that links actin fibres in the cortex of muscle cells to the extracellular basal lamina (connective tissue). It does this by forming a bridge between actin and a transmembrane glycoprotein complex called the dystrophin-associated sarcoglycan complex (DASC) located in the **sarcolemma**, a membranous sheath that surrounds muscle fibres. DASC is required for the function and maintenance of muscle cells. It provides a signalling pathway between the connective tissue and the cytoskeleton of muscle cells. In the absence of dystrophin this complex does not form properly. Mutations that affect proteins of the DASC complex also result in a form of muscular dystrophy, called limb girdle muscular dystrophy. In addition, some people suffering from congenital muscular dystrophy (CMD) harbour mutations affecting α-laminin-2, to which DASC is attached outside the cell.

5.4.3 Mutations affecting the DMD gene

DMD mutations arise at a rate of 1×10^{-4} per generation, two orders of magnitude higher than for other X-linked loci. This high rate of mutation is

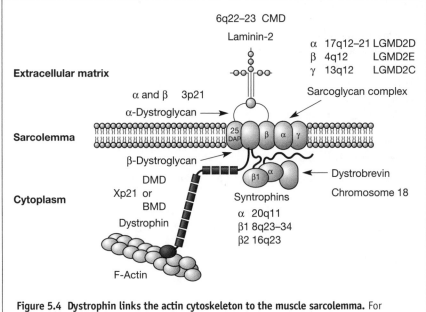

Figure 5.4 Dystrophin links the actin cytoskeleton to the muscle sarcolemma. For explanation see text.

probably due to the target presented by its extreme size. Deletions account for 60% of these mutations. The end points of the deletions are clustered in two particular introns. The reason for this is not clear, but it could be related to the presence of a mobile genetic element in both introns. Where deletions disrupt the reading frame of the protein, the severe Duchenne form of the disease results. Where the reading frame is preserved, the milder Becker form results. Surprisingly, one mutation has been identified that results in the Becker form, even though the protein is less than half its original size. A similar pattern is seen in small deletions and point mutations: where the mutation allows a full-length protein to form, the milder Becker form generally results, but nonsense and frameshift mutations producing truncated proteins cause the severe Duchenne form.

5.5 Trinucleotide repeat expansion mutations

We saw in Chapter 2 that microsatellites consist of tandemly repeated dinucleotide, trinucleotide or tetranucleotide sequences and that the number of repeats is polymorphic. The ability to clone genes and to analyse the mutations responsible for genetic disease led to the unexpected discovery that polymorphic trinucleotide repeats can dramatically increase in length to form a novel type of mutation called a **trinucleotide repeat expansion (TRE)**. A total of 11 such mutations has been identified (Table 5.1). In TRE mutations the number of repeats is unstable in meiosis, in some cases even

Table 5.1 Disorders caused by trinucleotide repeat expansions.

Fragile sites caused by CCG expansion in 5′ UTR
FRAXA (fragile-X syndrome)
FRAXE (mental retardation)
FRAXF
FRA16A
FRA11B (Jacobsen syndrome)

CTG expansion in 3′ UTR
Myotonic dystrophy (MD)

Neurodegenerative disorders caused by CAG expansions in coding regions
Huntington s disease (HD)
Spinal and bulbar muscular atrophy
Spinocerebellar ataxias SCA1, SCA2, SCA3, SCA6, SCA7, SCA17
Dentatorubral—pallidoluysian atrophy
Machado—Joseph disease

---

in mitosis, usually causing a further increase in repeat length. For this reason they are sometimes called **dynamic mutations**. Dynamic mutations are responsible for a phenomenon called **anticipation**, where either the severity or the degree of penetrance of a disease apparently increases with each succeeding generation or the age of onset decreases.

TREs can broadly be divided into two classes. In one class, exemplified by fragile-X syndrome, the trinucleotide concerned is CCG, which expands in the **untranslated regions** (UTRs) of genes. This typically results in chromosome fragile sites: cytogenetically visible, non-staining gaps in metaphase chromosome spreads. In the other class, the trinucleotide concerned is CAG, which occurs in a number of neurodegenerative disorders such as Huntington's disease (HD). In this class, the number of expanded repeats is much lower and occurs within the coding region of the gene affected. One TRE mutation that does not easily fit in this classification is responsible for Myotonic dystrophy (MD). In this case the trinucleotide concerned is CTG and the expansion site is located in the 3′ UTR of the gene that is probably affected.

5.5.1 Fragile sites caused by CCG expansions

In TRE mutations involving CCG, the normal repeat number of ~30 increases to several hundred or even thousands. It results in **fragile sites** in chromosomes. These are non-staining gaps in chromosomes visible in metaphase spreads of cells cultured under conditions such as folate deprivation or chemical inhibition of DNA synthesis. Five loci involved in fragile sites have been cloned (Table 5.1). All have been shown to involve expansions of a CCG repeat. Some of these fragile sites have no apparent phenotypic consequences. However, one of them, FRAXA, results in fragile-X syndrome, characterised by moderate to severe mental retardation.

Two other fragile sites, FRAXE and FRA11B, are also associated with mental retardation.

Fragile-X syndrome is the commonest cause of congenital mental retardation after Down's syndrome, affecting 1 in 2000 children. There are a number of unusual features in its pattern of inheritance.

1. It shows incomplete penetrance as, on average, 20% of boys who are known to carry the allele, because they transmit the disorder to subsequent generations, are not affected. These boys are classified as normal transmitting males. The degree of penetrance increases with each succeeding generation in a pedigree. This phenomenon is called anticipation and, in one form or another, is characteristic of TRE mutations.
2. It shows at least partial dominance because 30% of female heterozygotes are also affected. These affected females always inherit the fragile site from their mothers, never from their fathers.
3. In males the disorder is never caused by a new mutation during gametogenesis in their mothers.

FRAXA, the fragile site associated with fragile-X syndrome, is located at Xq27.3. The gene affected is called *FMR*1 (fragile-X mental retardation 1). The expansion affects a polymorphic site in which the number of CCG repeats normally ranges from 6 to 52. Individuals affected by the disease are said to carry a full mutation when the number of repeats increases to 230–1000 copies. The expansion site is located upstream of the coding region near to a CpG island. The full mutation induces **DNA methylation**, which extends into this CpG island and consequently inactivates the *FMR*1 gene. Some boys, who suffer from a severe form of the disorder, have been found to have deletions of the *FMR*1 gene. This confirms that the expansion does indeed cause loss of gene function and is not acting in some other way to change the activity of the *FMR*1 gene product. The role of the FMR1 protein has not yet been fully elucidated. It contains an amino acid sequence that is found within proteins known to bind RNA. The FMR1 protein binds specifically to approximately 4% of mRNA molecules found within brain cells, so it may have a role in controlling translation.

The number of repeats at the expansion site can show intergenerational instability, which accounts for the unusual pattern of inheritance. Normal transmitting males harbour **premutations** in which the number of repeats ranges from 60 to 230. These can expand into full mutations during meiosis in a carrier female but not in a normal transmitting male. The frequency with which the premutation expands into a full mutation is dependent upon its length, which can only increase in a female carrier. With each passage of a premutation through a carrier female its length may increase, consequently raising the chance of expansion into a full mutation in the next female carrier in the pedigree. This explains the phenomenon of anticipation. The requirement for a premutation to be expanded by passage through a carrier female explains why the disorder is never caused in boys by a new mutation during gametogenesis in their mothers. The observation that expansion can only occur in females explains why affected females always receive the mutant allele from their mothers. Both males and females bearing the full mutation show somatic instability where the repeat length and degree of methylation can show wide variation in different cells.

5.5.1.1 The role of FMRP

FMRP, the protein encoded by *FMR1*, acts as a negative regulator of protein synthesis at the tips of dendritic spines, where excitatory synaptic transmission occurs. Newly formed spines progressively become shorter as they mature. The spines of affected boys and *fmr1* knockout mice fail to mature and are longer than those of normal individuals. Such abnormalities in spine morphology are associated with many forms of mental retardation, such as Down's and other syndromes. The activity of the spines is affected by a surface receptor called AMPA. Internalisation of this receptor leads to long-term depression of synaptic activity (Figure 5.5). This internalisation is activated by local protein synthesis which in turn is stimulated by the glutamate receptor, mGluR5. Glutamate is the major excitatory neurotransmitter; when it binds to mGlu5 it elicits long-lasting intracellular effects (metabotrophic), which include local protein synthesis at the spine tip on pre-existing mRNA transcripts. It is this protein synthesis that is inhibited by FMRP. mGluR5 stimulation also results in increased *FMR1* transcription. Thus, the protein synthesis stimulated by mGluR5 is limited by a form of end product inhibition: so, in the absence of FMRP there will be an exaggerated response to mGluR5 stimulation, leading to excessive AMPA internalisation and consequent long-term depression. Elevated protein synthesis at the spine tip may also be responsible for the failure of the spines to mature in the absence of FMRP in affected boys.

FMRP is an RNA-binding protein, which binds to specific mRNA molecules through three domains: FH1, FH2 and an RGG box. The FH2 domain recognises a loop–loop pseudoknot, an intricate secondary structure

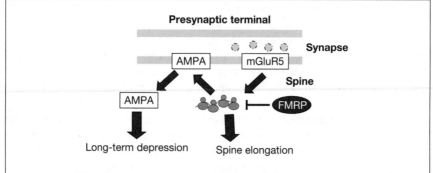

Figure 5.5 FMRP attentuates response to mGluR. Stimulation of the mGlu5 receptor leads to increased protein synthesis at the tip of dendritic spines. This has two effects: first, it leads to the internalisation of the AMPA receptor and thus long-term depression; second, it stimulates tip growth of the spine. FMRP inhibits mGlu5-stimulated protein synthesis.

within RNA molecules, to form a so-called kissing complex. The RGG box recognises an intramolecular G-quartet, a stem loop structure formed by four consecutive guanine bases. These interactions allow FMRP specifically to target transcripts encoding cytoskeletal proteins required for polarised growth of the spines and proteins involved in dendrite maturation and synaptic activity. The mRNA transcripts encoding these proteins are transported from the nucleus along microtubules to the spine tips. During this transport the transcripts are maintained in a translationally inactive state by the RISC complex that inhibits translation of target mRNAs (see Chapter 7). Stimulation of the mGluR5 receptor at the tip results in the destruction of the RISC components and so activates the translation of these mRNA molecules. It is this activation process which is inhibited by FMRP.

5.5.2 Neurodegenerative disorders caused by CAG expansions

At least nine neurodegenerative disorders have been shown to be caused by expansion of polymorphic CAG repeats located within the coding region of the gene affected (Table 5.1). All of the disorders are late onset, autosomal dominant and progressive in nature. They are all characterised by similar symptoms: ataxia, chorea, dementia and sometimes psychosis. The effect of each disease is limited to a different group of cells in the brain, yet each gene is expressed in many different classes of brain cells; there is currently no explanation for this apparently paradoxical observation. CAG encodes glutamine, so the expansion mutation will cause an increase in a polyglutamine tract in the encoded proteins. Since these are all dominant mutations this must in some way alter the activity of the protein rather than reduce its function.

The best-known and most common of this group of diseases is Huntington's disease (HD), which affects 1 in 10 000 Caucasians. The symptoms are generally manifested in mid-life, but it can show juvenile onset, in which case the course of the disease is more rapid and the symptoms more severe. A particularly distressing aspect of HD is that, as it is a late-onset, autosomal dominant disease, sufferers will normally have had children before the onset of the disease. Each of these children will have a 50% chance of being affected.

In 1983, HD was one of the first human genes to be mapped by linkage to polymorphic DNA markers. However, the gene responsible was only cloned in 1993 after an exceptionally prolonged and difficult gene hunt. Two strategies finally led to the gene:

1. Linkage disequilibrium showed that although multiple mutations have occurred to cause HD, one-third of HD chromosomes probably descend from the same ancestral chromosome. The haplotype analysis narrowed the location of the gene to a 500-kb region.
2. mRNA transcripts were mapped that arose from the region identified by the linkage disequilibrium analysis. In patients with HD the gene encoding one of these was affected by an expansion of a trinucleotide repeat $(CAG)_{\sim15}$.

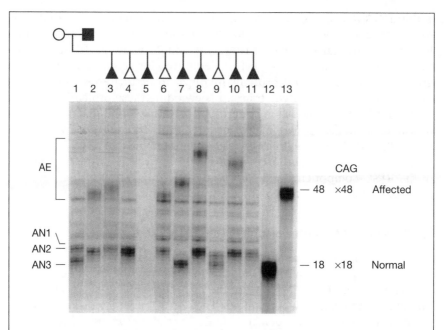

Figure 5.6 Trinucleotide repeat expansion in HD. The figure shows a large family in which the disease is segregating. The region surrounding the CAG repeat site has been amplified using PCR. All individuals in the pedigree have a normal allele with the CAG trinucleotide repeated about 18 times. Affected individuals also have a mutant allele where the repeat number has increased. Lanes 12 and 13 show reference cDNA clones where the repeat length is 18 (wild type) or 48 (mutant). The PCR reaction in lane 5 failed. Individual number 11 apparently only has a normal allele. Subsequent analysis showed that this individual carried an expansion so large that the PCR reaction failed. As a consequence of the very large repeat, the disease was particularly severe and showed juvenile onset. (Reproduced from McDonald *et al.* (1993) *Cell*, **72**, 971—83.)

The gene affected, known as *IT15*, is located near the telomere on chromosome 4. It encodes a protein called **huntingtin**. The coding region contains a polymorphic CAG repeat that expands in those affected by the disease (Figure 5.6). The normal range of repeat lengths is 15–35; in affected individuals this increases to 36–121. It is remarkable that the lowest number of repeats that causes the disease is only one more than the highest found in normal individuals, although it should be said that repeats in the 30–40 range are extremely rare. Apparently, increasing the polyglutamine tract from 35 to 36 residues is enough to trigger the onset of the disorder.

HD does not show anticipation in the formal sense, but there are some related phenomena. Longer repeat lengths are associated with juvenile onset and more severe symptoms. These are generally transmitted by affected fathers, suggesting that expansion takes place during male gametogenesis. This is reminiscent of fragile-X syndrome, where the expansion can only take place in the female germline. Age of onset is also inversely correlated with repeat length.

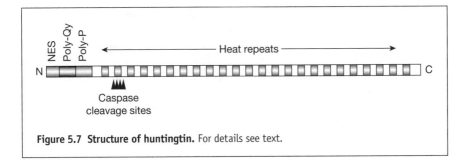

Figure 5.7 Structure of huntingtin. For details see text.

5.5.2.1 Huntingtin

Huntingtin, the protein encoded by *HTT*, is a large protein, ubiquitously expressed both within and outside the nervous system. A cartoon of its structure is shown in Figure 5.7. The polyglutamate (poly-Q) region is encoded by exon 1 and is located towards the N-terminus, immediately followed by a polyproline (poly-P) region. A characteristic of huntingtin is the presence of 36 repeats called HEAT repeats. Each HEAT repeat consists of a variable number of residues (about 50) that form a helix-turn-helix or hairpin structure with flat, hydrophobic sides. The hydrophobic sides from adjacent HEAT repeats associate with each other, so that the whole molecule forms an elongated superhelix with a continuous hydrophobic core. This structure is characteristic of other proteins with HEAT repeats that often act as scaffold proteins that interact with a number of other proteins: in fact, physical interactions have been documented between huntingtin and a total of 236 different proteins. This large number of interactions reflects the wide variety of roles that have been described for huntingtin in cell biology.

- Mice *htt⁻/htt⁻* embryos die as early embryonic lethals showing high levels of apoptosis. This suggests that huntingtin acts as an anti-apoptotic factor. A possible mechanism is that huntingtin sequesters HIP1 that promotes apoptosis by activating pro-caspase 8, the protease that initiates the apoptotic cascade. Huntingtin also associates with caspase 3, the protease activated by the cascade that actually causes the cellular destruction during apoptosis.

- It is associated with proteins of the endocytic machinery, particularly clathrin (a protein that coats endocytic pits) and dynamin (a protein that pinches of the endocytic vesicles from the plasma membrane).

Huntingtin associates with microtubular motor proteins such as dynein and dynactin, which form a minus-end directed motor complex, and kinesin, which is a plus-end directed motor. Knockdown of huntingtin expression using RNAi inhibits the movement of vesicles and mitochondria within neurones.

5.5.2.2 The molecular pathology of Huntington's disease

As noted above, mice lacking a functional *HTT* gene die as early embryonic lethals, indicating that wild type huntingtin performs an essential function. However, heterozygous *HTT/htt⁻* mice develop normally, so two copies of the gene are not required. This makes it unlikely that the Huntington's phenotype arises from a loss of function of the mutated HTT protein, so the expansion of the poly-Q tract must be conferring a gain of function. In normal cells, huntingtin is largely located in the cytoplasm, with only a small amount in the nucleus. Mutant huntingtin is cleaved at the caspase sites located downstream of the poly-Q tract (Figure 5.7) and the resulting N-terminal fragments accumulate in the nucleus as insoluble aggregates called inclusion bodies. Caspase cleavage is essential for its toxicity, since removal of the sites ameliorates the harmful effects of the poly-Q expansion. A transgenic mouse, in which just exon 1 containing the CAG expansion has been introduced, shows a remarkably similar phenotype to human Huntington's disease. In these mice, the N-terminal fragment also accumulates in nuclear inclusion bodies. So these insoluble nuclear aggregations of the N-terminal fragment containing expanded poly-Q tract are apparently responsible for the toxicity of the mutant protein. However, cells expressing mutant huntingtin are perturbed in a variety of different ways, described below. It is not clear to date whether one of these on its own is responsible for the eventual death of the neurones, or whether this is a result of the combined effect of the multiple system failures.

1. The Huntington's inclusion bodies associate with a variety of transcription factors and associated proteins:

    a) CBP (CREB-binding protein), as its name implies, is a transcriptional co-activator that associates with CREB (cAMP response element binding protein). CBP is a major transcriptional regulator whose action mediates the response to many signals essential for the long-term survival of neurones. CBP has histone acetyltransferase activity, which activates transcription by opening up chromatin structure. CBP is found in intranuclear aggregates of mutant huntingtin along with p/CAF, another protein with histone acetyltransferase activity. The expanded poly-Q tract of mutant huntingtin interacts with a poly-Q tract in CBP and inhibits its acetyltransferase activity, thus interfering with its ability to promote transcription. The action of histone acetylases is reversed by histone deacetylases (HDAC). HDAC inhibitors have been shown to reduce the toxic effects of mutant huntingtin in model organisms, including the mouse model. This provides a promising therapeutic approach for the treatment of Huntington's disease.

    b) BDNF (brain-derived neurotrophic factor) is a growth factor that is essential for the survival of striatal neurones that specifically die in Huntington's disease. Transcription of *BDNF* is negatively regulated by a transcriptional regulator variously called either NRSF (neurone-restrictive silencing factor) or REST (RE1-silencing transcription

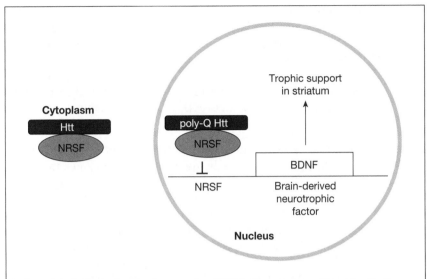

Figure 5.8 Wild type huntingtin sequesters NRSF in the cytoplasm. Mutant huntingtin allows NRSF to enter the nucleus where it inhibits the expression of BDNF required for trophic support in the striatum.

factor) which binds to a specific element in the *BDNF* promoter called NRSE (neurone-restrictive silencing element) (Figure 5.8). Wild type huntingtin promotes the transcription of *BDNF* by binding to NRSF/REST and sequestering it in the cytoplasm, thus preventing it from repressing the transcription of *BDNF* in the nucleus. Mutant huntingtin can no longer sequester NRSF/REST in the cytoplasm, so it can now enter the nucleus and repress the expression of *BDNF* (Figure 5.8). As well as promoting the transcription of *BDNF*, wild type huntingtin facilitates the transport of BDNF-containing vesicles along microtubules to the sites of its secretion at axon terminals:

c) SP1 (specificity protein 1) is a transcriptional activator that binds to upstream GC tracts and interacts with the general transcription initiation complex. Mutant huntingtin also interacts with SP1. However, there are conflicting reports as to whether this contributes to the pathology of HD or whether in fact it is protective.

2. Mitochondrial dysfunction has been suspected to be involved in a number of neurodegenerative diseases, and there are several indications that this is the case in Huntington's disease:

a) Mitochondria isolated from post-mortem samples from patients with HD and from the mouse model show a decreased membrane potential and were depolarised in response to Ca^{2+} loading. In addition to nuclear aggregates, mutant huntingtin is found associated with the outer mitochondrial membrane. The addition of mutant huntingtin to normal mitochondria induced the same effects.

b) Production of ATP is impaired in mitochondria isolated from the mouse model and this is associated with a decreased rate of ADP uptake from the cytoplasm.

c) PGC-1α (peroxisome-activated receptor gamma-1α) is a transcription factor that programmes the expression of genes required for mitochondrial biogenesis. Expression of genes controlled by PGC-1α is reduced in the striatum of HD mice and human patients.

Protein degradation is an essential cellular function. It acts to remove proteins that fail to fold properly after translation and ensures the irreversible passage of the cell through its cycle by degrading proteins once they have acted to passage the cell to the next stage. Proteins destined for degradation are marked by conjugation with ubiquitin. These marked proteins are then degraded by the 26S proteosome complex. The whole system is known as the UPS (ubiquitin proteosome system). There are conflicting reports that the UPS is upregulated in cells expressing mutant huntingtin. One hypothesis is that mutant huntingtin aggregates are resistant to degradation in the 26S proteosome complex, which thus becomes blocked. This prevents the quality surveillance on newly synthesised proteins from operating normally, resulting in the accumulation of inactive misfolded proteins in the cell. An interesting test of this hypothesis was based on the expression of a poly-Q peptide in the worm *Caenorhabditis elegans*, which models some of the HD phenotypes. Many temperature-sensitive mutations have been isolated in this worm. Although temperature-sensitive proteins may function at the permissive temperature, the altered polypeptide may not fold completely as efficiently, placing an extra reliance on the UPS to maintain protein quality. When worms expressing the poly-Q in muscle cell tracts were crossed with worms carrying temperature-sensitive mutations, the mutant phenotype was expressed at the temperature that is normally permissive. Thus, the load placed on protein quality surveillance by the poly-Q tract may affect the folding of other proteins and contribute to pathogenesis. Furthermore, in cells carrying the temperature-sensitive mutation and the poly-Q tract, the formation of poly-Q aggregates was markedly increased. This may explain how genetic modifiers act to influence the severity and age of onset of HD mutations since alleles of other genes which hinder the folding of the encoded protein may also have an effect on the formation of huntingtin aggregates.

5.5.3 Myotonic dystrophy or dystrophia myotinica (DM)

DM (also known as **dystrophia myotonica**) is an autosomal dominant neuromuscular disorder. It affects 1 in 8500 individuals, making it the most common neuromuscular disorder showing adult onset. It also shows a range of symptoms, from cataracts and frontal baldness in mildly affected individuals to multisystem failure in severely affected individuals with symptoms including muscular dystrophy, myotonia (an inability to relax muscles after voluntary contraction), endocrine dysfunction, mental retardation, cardiac abnormalities and testicular atrophy in males. It also shows a variable pattern of onset, ranging from a congenital condition with severe symptoms to a classic adult-onset form and a mild late-onset form. The disorder shows anticipation, with increasing severity and decreasing age of onset with each succeeding generation. Often the symptoms are so mild in the grandparents or parents that the disease had not been diagnosed before its occurrence in a severe form in the grandchildren.

The disorder is caused by the expansion of a CTG repeat located in the 3′ UTR of a gene known as the **dystrophia myotonica protein kinase (*DMPK*)**. The repeat is polymorphic in normal populations with repeat numbers in the range 5–35. This expands in affected individuals to repeat numbers in the range 50 to >3000. There is an approximate correlation between repeat length and severity and age of onset. The number of repeats increases with each succeeding generation, explaining the observed anticipation. Severely affected congenital cases show the highest number of repeats. Mildly affected or asymptomatic forebears of more severely affected individuals will have between 50 and 80 repeats. The expansion is not sex-limited as it has been observed in parents of both sexes.

5.5.3.1 The molecular pathology of myotonic dystrophy

At first the way in which the expansion caused the diverse symptoms of DM was puzzling. Since *DM1* is an autosomal dominant disease, loss of *DMPK* function could have its effect through haploinsufficiency; that is, one copy does not express sufficient protein for normal function. However, *DMPK* knockout mice did not show the symptoms of the disease, which makes haploinsufficiency unlikely. Another hypothesis was that the expansion altered local chromatin structure, affecting the expression of a nearby gene, *SIX5*, which was an attractive candidate because it encodes a homeobox transcription factor, a class of protein typically involved in development. However, while *six5* knockout mice did show a form of cataract, it was not exactly the same as the cataracts that form in patients with DM, and the mice showed none of the other DM symptoms.

The first clue as to what was happening in patients with DM was the observation that transcripts bearing the expansion were retained in the nucleus in discrete foci. Then a second gene was identified, *DM2*, where mutations caused a similar disease to *DM1*. Like *DM1*, the mutation in *DM2* was an expansion, this time of a tetranucleotide CCTG, which is related to the CTG expanded in *DM1* patients. *DM2* encodes a zinc finger transcription factor. The different types of protein encoded by *DM1* and *DM2* – a kinase and a transcription factor – make it unlikely that the loss of a specific protein function is responsible for the disease, especially as the *DM2* expansion occurs in a non-coding intron. Significantly, *DM2* mRNA molecules carrying the expansion are also retained in the nucleus in discrete foci. This suggested that, rather than affecting the expression of the gene in which they were located, transcripts carrying the expansion sequester a protein required for processing and/or export of mRNA from the nucleus, and this general effect of transcription is responsible for the diverse aspects of the

DM syndrome. We now know that the expansion affects the balance of two proteins in the nucleus called muscle blind-like (MBNL-1) and CUG-BP1: when the expansion is present the level of MBNL-1 is reduced compared with CUG-BP1. The balance of these two proteins controls a developmental switch in alternative splicing in 13 genes. When MBNL-1 level is lower than CUG-BP1, the fetal form of transcript persists into adulthood. The identity of these 13 genes is highly significant. They encode proteins involved in exactly those functions that are defective in patients with DM. A striking example is the splicing of the mRNA encoding the insulin receptor, IR. The fetal form of the spliced mRNA lacks exon 11, which is present in the adult form. The resulting altered structure of the insulin receptor explains the insulin-resistant diabetes typically suffered by patients with DM. Other examples include the mRNA for a chloride channel which may be responsible for the myotonia and cardiac troponin T, which explains the cardiac abnormalities.

5.6 Haemoglobinopathies

Haemoglobin is the oxygen-carrying molecule in the blood. It forms 90% of the protein content of red blood cells and is by far the most abundant blood protein. It is a tetramer consisting of two α-globin and two β-globin polypeptide chains (α_2 or β_2) each complexed with a molecule of iron-containing haem. It binds oxygen cooperatively, which is important to its physiological function of binding oxygen at the high levels of dissolved oxygen found in the lungs and releasing it at the lower levels found in body tissues. Homotetramers of the same globin type (α_4 or β_4) show hyperbolic oxygen binding and are unable to release oxygen under physiological conditions.

The haemoglobinopathies are a group of diseases that affect haemo-globin production or function, leading to anaemia. Genetic defects directly affecting the globin genes can be divided into two classes of almost equal prevalence: sickle cell disease and the thalassaemias. As well as sickle cell disease and the thalassaemias, glucose-6-phosphate dehydrogenase deficiency is another common hereditary disease that can affect red blood cells. This results in favism (haemolysis after eating beans), anaemia in response to the anti-malarial drug primaquine, and neonatal jaundice.

In sickle cell disease, β-globin is produced in normal quantities but an amino acid substitution causes it to precipitate into inactive aggregates within red blood cells, resulting in the characteristic sickle morphology. In the thalassaemias, one of the globin chains is reduced or absent. This results in an excess of the other type, which is responsible for the clinical manifestations. β Thalassaemias describe disorders in which β-globin is reduced or absent; α thalassaemias describe corresponding disorders in which α-globin is reduced or absent. A complete absence of α-globin (α^0 thalassaemia) causes **hydrops fetalis**, which is fatal at birth due to oxygen starvation. Sickle cell anaemia, milder forms of α thalassaemia and the β thalassaemias show a wide range of clinical severity that may be treatable by regular blood transfusions.

Collectively, the haemoglobinopathies are the most common form of disease traceable to a monogenic defect. Approximately 7% of the world's population are thought to be carriers for one of the disorders. Sickle cell anaemia is most common in Africa, where it affects 100 000 infants each year. There are also significant numbers of cases in other regions where there are populations of African descent, such as the USA (1500 new cases annually), the Caribbean (700 new cases annually) and the UK (140 cases annually). The thalassaemias are especially prevalent in South East Asia. In Thailand, for example, 500 000 people suffer from some form of hereditary anaemia. β Thalassaemia is also common in Mediterranean peoples. Some Mediterranean islands have a particularly high rate: the carrier frequency is 20% in parts of Sardinia and Cyprus, resulting in a population incidence of 1%.

High frequencies of the haemoglobinopathies correlate closely with the worldwide incidence of malaria, indicating how they have arisen. Heterozygotes are more resistant to infection by *Plasmodium falciparum*, the causative agent of the most severe form of malaria. This leads to a polymorphism, where selection for mutant alleles when carried by heterozygotes is balanced by selection against the same alleles when carried by homozygotes.

5.6.1 Globin genes

Both α-globin and β-globin are encoded by a family of related genes that are expressed at different times during development. The α-globin gene cluster occupies 30 kb of chromosome 16 (Figure 5.9). There are two α-globin genes, α_1 and α_2, encoding identical polypeptides and differing only

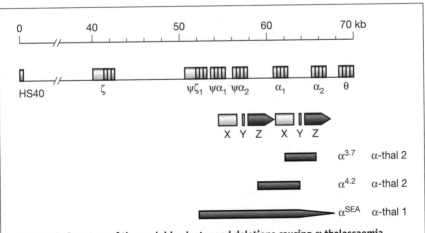

Figure 5.9 Structure of the α-globin cluster and deletions causing α thalassaemia.
X, Y and Z are blocks of sequence homology repeated in the regions surrounding α_1 and α_2 genes. The blue bars show the extent of deletions resulting in α thalassaemia. The figures in superscripts show the size of the deleted regions. SEA, South East Asia.

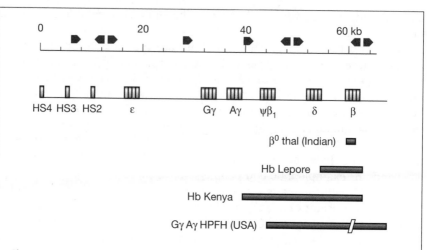

Figure 5.10 Structure of the β-globin gene cluster and deletions causing β thalassaemia. The solid bars indicate the extent of deletions found in particular deletion mutations. HS4, HS3 and HS2 are locus control elements. The arrows above the scale indicate the location and orientation of *Alu* elements.

in small changes in non-coding parts of the gene, such as the 3′ UTR and introns. The two genes are thought to have arisen by duplication and this is reflected by blocks of homology in the surrounding regions. As well as the α-globin genes, there is a related α-globin-like gene called ζ_2-globin which is expressed in early embryonic development (see below). In addition, there are pseudogenes (see Chapter 2) of both α-globin and ζ-globin called $\psi\alpha_1$ and $\psi\zeta_2$ respectively. Finally, there is a gene that resembles α-globin called θ-globin, the status of which is uncertain. It is transcribed, but the resulting protein is apparently not incorporated into haemoglobin. It may be a pseudogene in the early stages of evolution.

The β-globin gene cluster is 70 kb in size and located on chromosome 11 (Figure 5.10). It consists of five genes and one pseudogene. There are two similar adult forms, β-globin itself and δ-globin, which constitutes about 6% of adult β-globin-like chains. During the fetal stage of development, Aγ and Gγ are expressed; they encode β-globin-like proteins that differ only at position 136, which is alanine in Aγ and glycine in Gγ. Finally, ε-globin is expressed in early embryonic development.

As a result of the changing pattern of expression during development, a variety of haemoglobin molecules appear that differ in the composition of the haemoglobin subunits (Table 5.2). These different haemoglobins meet the changing oxygen transport requirements at different developmental stages. During development, the pattern of expression of both the α-globin and β-globin clusters is under the control of locus control regions (LCRs). HS2–HS4 control the expression of β-globin and HS40 controls the expression of α-globin (see Figures 5.9 and 5.10). LCRs are enhancers that bind erythroid-specifc transcription factors. Without the LCR there is very little

Table 5.2 Haemoglobins made at different times during development.

	β-Globins				
	Embryonic		Fetal	Adult	
α-Globins	ε	γ	γ	δ	β
ζ	$\zeta_2\varepsilon_2$ Hb Gower1	$\zeta_2\gamma_2$ Hb Portland			
α			$\alpha_2\gamma_2$ Hb F	$\alpha_2\delta_2$ Hb A$_2$	$\alpha_2\beta_2$ Hb A

expression of the genes, but exactly how they act is still unclear. One striking fact about both clusters is that the physical arrangements of the genes reflect the order in which they are expressed during development, but the significance of this is still unknown. The LCRs map a considerable distance upstream of the clusters. For example, HS40 is located over 60 kb upstream of the α_1 and α_2 genes. Thus they are an extreme example of the way that enhancers can act at distance to activate transcription of the gene they control.

5.6.2 α Thalassaemia

α Thalassaemias are caused by deletions of one or both α-globin genes. An α-thalassaemia 1 (α-thal 1) chromosome has one remaining gene while an α-thalassaemia 2 (α-thal 2) chromosome has both genes deleted. These chromosomes can combine in different ways to result in the presence of zero, one, two or three α-globin genes (Figure 5.11). The severity of the resulting thalassaemia is determined by the number of remaining copies.

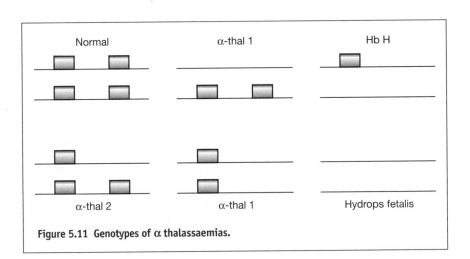

Figure 5.11 Genotypes of α thalassaemias.

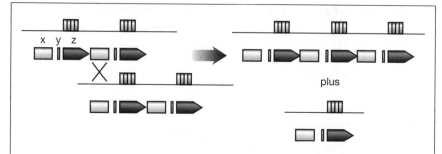

Figure 5.12 Unequal crossing-over results in a deletion of an α-globin gene. Unequal crossing-over between X regions produces one chromosome with a triplicated α-globin locus and a chromosome containing only one. In this example, the resulting deletion would be 4.2 kb in size, as observed in some α-thal 2 chromosomes (see Figure 5.9). Chromosomes carrying three α-globin loci have been observed in populations where α-thal 1 is common. (Redrawn with permission from Roses, A.D. (2000) *Nature*, **405**, 857—865.)

- Individuals with three α-globin genes (α-thal 2) are asymptomatic carriers.

- Two functioning α-globin genes (α-thal 1) results in mild anaemia.

- One functioning α-globin gene results in moderate anaemia with inclusions in red cells consisting of β_4 tetramers known as Hb H.

- Individuals with no α-globin genes die shortly after birth from a condition known as hydrops fetalis. The primary haemoglobin is Hb Barts (γ_4), which has such a high oxygen affinity that it fails to release any oxygen in the tissues.

The deletions apparently result from unequal crossing-over between the duplicated segments which originally gave rise to the two α-globin genes (Figure 5.12). Reciprocal chromosomes with three α-globin genes have been observed.

5.6.3 β Thalassaemia

The severest form of β thalassaemia is β^0 thalassaemia where there is a complete lack of β-globin chains in adult life. The resulting excess of α-globin chains precipitate, leading to red blood cell damage and also inhibiting the production of new red blood cells (**erythropoiesis**). In β^+ thalassaemia there is a large reduction but not complete absence of β-globin synthesis. A relatively benign form of β thalassaemia is hereditary persistence of fetal haemoglobin (HPFH), which occurs when the absence of β-globin is compensated by the continued synthesis of the fetal form. Sometimes, synthesis of fetal haemoglobin persists, but at insufficient levels to compensate for the loss of β-globin; this condition is known as δβ thalassaemia because there is also an absence of δ-globin.

Virtually every type of point mutation that could lead to a reduction or absence of β-globin synthesis has actually been observed in cases of β^0 and β^+ thalassaemia. These include:

- a deletion of the whole β-globin gene;

- promoter mutations preventing or reducing transcription;

- mutations of splice sites preventing the removal of introns;

- mutation of the poly-A acceptor site affecting RNA processing;

- nonsense and frameshift mutations causing termination of translation.

Many deletions affecting the β-globin cluster have been analysed; these are represented in Figure 5.10. One of them, Hb Indian, is the deletion affecting the β-globin gene that results in β^0 thalassaemia. Some of the others are apparently the result of unequal crossing-over between two β-globin genes, resulting in their fusion and the deletion of intervening DNA. For example, Hb Lepore results from a deletion that fuses the 5′ portion of δ-globin with the 3′ portion of β-globin. The reciprocal product of this recombination event, known as Hb anti-Lepore, has been observed.

5.7 Inherited predisposition to cancer

All cancers can be said to be genetic diseases, in that they arise through damage to 100 or so cellular **oncogenes** which, in their unmutated state, normally cooperate to maintain control over cell proliferation (see Box 5.3). The controls are complex so that, for the most part, failure of one component is compensated by the action of a different part of the control network. Before a cell is completely liberated from proliferation controls, multiple parts of the control system must be damaged. The origin of a tumour, and its progression to an invasive malignant state, is thus a multistage process: successive mutations to different oncogenes are necessary. For this reason, the risk of cancer increases progressively with age as the mutations accumulate.

The mutations may be caused by endogenous cellular processes or they may be inflicted by environmental mutagens. Rarely, individuals are born with germline mutations affecting key components of the regulatory network. These are responsible for hereditary predispositions to cancer. Many of the products of these genes have also been found to be mutated in sporadic cancers, and the study of the genes affected by these mutations has provided critical insights in understanding the mechanisms that normally control cell division as well as allowing diagnostic tests for families at risk.

Products of oncogenes operate at different levels. Some function in signal transduction pathways, ensuring that cells divide only when positively stimulated to do so by external **growth factors** collectively known as **mitogens**. Others may encode cellular receptors of the external signal, components of the intracellular signalling apparatus or nuclear transcription

BOX 5.3: THE CONTROL OF CELL DIVISION

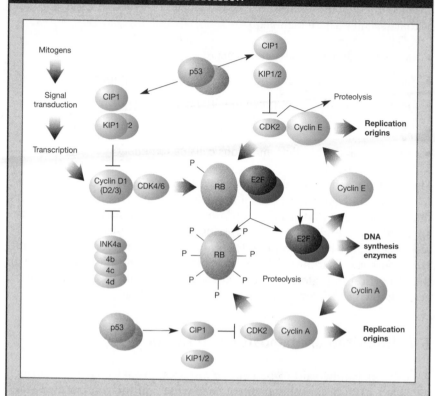

Progress through the cell cycle is controlled by the **cell cycle engine**, which consists of a family of **protein kinases** called cyclin-dependent protein kinases
(CDKs). As their name implies, these kinases must be associated with a cyclin
partner to be active. The periodic appearance and disappearance of different
cyclin families controls the action of the different CDKs, which in turn regulates
the passage of cells to division. Cells are normally only able to divide in the presence of mitogens; in the absence of mitogens cells remain in a quiescent state
with unreplicated DNA (a state sometimes called G_0). Stimulation of a quiescent
cell results in the rise of the cyclin D family of proteins, of which there are three
members, cyclin D1, D2 and D3, which differentially respond to various signals in
different tissues. Cyclin D molecules associate with CDK4 or CDK6. This complex
phosphorylates the retinoblastoma protein (RB), causing it to release a hetero-
dynamic transcription factor called E2F. Free E2F activates transcription of genes
required for DNA synthesis. E2F may also activate its own transcription and that of
cyclin E and cyclin A. Cyclin E expression is periodic, normally peaking in late G_1.
The catalytic partner of cyclin E is CDK2, which also phosphorylates RB, making
this process independent of cyclin D/CDK4/6 and hence the presence of mitogens.
Cyclin A normally appears in **S phase** and levels remain high throughout the rest
of the cycle. It associates with CDK2 in early S phase and in G_2 with a different
cyclin-dependent kinase called CDC2. Cyclin E/CDK2 and cyclin A/CDK2 carry out

key phosphorylation steps that activate replication origins and so trigger DNA replication. Once this occurs, E2F is inactivated through phosphorylation by cyclin A/CDK2, and cyclin E is destroyed by proteolysis to prevent further rounds of DNA synthesis. Cyclin E may be targeted for proteolysis by phosphorylation by its own CDK2 partner.

A key part in the regulation of this process is played by two families of CDK inhibitors called the INK4 family (four members) and the CIP1/KIP family. The INK4 proteins specifically inhibit cyclin D/CDK4. Mutations that inactivate INK4 or cause amplification of the cyclin D gene have been found in a wide variety of commonly occurring cancers. CIP1 is induced by p53, which mediates the DNA damage checkpoint pathway, pausing the cell cycle to allow DNA repair. In the presence of severe damage p53 provokes apoptosis. DNA damage is sensed by p53 through a pathway that includes the AT protein, which is inactivated in ataxia telangiectasia.

factors that change the pattern of gene expression. Mutations in these pathways are often dominant at a cellular level: they change the pattern of activity or result in hyperactivity of the gene product so that the signalling pathway is activated inappropriately. Some of the most commonly occurring types of mutation in this category are mutations to members of the **ras** family, which act at the interface of growth factor receptors and intracellular signalling pathways.

Other types of cell division regulators act to inhibit cell cycle progression. They provide mechanisms, called **checkpoints**, that prevent cell division when it would be inappropriate; for example, they may halt the cell division cycle to allow damage to DNA to be repaired. In the presence of severe damage to DNA they even direct the cell to commit suicide, a process known as **apoptosis**. The way in which they operate is described in Box 5.3. Finally, other types of gene product are involved in the maintenance of genome integrity and DNA repair. Failure of checkpoint and DNA repair functions allows other genetic changes to accumulate, which may further incapacitate the mechanisms that control cell proliferation. Oncogenes therefore occupy a key position in the regulatory network, and their failure has been implicated in virtually every type of cancer.

Mutations to genes encoding checkpoint and repair functions are generally recessive at a cellular level because inactivation of one copy of the gene can be compensated by the function of the other copy. Such genes are known as **tumour suppressors** or **anti-oncogenes** because they act to prevent tumours from occurring. Many of the germline mutations that give rise to hereditary predisposition to cancer are tumour suppressors. Although they are recessive at the cellular level, germline mutations to these genes result in a predisposition to cancer that is inherited in an autosomal dominant pattern. This comes about because a cell in which one copy of the gene is inactivated by a germline mutation requires one further mutation event to inactivate the other. The chances that this will occur during somatic development are high and tumours arise at a high frequency. In contrast, two independent mutation events are required in somatic cells not carrying a germline mutation; such events will occur and result in sporadic cancers, but at a lower frequency.

Germline mutations to tumour suppressor genes commonly result in predisposition to tumours in diverse types of tissue. Sometimes, as in the case of retinoblastoma, one tissue predominates, presumably because of additional controls that protect cells in other tissues. However, in such cases tumours do occur in other tissues. In other cases, such as Li–Fraumeni syndrome, the predisposition is to multiple tumour types.

5.7.1 Retinoblastoma

The study of **retinoblastoma** allowed Alfred Knudson to formulate the tumour suppressor paradigm described above, which has been central to the study of hereditary cancers. Retinoblastoma is a rare childhood cancer of the retina, affecting about 5 in 100 000. About 40% of the cases show an autosomal dominant pattern of inheritance, while the remaining 60% are sporadic. Several features of the inherited form distinguish it from the sporadic form. The penetrance of the hereditary form is 95%, the mean number of tumours is three and tumours in both eyes are common. Survivors are susceptible to other tumours, particularly osteosarcoma and soft-tissue sarcomas, at a rate of about 15% by age 30. In contrast, the children with the sporadic form will normally suffer from a single tumour in one eye and are not susceptible to other types of cancer.

Knudson proposed that the hereditary form occurs when one copy of the gene involved, called *RB1*, is inactivated in the germline and the other is inactivated by mutation during somatic development. He showed that the known rate of mutation at the *RB* locus, the number of target embryonic **retinoblasts** (the embryonic precursor cells), the penetrance of the hereditary form and the population frequency of the sporadic form were all consistent with such a hypothesis. Multiple tumours occur in the hereditary form because of the high probability that more than one retinoblast will be affected by a second mutation hit. The increased occurrence of other tumour types reflects a common role for the *RB1* gene product in proliferation control in other tissues. In contrast to the hereditary form, multiple tumours are rare in sporadic cases because it is extremely unlikely that more than one retinoblast will be independently affected by two hits.

This model predicts that when the second copy of the *RB1* gene is examined it will be found to have been inactivated. This can happen in a variety of ways: (i) point mutation; (ii) deletion; (iii) chromosome non-disjunction; or (iv) mitotic crossing-over. Three of these mechanisms will result in tumour cells becoming homozygous for markers near the *RB1* locus on the chromosome bearing the germline mutation (Figure 5.13). Such a situation is known as **loss of heterozygosity** (LOH).

About 5–10% of hereditary retinoblastoma cases have a deletion of part or all of chromosome band 13q14. This observation identified the location

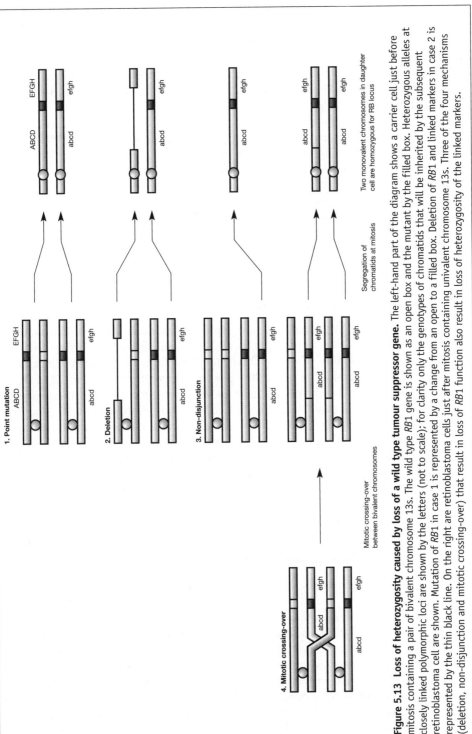

Figure 5.13 Loss of heterozygosity caused by loss of a wild type tumour suppressor gene. The left-hand part of the diagram shows a carrier cell just before mitosis containing a pair of bivalent chromosome 13s. The wild type *RB1* gene is shown as an open box and the mutant by the filled box. Heterozygous alleles at closely linked polymorphic loci are shown by the letters (not to scale); for clarity only the genotypes of chromatids that will be inherited by the subsequent retinoblastoma cell are shown. Mutation of *RB1* in case 1 is represented by a change from an open to a filled box. Deletion of *RB1* and linked markers in case 2 is represented by the thin black line. On the right are retinoblastoma cells just after mitosis containing univalent chromosome 13s. Three of the four mechanisms (deletion, non-disjunction and mitotic crossing-over) that result in loss of *RB1* function also result in loss of heterozygosity of the linked markers.

of *RB*1. The isolation of *RB*1 allowed these predictions to be tested. The use of RFLP markers showed that LOH was common in tumour cells from hereditary cases. Moreover, the introduction of the wild type *RB*1 gene into such tumour cells suppressed their tumour characteristics, providing further experimental verification of the tumour suppressor hypothesis. LOH is now used as a key diagnostic test of whether a gene is acting as a tumour suppressor.

Although the *RB*1 gene was first identified through a rare form of cancer, subsequent research has shown that its product (RB) plays a key role in regulating the cell cycle (see Box 5.3). It acts to block cells from initiating a round of mitotic cell division unless stimulated by mitogens. Other mutations that affect components in the same pathway as RB have also been found in hereditary cancers. For example, INK4a, which encodes one of the p16 family of CDK inhibitors (see Box 5.3), is mutated in familial melanoma. Many sporadic cancers have a high frequency of mutations in components of the RB pathway. For example, homozygous loss of *RB*1 is found in nearly all small-cell lung carcinomas.

5.7.2 Li–Fraumeni syndrome

Li–Fraumeni syndrome is a rare autosomal dominant disease characterised by predisposition to multiple forms of cancer. In women, breast cancer is the most common of these. The gene affected by Li–Fraumeni syndrome is *TP*53, which encodes p53, a key cell cycle regulator (see Box 5.3). p53 provides a checkpoint function in cells with damaged DNA. If the damage is repairable, p53 stalls the cell cycle by stimulating inhibitors of CDK4/cyclin D and CDK2/cyclin E, which are protein kinase/cyclin complexes responsible for triggering the start of the cell cycle (see Box 5.3). If the damage is severe, p53 triggers cell death by apoptosis. The removal of this function allows cells with damaged DNA to divide, increasing genome instability and increasing the chances of damage to another component of the control machinery. Inactivation of p53 is probably the most frequent mutational event in sporadic cancers.

5.7.3 Hereditary breast cancer

Breast cancer is the most common form of malignant cancer among women. Each year in the United States an estimated 182 000 women will develop breast cancer and 46 000 women will die from the disease. Estimates of the cumulative lifetime risk vary from 1 in 12 to 1 in 8. The risk of breast cancer is two-fold higher in women with first-degree relatives (mother or sister) with the disease, with higher risk associated with an earlier age of diagnosis in the relative (<40 years of age). Despite this, no specific genetic factors have been identified in the majority of these cases. However, a subset of between 5% and 10% of cases, known as hereditary breast cancer (HBC), are due to germline mutations. HBC shows a number of characteristics which distinguishes it from the sporadic form:

- HBC clusters in families, showing a pattern of inheritance that is characteristic of an autosomal dominant trait.

- As well as breast cancer there is an increase in other forms of cancer, particularly ovarian cancer and male breast cancer, otherwise a rare disease.

- Age of onset in HBC is much lower: the mean age of diagnosis in HBC is 47 years compared with 62 years in the sporadic form.

- In HBC there is a much higher incidence of bilateral cancers; that is, tumours in both breasts.

5.7.3.1 *BRCA*1 and *BRCA*2

Breast cancer is twice as common in first-degree relatives of women with breast cancer, but generally the disease does not show a Mendelian pattern of inheritance. However, in some extended family pedigrees there are multiple cases and the pattern of inheritance is consistent with an autosomal dominant mutation. Because breast cancer is such a common disease, it is possible that this pattern simply occurs by chance. So, first it was necessary to demonstrate that there really was a mutation segregating. This was done by a survey of the family history of breast cancer in 1579 women diagnosed with breast cancer in a San Francisco clinic. Compared with a random pattern, the data were statistically a much better fit to a model where an autosomal dominant allele was segregating in about 5% of the families. Once it was established that it was likely that there was a subset of families with Mendelian inheritance of breast cancer, families where there were multiple cases of breast cancer were collected and the gene was mapped by testing linkage between mapping markers and the disease. Linkage was found to a marker on 17q21, but to see the linkage it was necessary to rank the families according to age of onset. Only in families where the mean age of onset was less than 47 years was linkage observed; breast cancer in the other families was therefore not caused by an allele of this gene. Mapping the gene allowed it to be cloned using positional cloning methods and it was given the designation *BRCA*1 (breast cancer-associated 1). In other families, although breast cancer was segregating as an autosomal dominant, there was no linkage to *BRCA*1. This led to the identification and cloning of a second gene, *BRCA*2, at chromosome 13q12–13, where autosomal dominant mutation caused a high risk of cancer. A woman carrying either of these mutations has about a 40% chance of suffering from breast cancer by the age of 40, a 66% chance by age 55 and an overall lifetime risk of 82%. *BRCA*1 mutations also result in a high risk of ovarian cancer (40% lifetime risk); *BRCA*2 mutations also elevated the risk of ovarian cancer to a lesser extent (20% lifetime risk), and *BRCA*2 mutations also result in an elevated risk of male breast cancer, an otherwise rare disease.

An accurate estimation of the penetrance of *BRCA* mutations is important for counselling women who test positive for such a mutation. However, there is still considerable uncertainty concerning the penetrance of such mutations. This is because the estimates of penetrance are derived from high-risk families – there may be other factors contributing to the risk in these families that are not present in normal families. For example, the penetrance of a BRCA1 or BRCA2 mutation may be modified by alleles at other loci. In the breast cancer families there may be a higher frequency of unfavourable modifying alleles leading to a higher penetrance of BRCA1 and BRCA2 mutations. In other words, there may have been a bias of ascertainment that led to unusual families being selected because in these families the BRCA1/2 alleles behaved as simple autosomal dominants. BRCA1 and BRCA2 mutations may have a lower penetrance in genetic backgrounds that are more common. If this is the case, the family history of a woman testing positive for a BRCA mutation may be an important factor in estimating the risk.

Cloning BRCA1 and BRCA2 allowed mutations which predispose to breast cancer to be identified. A very large number of different mutations have now been characterised, which, together with the large size of both genes, will make testing for mutations very difficult. In most part, the mutations that have been identified are frameshift and nonsense mutations which would be expected to destroy gene function and thus be recessive at a cellular level. However, this excess of frameshift and nonsense mutations that have been recognised may be an artefact, because it is much easier to decide that such mutations are inactivating compared with mis-sense mutations which may also be harmless variants.

Since the mutations in both genes are inherited as autosomal dominants, it was expected that both genes would be tumour suppressors following the paradigm established for retinoblastoma (see above). This was confirmed when it was found that LOH occurred in nearly all cases examined. Thus HBC comes about when one copy of the gene is inactivated by a germline mutation and the other is lost by an event which led to the inactivation of the remaining functioning copy in somatic cells. Furthermore, BRCA1 has been shown to function as a tumour suppressor *in vitro* as introduction of the wild type gene corrects the phenotypes of cells lacking BRCA1 function.

5.7.3.2 Population distributions of BRCA1 and BRCA2 mutations

The size of the BRCA1 and BRCA2 genes and the large number of different mutations that can occur make it difficult to undertake population screens to measure the mutation frequencies of these genes. The allele frequency of inactivating alleles affecting each BRCA gene is about 1 in 800. Thus the carrier frequency for all HBC mutations may be as high as 1 in 200–300. Although such mutations account for only 5–10% of all breast cancer cases, the frequencies are very high for germline mutations, and breast cancers due to BRCA1/2 mutations may be said to be one of the most common monogenic disorders.

Two particular mutations, BRCA1 185delT and BRCA2 6174delT, have each been found to present at a frequency of about 2% in the US Ashkenazi Jewish population. The BRCA1 185delT was found to be responsible for about 28% of early-onset breast cancers in this population, but the BRCA2 mutation was only found to be responsible for 8% of such cancers. This may indicate that the penetrance of the *BRCA2* mutation is lower.

5.7.3.3 Function of *BRCA1* and *BRCA2* proteins

BRCA1 and BRCA2 proteins are essential for viability because mice in which either *BRCA1* or *BRCA2* genes have been deleted die during early embryogenesis. Cell lines established from tumours resulting from null mutations in either *BRCA1* or *BRCA2* are very sensitive to DNA damaging agents and are characterised by chromosomal abnormalities. Thus, BRCA1 and BRCA2 play an essential role in DNA repair and genome stability. Before describing what is known about their role in these processes, it is necessary to summarise the ways in which DNA can be damaged and the complex cellular responses to this damage.

Damage to DNA can take various forms. Double-strand breaks (DSBs) are caused by external agents such as ionising radiation. Dimerisation of adjacent thymine bases is caused by UV irradiation. Chemical mutagens can cause mismatches between the two strains of the DNA molecule. During DNA replication the replication forks can stall or collapse, which can be induced by treatment with drugs such as hydroxyurea that interrupts the supply of nucleotide substrates or by chemical agents that form covalent crosslinks between the DNA strands. A variety of cellular mechanisms respond to DNA damage as follows:

- DNA damage is recognised by as yet uncharacterised sensors which activate two kinases called ATM (ataxia telangiectasia mutated) and ATR (ATM and Rad3-related). ATM predominantly responds to double-strand breaks while ATR responds to other forms of damage such as stalled replication forks, but there is clearly overlap. These kinases target a variety of downstream effectors which are responsible for the damage responses listed below and illustrated in Figure 5.14. A third kinase, CHEK2, is also a target of ATM/ATR and participates in the phosphorylation of downstream targets. One target of ATR/ATM is histone H2AX. Phosphorylated H2AX (γH2AX) serves as a landmark for localisation of protein complexes that can carry out the responses to DNA damage. Chromatin reorganisation mediated by SWI/SNF proteins at these sites is required to allow access of the repair complexes.

- Checkpoint pathways operating to pause the cell cycle in S phase when DNA is damaged or in G_2 when replication is not complete. Loss of these checkpoints allows the cell to attempt nuclear and cell division with damaged or incompletely replicated DNA. The additional damage this will cause in surviving daughter cells may cause the loss of other controls over cell proliferation, taking the cell further down the path to malignancy.

- Double-strand breaks are a serious form of DNA damage which must be repaired for cell viability. Two pathways can operate to repair double-strand breaks:

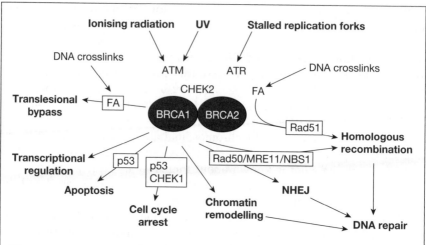

Figure 5.14 The *BRCA*1/2 DNA surveillance and repair network. For explanation see text. FA, Fanconi anaemia pathway.

○ In the homologous recombination pathway, one strand of the DSB is resected in a 5′ → 3′ direction to leave a single-stranded DNA end which invades the DNA duplex of a sister chromatid and uses it as a template to synthesise new DNA to replace the resected sequences. A complex of three proteins, MRE11 RAD51 and NBS1, carries the resection while the RAD50 protein promotes the strand invasion. As result of the strand invasion a complex is formed between the two molecules called a Holiday junction. Holiday junctions are resolved to restore the two DNA duplexes. In meiosis, DSB formation initiates recombination as strand invasion is targeted to the homologous chromosome, resulting in chiasmata and crossing-over. The homologous recombination pathway is an accurate process which repairs the DSB in an error-free manner.

○ Non-homologous end joining (NHEJ) is a process that joins the free ends of two DNA molecules by resecting one strand of each end until a short stretch of homology in each is uncovered; this is then used to join the two ends together. NHEJ repair of the DSB can result in deletion of the resected sequences and can even join two ends together that originated from different DNA molecules, causing chromosomal rearrangements. NHEJ is therefore an error-prone form of repair. It is presumably used as an emergency backup to the error-free homologous recombination pathway.

● Mismatch repair, as its name implies, repairs mismatched base pairs by excising one strand of the DNA and using the remaining strand as a template to fill in the gap. This process can be activated during transcription when the RNA polymerase encounters a mismatch – a process known as transcription-coupled repair. In transcription-coupled repair the transcribed strand is preferentially used as the template. The mechanism of mismatch repair was first elucidated in bacteria and yeast, but it is conserved in humans. Mutations in the human homologs are responsible for a form of colon cancer (HNPCC – see Section 5.7.4.2).

• If the damage to DNA is too severe to be repaired, apoptosis is induced to kill the cell and prevent it from proliferating. If cells with damaged DNA escape apoptosis, they may proliferate in an uncontrolled fashion as the damage inactivates mechanisms that normally control proliferation. A key protein is p53, which makes the decision of whether to halt the cell cycle if the damage to the DNA is repairable or initiate apoptosis if it is not. Germline mutations to *TP53*, which encodes p53, cause Li–Fraumeni syndrome which is characterised by a predisposition to a variety of cancers (see Section 5.7.6). Somatic mutations to p53 are found in a high proportion of spontaneous cancers.

5.7.3.4 *BRCA*1 and *BRCA*2 proteins

BRCA1 plays a central role in mediating many of the different DNA damage responses described above (Figure 5.14), while BRCA2 is thought to be mainly involved in promoting homologous recombination to repair DSBs. Both BRCA1 and BRCA2 are phosphorylated by ATM/ATR in response to DNA damage and BRCA1 facilitates ATM/ATR to phosphorylate many of their other targets. Both BRCA1 and BRCA2 co-localise in nuclear foci together with other proteins involved in DNA repair, presumably at sites of DNA damage. In their absence, the repair complexes do not form. Both BRCA1 and BRCA2 are large proteins – 1863 and 3418 amino acids respectively) that interact with each other and with many other proteins, acting as a scaffold for the assembly of the protein complexes that repair DNA (Figure 5.15).

At its N-terminal end, BRCA1 contains a ring finger domain that interacts with BARD1 to form a heterodimer which has ubiquitin-ligase activity that may target other proteins for degradation, although the nature of the targets remains unclear. The central region of BRCA1 binds to DNA and also interacts with a large protein super-complex called BASC (BRCA1-associated genome surveillance complex) which includes:

• the upstream ATR signalling kinase;

• the RAD50–MRE–NBS1 complex that mediates homologous recombination;

• the MSH2, MSH6 and MLH1 proteins that mediate mismatch repair;

• replication factor C, which loads DNA polymerase onto DNA during repair synthesis;

• a helicase called BLM, encoded by the gene mutated in Bloom's syndrome, which is characterised by radiation sensitivity.

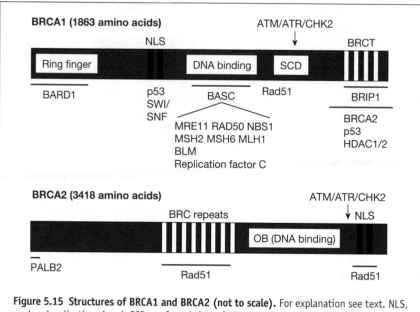

Figure 5.15 Structures of BRCA1 and BRCA2 (not to scale). For explanation see text. NLS, nuclear localisation signal; SCD, preferred sites of ATM phosphorylation; HDAC, histone deacetylase. SWI/SNF are involved in chromatin remodelling.

At its C-terminus BRCA1 contains a series of repeats called BRCT repeats that are found in other proteins involved in DNA repair. BRIP1 (BRCA1-interacting protein) binds to this region. BRIP1 has helicase activity and may be important in translesion DNA synthesis that allows the DNA replication machinery to bypass a site of DNA damage. BRCA1 associates with BRCA2 in the C-terminal region but is at least partly independent of the BRCT repeats.

BRCA2 is characterised by a central region that contains a series of repeats called BRC repeats (Figure 5.15) where Rad51 binds to facilitate homologous recombination. Rad51 binds to a second site near the C-terminus. The BRC repeat region is also where BRCA1 binds. BRCA2 binds to DNA through a region called the OB fold. A protein called PALB2 binds BRCA2 in its N-terminal region, which is required to localise BRCA2 to sites of DNA damage.

5.7.3.5 Fanconi's anaemia and the BRCA network
Fanconi's anaemia (FA) is a rare disease characterised by bone marrow failure, developmental abnormalities and predisposition to cancer. Cell lines derived from patients with FA are characterised by a high sensitivity to DNA crosslinking agents which prevent the DNA strands from separating during replication and transcription. There are 13 separate FA genes or complementation groups. Three of these genes are part of the *BRCA1/BRCA2* genome surveillance network: *FANCD1* is actually the same gene as *BRCA2*, while *FANCN* is the same gene as *PALB2*, whose product interacts

with *BRCA2* to locate it at sites of DNA damage. Thus the FA pathway apparently targets *BRCA2* to guide to sites of DNA crosslinks to promote repair by homologous recombination. *FANCJ* is the same gene as *BRIP1*, whose helicase activity will allow it to disentangle collapsed and stalled replication forks to allow translesion DNA synthesis.

5.7.3.6 Alleles of *BRCA1*-associated and *BRCA2*-associated genes elevate breast cancer risk

BRCA1 and *BRCA2* mutations account for about 15% of the two-fold excess risk to the first-degree relative of a patient with breast cancer. Despite extensive searches, no other genes have been identified conferring a similar risk. However, alleles in the genes encoding ATM, CHEK2, PALP2 and BRIP1 have been found to elevate the risk of breast cancer two- to three-fold. While this is much lower than the risk associated with mutations in the *BRCA* genes, they constitute moderately high risks compared with most of the risk alleles of common complex diseases discussed in Chapter 6, which typically confer a risk of 1.2 or lower. They are present at a population frequency of 0.1–1%, which is less common than risk alleles of common complex diseases, the frequency of which is typically over 10%. Taken together, the combined effect of known risk alleles of breast cancer accounts for about 25% of the total familial risk. The remaining risk is probably polygenic; that is, due to the combined effects of a large number of alleles which each confer a low risk. Efforts to identify these alleles are described in Chapter 6.

5.7.4 Colorectal cancer

Colorectal cancer is the second or third most common cancer of the western world, with a lifetime risk of between 5% and 6%. It is rarer in the developing world and Japan. About 90% of cases are sporadic. However, there are two well characterised hereditary forms: **familial adenomatous polyposis** and **hereditary non-polyposis colorectal cancer**. Both forms fulfil the classic criteria of tumour suppressors.

5.7.4.1 Familial adenomatous polyposis

Familial adenomatous polyposis (FAP) affects between 1 in 15 000 and 1 in 8000 of the population. It is characterised by the presence of hundreds to thousands of benign adenomas or polyps covering the colon and rectum. These polyps are clonal in origin. Some of these polyps are larger than others and from these malignant carcinomas arise. FAP is inherited as an autosomal dominant disease. The gene involved is called *APC* (adenomatous polyposis coli). It was mapped by linkage studies to 5q21–22 and positionally cloned in 1991. *APC* is composed of 15 exons, producing a 9-kb mRNA. The cellular role of the gene is currently unknown but it is thought to be involved in cell-to-cell signalling or adhesion. Most cells in polyps are homozygous for mutations or deletions of the *APC* gene. Occasional polyps are found in normal individuals and these also bear mutations in the *APC* gene. However, loss of the *APC* gene is not sufficient for carcinomas to develop. Mutations in other genes, such as *TP53* encoding p53, *KRAS* (a member of the *ras* family), and another tumour suppressor gene known as *DCC* (deleted in colon cancer) are found in full-blown carcinomas.

5.7.4.2 Hereditary non-polyposis colorectal cancer

Hereditary non-polyposis colorectal cancer (HNPCC), also known as Lynch syndrome, is a hereditary cancer that is not associated with the formation of polyps. It is characterised by an earlier age of onset than sporadic colorectal cancer and frequently involves cancer in other organs. Estimates of its incidence vary widely because of difficulties in distinguishing it from the sporadic form. The classical form of the disease is defined by its occurrence in three members of a family in successive generations, with one affected individual diagnosed before age 50. Some estimates are as low as 1 in 10 000, while others are as high as 1 in 200. Estimates of what proportion of all colorectal cancers are caused by HNPCC range from 0.7% to 13%. A mean of these figures would indicate a population incidence of 1 in 400, making it one of the most common of monogenic disorders.

HNPCC is characterised by instability of multiple microsatellite loci located throughout the genome. This suggests a defect in mismatch repair of DNA. Mismatch repair operates by excising one strand of a stretch of DNA that contains a base pair mismatch. The missing strand is then replaced by the action of DNA polymerase, using the remaining strand as a template. Mutations in genes mediating this process characteristically result in a decrease in the stability of repeated DNA sequences.

After an HNPCC susceptibility locus was located to chromosome 2p by linkage analysis, a search was undertaken for genes that might provide this function. A candidate gene, human *MSH2*, was cloned that is a homolog of the yeast *MSH2* gene and the bacterial *mutS*, both of which are known to mediate mismatch repair. It was shown that mutations affecting *MSH2* co-segregated with the disease in several HNPCC kindreds. Subsequent to the identification of *MSH2*, three other genes have been cloned that are also involved in mismatch repair and in which germline mutations co-segregate with the disease in different HNPCC kindreds. These genes are known as *MLH1*, *PMS1* and *PMS2*. Most cases (~80%) of HNPCC are caused by mutations in *MLH1* or *MSH2*. Most mutations affecting these genes result in truncated proteins or substitutions at highly conserved residues. In both cases an inactive protein would result. Thus it seems likely that the two-hit tumour suppressor model applies to HNPCC. Individuals with germline mutations inactivating one copy of the gene develop tumours when the other copy is lost because of somatic mutation.

5.7.5 Neurofibromatosis

Neurofibromatosis (NF) is characterised by so-called *café-au-lait* pigmented skin patches and disfiguring benign tumours called neurofibromas, which are outgrowths of **neural crests**. It exists in two forms, the more common, NF type 1 (NF1), being one of the most widespread autosomal dominant diseases, affecting 1 in 3000 children. Mutations in the *NF1* gene predispose to a number of tumours that originate in cells of the nervous system.

The *NF1* gene is located at chromosome 17q11 and encodes a protein known as neurofibromin. Neurofibromin is an inhibitor of the *ras* oncogene, acting in the same manner as another protein, GTPase-activating protein (GAP; Figure 5.16). The tissue specificity of *NF1* mutations may reflect an absence or lowering of GAP function in neural crest cells, making them more dependent on neurofibromin. Tumours arising in patients with *NF1* frequently have inactivating mutations in the remaining copy of the *NF1* gene. Therefore the *NF1* gene behaves as a classic tumour suppressor gene, in terms of both its cellular function as an inhibitor of an oncogene and its genetic properties.

5.7.6 Ataxia telangiectasia

Ataxia telangiectasia (AT) is another disease, like Li–Fraumeni syndrome and retinoblastoma, where rare germline mutations help to identify an important part of the normal machinery that protects against cancer. AT is a rare autosomal recessive disease, affecting between 1 in 100 000 and 1 in 40 000

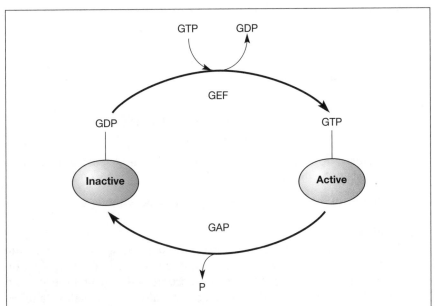

Figure 5.16 The *ras* cycle. When bound to a molecule of GTP, *ras* is inactive. Hydrolysis of GTP mediated by GTPase-activating protein (GAP) converts it to the inactive state. Exchange of GDP for GTP, mediated by guanosine exchange factor (GEF), converts it back into the active state. Oncogenic mutations lock *ras* into the active form.

individuals worldwide. It is a complex disease, characterised by a variety of apparently unconnected symptoms that first become apparent at the age of 3 years and lead to death in the second or third decade:

- **Ataxia** caused by cerebellar degeneration, which leads to progressive neuromotor deterioration.

- Dilated blood vessels in the eye and face (telangiectasia).

- Severe immune deficiency and absence of the thymus.

- Acute sensitivity to ionising radiation.

- Predisposition to multiple cancers.

- Heterozygotes are mildly affected but exhibit predisposition to cancer and radiation sensitivity.

Genetic studies were confused by there being apparently four complementation groups, each manifesting distinct phenotypic characteristics. A single gene was located by linkage mapping, so the complementation observed is apparently an example of intragenic complementation. At a cellular level, AT cells behave as if they have a defect in the p53-mediated checkpoint pathway. The gene encodes a large protein with similarities to phosphoinositol 3-kinases, which have already been shown to be part of intracellular signalling pathways and to be involved in the control of DNA synthesis and meiosis in yeast. The checkpoint defect explains the predisposition to cancer and radiation sensitivity.

5.8 Summary

Benefits of cloning genes
- It tells us about the role of the affected protein within the cell, allowing research into new drugs to treat the disease.

- Characterisation of mutations that result in the disease allows diagnostic tests to be developed.

- Availability of the wild type gene may lead to gene therapy.

Positional cloning
- The gene is mapped by looking for linkage between polymorphic genetic markers and the occurrence of the disease. Key recombination events can define the limits on a chromosome within which the affected gene is located. Cytogenetic rearrangements also provide valuable information about a gene's location.

- Detailed annotation of the human genome identifies all the genes in the region identified by genetic mapping. Often a candidate gene can be identified because the likely function of one of the genes suggests it could be plausibly involved in the disease.

- Co-segregation of mutations with the disease confirms that the correct gene has been identified. However, it is necessary to distinguish between neutral variants and disease-causing mutations. Nonsense and frameshift mutations or expansions of tandem repeats that exclusively segregate with the gene are strong indicators that a mutation is causal. Other criteria include a substitution of a highly conserved amino acid or an amino acid substitution that is likely to disturb protein function, such as a polar residue in a transmembrane domain.

Cystic fibrosis
- CF is the most common autosomal recessive disease that affects Europeans.

- The gene affected is called *CFTR*.

- The CFTR protein transports chloride ions across the cell membrane. Loss of function disrupts the flow of water onto the epithelial surface, leading to a thick sticky mucus that cannot be cleared by the action of cilia. As a result, bacteria are not cleared, leading to chronic and recurrent infections.

- Two-thirds of mutant alleles are the same mutation, called ΔF508.

- As well as acting as a chloride channel, it represses the activity of the ENaC sodium channel. Loss of this regulation may also be an important aspect in the loss of epithelial surface hydration.

Duchenne muscular dystrophy
- DMD is a sex-linked disease.

- The chromosomal locus is exceptionally large (2.5 Mb).

- The gene affected encodes a protein called dystrophin that links the actin cytoskeleton to the extracellular cortex via a complex called DASC.

Trinucleotide repeat expansions
- Diseases that result from trinucleotide repeat expansions show anticipation. This is where the penetrance, severity or age of onset of a disease increases through successive generations.

- Expansions involving the trinucleotide CCG are responsible for fragile sites in chromosomes, including fragile-X syndrome. These occur in the 5′ non-coding area of genes and result in their inactivation. FMRP, the protein depleted in fragile-X syndrome, binds mRNA and inhibits their translation at the tips of dendritic spines. Loss of this regulation results in long immature spines and long-term depression of spine activity.

- Expansions involving CAG in the coding region are responsible for neurodegenerative disorders, including HD. The CAG expansion results in the expansion of the polyglutamate tract near the N-terminal of the encoded protein which is called huntingtin. Huntingtin has multiple roles in cell biology, and the molecular pathology is complex and still not fully understood.

- Myotonic dystrophy 1 (DM1) is caused by an expansion of a CTG in the 3′ UTR of a protein kinase. A similar disease, DM2, is caused by a CCTG expansion in a gene encoding a zinc finger transcription factor. In both cases the pathology is likely to be caused by mRNAs carrying the expansion affecting the splicing of other genes. This results in a multiple loss of gene function which is responsible for the diverse defects in myotonic dystrophy.

Haemoglobinopathies

- These are diseases affecting haemoglobin that result in severe anaemias.

- Sickle cell disease results from a single amino acid substitution.

- Thalassaemia results in a reduction of α-globin or β-globin levels.

- β^0 Thalassaemia results from mutations that prevent any β-globin expression. Milder forms of α thalassaemia can arise when deletions remove the β-globin gene but there is a compensatory persistence of fetal haemoglobin.

- There are two copies of the α-globin gene in the cluster so there are four copies in each diploid individual. The severity of α thalassaemias correlates with the number of remaining α-globin genes. The most severe is hydrops fetalis in which there are no remaining copies.

Hereditary cancers

- A complex network controls cell proliferation. Cancer results from the accumulation of somatic mutations that disrupt this network.

- Tumour suppressors are genes that normally suppress cancers; cancers result when both copies are knocked out by mutation.

- Many hereditary cancers result when one copy of a tumour suppressor gene is inactivated by a germline mutation and the other copy is lost by a somatic event.

- Retinoblastoma is a rare form of cancer, the study of which led to the identification of a key tumour suppressor gene that is mutated in many sporadic cancers.

- About 5–10% of breast cancer is caused by mutations in two genes, *BRCA*1 and *BRCA*2, that act as tumour suppressor genes.

- The BRCA1 and BRCA2 proteins act in the DNA repair and genome surveillance pathways and show multiple interactions with proteins involved in DNA damage and checkpoint control.

- Hereditary colon cancers can also be caused by mutations in tumour suppressor genes:
  - ○ FAP is caused by mutations in the *APC* gene.
  - ○ HNPCC is caused by mutations in genes responsible for mismatch repair.
- NF is caused by mutations in a gene that is a negative regulator of the *ras* oncogene that is commonly mutated in sporadic cancers.

Further reading

Positional cloning

ROCK, M.J., PRENEN, J., FUNARI, V.A. *et al.* (2008) Gain-of-function mutations in *TRPV*4 cause autosomal dominant brachyolmia. *Nature Genetics*, **40**, 999–1003.

The example of positional cloning used in Box 5.1

Cystic fibrosis

BOUCHER, R.C. (2007) Cystic fibrosis: a disease of vulnerability to airway surface dehydration. *Trends in Molecular Medicine*, **13**, 231–240.

Reviews the action of CFTR in maintaining hydration of lung airway surfaces and pathology caused by the absence of CFTR in patients with CF.

COLLINS, F.S. (1992) Cystic fibrosis: molecular biology and therapeutic implications. *Science*, **256**, 774–779.

Summarises the structure of the *CFTR* gene and the protein it encodes.

Duchenne muscular dystrophy

AHN, A. and KUNKEL, L.M. (1993) The structural and functional diversity of dystrophin. *Nature Genetics*, **3**, 283–291.

BROWN, R.H. (1997) Dystrophin-related proteins and the muscular dystrophies. *Annual Review of Medicine*, **48**, 457–466.

KOENIG, M., HOFFMAN, E.P., BERTELSON, C.J. *et al.* (1987) Complete cloning of the Duchenne muscular dystrophy (DMD) cDNA and preliminary genomic organisation of the DMD gene in normal and affected individuals. *Cell*, **50**, 509–517.

MONACO, A.P. and KUNKEL, L.M. (1987) A giant locus for the Duchenne and Becker muscular dystrophy gene. *Trends in Genetics*, **3**, 33–37.

ROBERTS, R.G. (1995) Dystrophin, its gene, and the dystrophinopathies. *Advances in Genetics*, **33**, 177–231.

Trinucleotide repeat expansions

ORR, H.T. and ZOGHBI, H.Y. (2007) Trinucleotide repeat disorders. *Annual Review of Neuroscience*, **30**, 575–621.

A comprehensive review of all the trinucleotide repeat expansion disorders.

Fragile-X syndrome

BEAR, M.F., HUBER, K.M. and WARREN, S.T. (2004) The mGluR theory of fragile X mental retardation. *Trends in Neurosciences*, **27**, 370–377.

Describes in detail the theory that FMRP acts to inhibit the long depression that is triggered by stimulation of the mGluR receptor.

GARBER, K.S., SMITH, K.T., REINES, D. and WARREN, S.T. (2006) Transcription, translation and fragile X syndrome. *Current Opinion in Genetics and Development*, **16**, 270–275.

Describes how the FMRP protein is an RNA-binding protein that interacts with the RISC machinery to inhibit translation at the tip of dendrite spines.

GARBER, K.B., VISOOTSAK, J. and WARREN, S.T. (2008) Fragile X syndrome. *European Journal of Human Genetics*, **16**, 666–672.

General review of fragile-X syndrome, including its symptoms, molecular pathology and possible treatments.

O'DONNELL, W.T. and WARREN, S.T. (2002) A decade of molecular studies of fragile X syndrome. *Annual Review of Neuroscience*, **25**, 315–338.

Reviews the unusual genetics of fragile-X syndrome.

Huntington's disease

IMARISIO, S., CARMICHAEL, J., KORULCHUK, V. *et al.* (2008) Huntington's disease: from pathology and genetics to potential therapies. *Biochemical Journal*, **412**, 191–209.

Comprehensive general review of Huntington's disease.

BATES, G.P. (2001) Exploiting expression. *Nature*, **413**, 691–694.

Brief review of the way that mutant huntingtin may inhibit histone transacetylase activity of CBP and how this could be potentially reversed by HDAC inhibitors.

BATES, G.P. (2006) One misfolded protein allows others to sneak by. *Science*, **311**, 1385–1386.

Brief review of the experiments in the worm that demonstrate that misfolded aggregates of poly-Q peptides impair protein quality surveillance mechanisms.

Globin gene clusters

COLLINS, F.S. and WEISSMAN, S.M. (1984) The molecular genetics of human haemoglobin. *Progress in Nucleic Acid Research and Molecular Biology*, **31**, 314–462.

GROSVELD, F., ANTONIOU, M., BERRY, M. *et al.* (1993) The regulation of human globin gene switching. *Transactions of the Royal Society of London Series B*, **339**, 183–191.

Inherited cancers: general

BROWN, M.A. and SOLOMON, E. (1997) Studies on inherited cancers: outcomes and challenges of 25 years. *Trends in Genetics*, **13**, 202–207.

KNUDSON, A.G. (1993) Anti-oncogenes and human cancer. *Proceedings of the National Academy of Sciences USA*, **90**, 10914–10921.

SHER, C.J. (1996) Cancer cell cycles. *Science*, **274**, 1672–1677.

Breast cancer

NAROD, S.A. and FOULKES, W.D. (2004) *BRCA1* and *BRCA2*: 1994 and beyond. *Nature Reviews Cancer*, **4**, 665–676.

General review of the structure of *BRCA1* and *BRCA2* and their function in DNA repair and the maintenance of genome integrity.

WANG, W.D. (2007) Emergence of a DNA-damage response network consisting of Fanconi anaemia and BRCA proteins. *Nature Review Genetics*, **8**, 735–748.

Reviews the interaction between *BRCA1/2* and the Fanconi's anaemia pathway.

STRATTON, M.R. and RAHMAN, N. (2008) The emerging landscape of breast cancer susceptibility. *Nature Genetics*, **40**, 17–22.

Reviews the different mutations and alleles that have been identified to increase the risk of breast cancer, including alleles affecting genes in the DNA repair pathway that physically interact with *BRCA1/2*.

Hereditary non-polyposis colorectal cancer

De la CHAPPELLE, A. and PELTOMÄKI, P. (1995) Genetics of hereditary colon cancer. *Annual Review of Genetics*, **29**, 329–348.

Neurofibromatosis type 1

SHEN, M.H., HARPER, P.S. and UPADHYAYA, M. (1996) Molecular genetics of neurofibromatosis type-1 (NF1). *Journal of Medical Genetics*, **33**, 2–17.

Ataxia telangiectasia

SAVITSKY, K., BARSHIRA, A., GILAD, S. *et al.* (1995) A single ataxia gene with a product similar to PI-3 kinase. *Science*, **268**, 1749–1753.

CHAPTER SIX

The genetic components of complex diseases

Key topics

6.1 Introduction

Most of the genes involved in the common monogenic disorders have now been isolated by positional cloning. Monogenic disorders are often devastating in their consequences, but their population incidence is low so their overall contribution to public health is relatively minor. Diseases such as cancer, asthma, migraine, types 1 and 2 diabetes, rheumatoid arthritis, multiple sclerosis, hypertension, cardiac disease, obesity, Crohn's disease and psychiatric disorders such as schizophrenia, manic depression and autism are much more common. Such diseases do not show any clear pattern of Mendelian inheritance. They do, however, cluster in families, indicating that a genetic component is likely to be operating. The aetiology of such diseases is complex, being a mix of environmental and genetic causes. For this reason they are often referred to as complex or multifactorial diseases. Research in human molecular genetics is now focused on identifying the genetic factors involved in complex diseases.

The results from the ENCODE and HapMap projects described in Chapter 4 suggest that there are between 11 and 15 million common single nucleotide polymorphisms (SNPs) in the human gene pool, together with about 5000 **copy number variants** (**CNVs**). In this context 'common' is taken to mean a **minor allele frequency** (**MAF**) of greater than 5%. There is at least the same number of less common alleles. Any of these SNPs and CNVs could contribute to the risk of a complex disease; however, the risk conferred by each may be small. Moreover, there may not be a strong correlation between genotype and phenotype – some individuals may suffer from a disease but not carry a particular risk and vice versa. Thus, the identification of alleles that increase risk may be recognised only by studies that involve thousands of individuals and advanced statistical analysis. It is not surprising that identifying these risk alleles has proved an immensely difficult task and there has been a long history of frustration where reports that an allele contributes to the risk of a disease have not been confirmed by subsequent studies. As recently as 2003 only a small number of alleles could be confidently stated to increase the risk of a complex disease. However, since then a methodology called **genome-wide association study** (**GWAS**) has been developed and is systematically identifying these risk alleles and doing so in a way that is reproduced from study to study.

6.1.1 Why is it important to identify the genetic components of complex disease?

Although the identification of risk alleles is difficult, there are three reasons why it is important to undertake the research:

- Identification of individuals with an increased susceptibility to a disease will allow lifestyle changes to lower the risk, and more intensive medical surveillance to detect the first signs of its onset should it occur. Early diagnosis is nearly always a strong factor in successful treatment.

- Many diseases with apparently similar clinical courses may be heterogeneous in their molecular aetiology (causes). This may affect the prognosis and be the reason why some patients respond well to one particular treatment and others do not. If the different types of defect that can lead to the same disease can be elucidated and suitable diagnostic tests developed, then treatments can be better fitted to the actual disease.

- Cloning the genes contributing to the risk will lead to the characterisation of the encoded proteins. These proteins will be components of the cellular processes and physiological pathways that are malfunctioning in the disease. This will clearly lead to a better understanding of the normal state and what is going awry in the pathogenic state. Such knowledge will be the starting point for the development of more rational therapies.

6.1.2 Overview of methods to find risk alleles

This chapter will describe the different strategies that have been used to identify the alleles that confer a risk of complex disease. We shall see that these can be divided into strategies that look for linkage between mapping markers and the occurrence of the disease in family-based studies, and strategies that search for an association between a genotype and the disease in a survey of a population; that is, alleles that have a significantly higher frequency in affected individuals (Figure 6.1).

Population association studies are inherently more powerful at detecting alleles that contribute a small risk. The recent advances have come about through the development of a methodology called GWAS that combines the power of population-based association studies with the ability to scan systematically the whole of the genome of each individual in the population for alleles that might be contributing to the risk. An important discovery that made this possible is that the human genome is composed of so-called haplotype blocks. This means that the genotype of a small number of SNPs in each block predicts the genotype of all the others in the block. This greatly reduces the number of tests that need to be carried out to discover which of the total 11 million common SNPs in the human genome contribute to the risk of complex disease.

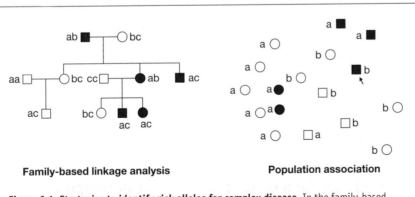

Family-based linkage analysis   **Population association**

Figure 6.1 Strategies to identify risk alleles for complex disease. In the family-based linkage analysis on the left, the disease is co-segregating with the mapping marker 'a'. In the population-based association study on the right, the frequency of allele 'a' is higher in affected compared with unaffected individuals. However, note that this allele is still carried by many unaffected individuals and that some affected individuals will not carry this risk allele (arrow).

Table 6.1 Abbreviations used in Chapter 6.

α	Threshold of statistical significance
AD	Alzheimer's disease
APM	Affected pedigree member. Members of a pedigree affected by a complex disease
ASP	Affected sib pair. Two siblings that are both affected by a disease
CDCV	Common disease common variant. A model of complex disease that considers the risk is caused by the combined action of many alleles, each of which is common in a population
CNV	Copy number variant. A region of the genome that departs from the normal diploid copy number
GRR	Genotype relative risk. The risk of a complex disease owing to a particular genotype compared with the general population
GWAS	Genome-wide association study. Scans whole genomes in many individuals for alleles that are associated with a complex disease
HLA	Complex locus that encodes human leucocyte antigens that form the major histocompatibility antigens (see Box 6.2)
LD	Linkage disequilibrium (see Box 6.3)
MAF	Minor allele frequency
MLS	Maximum likelihood score (see Chapter 3)
OR	Odds ratio. Used in epidemiological studies to compare the risks of a complex disease in two population groups with different genotypes
PAR	Population-attributable risk. The fraction of the overall population incidence of a disease that would be prevented if the allele were not present in the population
SNPs	Single nucleotide polymorphisms
T1D	Type 1 diabetes (see Box 6.1 for description)
T2D	Type 2 diabetes (see Box 6.1 for description)
WTCCC	Wellcome Trust Case Control Consortium
λ	Familial recurrence risk. Risk to a relative of an affected individual compared to the general population. λ_s specifies the relative is a sibling. Locus-specific λ refers to the recurrence risk due to a particular allele

This chapter will make extensive use of abbreviations. These will be introduced formally at the appropriate point; however, to aid overall clarity a summary is presented in Table 6.1.

6.2 Evidence for a genetic component in complex disorders and phenotypic traits

6.2.1 Heritability

Most phenotypic traits (such as height and weight) show continuous variation in a population. Susceptibility to disease also shows population variation, consisting of the sum of the variation in both environmental factors and genetic factors that together constitute an underlying **liability**. The proportion of variation in a population that is due to genetic variation is called **heritability** and is represented on a scale of 0 to 1. It is important to realise that heritability partitions the relative contributions of genetic and environmental variation in the particular population in which it was measured and at the particular time it was measured. Thus, heritability is not a fixed parameter but may differ in different populations, particularly if they are exposed to different environments. In terms of medical genetics, heritability is an important parameter because a high value implies that there are alleles that contribute to the risk that in principle can be identified and mapped. A high value for heritability does not mean that there is no environmental influence. A good example of this is height, which many studies have shown has a heritability of about 0.8 in both humans and animals. Nevertheless, the mean height of many western populations has increased significantly over the past 100 years. Since this change is too rapid to be due to genetic changes, it must be due to changing environments in the form of better nutrition. So heritability for height is high but is subject to environmental alteration. This is not a paradox; it implies that the environmental changes that led to an increase in height have applied uniformly across the population. An important conclusion that can be made from this example is that a high value for heritability does not mean genetic determination. Even though an individual has a genotype predisposing to disease, the onset of the disease may still be averted by appropriate environmental changes, such as diet or exercise.

Since family members are likely to share the same environment as well as the same genes, it is not easy to disentangle the relative contributions of genetic and environmental factors. For the most part, susceptibility to common diseases and phenotypic characteristics does not follow simple patterns of inheritance, so neither is likely to be determined by alleles of a single gene. Nevertheless, family, adoption and twin studies clearly show a high heritability value for many common diseases.

6.2.2 Twin studies

Twin studies provide a natural experiment with which to examine the relative contributions of nature and nurture. The parameter that is usually measured in twin studies is **concordance**. This is defined as the percentage of identical or non-identical twins that both suffer from a disease or exhibit a phenotypic trait when that disease or trait occurs in one member of the pair. Monozygotic or identical twins are genetically identical; in dizygotic twins only 50% of the genes are identical. Since both types are born at the same time, in the same family, the environment in which they grow up is likely to be similar. So the difference in concordance between monozygotic and dizygotic twins is a measure of the contribution that 50% of their genes makes towards the population variability. This allows the heritability to be estimated – the bigger the difference, the higher the heritability. The assumption that twins experience identical environments is subject to a number of caveats. In practice, there will be differences in the actual environment they experience. There may be differences in the position of the fetus, difficulties during birth, they may contract different diseases, their interaction with other family members, friends, teachers, etc. may be different, they will experience different life events such as accidents and so on. Some differences may come about simply through the action of chance and some perceived differences may be the result of error in experimental measurement. All these diverse effects are collected together under the umbrella term non-shared environment.

Table 6.2 shows some examples of twin concordance studies. The difference in concordance between monozygotic and dizygotic twins shows that there is a strong genetic component to many common diseases. Clearly, lifetime health is strongly influenced by genetic constitution. It is interesting to note that a disease such as tuberculosis which may have been assumed to be wholly environmental because it is caused by a bacterial infection shows a high concordance between identical twins. This illustrates how resistance to infection is influenced by genetic factors. For most of the human race, for most of its history, infectious diseases were a common cause of death. It is hardly surprising that natural selection should operate to favour genes that increase resistance.

Table 6.2 Examples of twin concordance values for some common diseases.

Disorder	Concordance (%)	
	Monozygotic	Dizygotic
Breast cancer	6.5	5.5
Cleft lip	35	5
Type 1 diabetes	50	5
Type 2 diabetes	100	10
Multiple sclerosis	20	6
Peptic ulcer	64	44
Rheumatoid arthritis	50	8
Tuberculosis	51	22

Twin studies also show that some behavioural disorders (such as schizophrenia) have at least in part a genetic basis (Table 6.3). One disorder that seems to be strongly influenced by genetic factors is autism. This disorder was previously considered to be entirely environmental; it was said to be caused by cold and aloof parents. Sparing such parents the additional trauma of such a label is a powerful benefit of such studies.

Table 6.3 Some examples for twin concordance values for common psychiatric and behavioural disorders.

Disorder	Concordance (%)	
	Monozygotic	Dizygotic
Alcoholism	40	20
Autism	60	7
Schizophrenia	44	16
Alzheimer's disease	58	26
Major affective disorder	60	20
Reading disability (dyslexia)	64	40

The application of twin and adoption studies to the study of normal phenotypic variation has revealed a strong genetic contribution to the observed phenotypic variation of many fundamental human attributes. Table 6.4 shows that twin concordance is high for general intelligence, personality (extroversion/neuroticism) and even perceived happiness. These studies on the contribution of genetic factors to innate abilities and personality are controversial. In part the controversy is based on technical factors such as the design of the experiments and what is actually being measured. However, the controversy also arises out of a fundamental misconception as to how these studies can be interpreted. They do not show that health, intelligence or happiness are unalterably predetermined by genetics; rather, they allow the observed phenotypic variation in a particular population, in a particular environment, to be partitioned into genetic and environmental variation.

Table 6.4 Concordance studies of general traits.

Trait	Concordance (%)	
	Monozygotic	Dizygotic
General intelligence	80	56
Perceived happiness	50	8
Neuroticism	44	18
Extroversion	50	18

6.2.3 Family studies

Members of a family share a proportion of alleles in common that may be simply predicted from their degree of relatedness. For example, 50% of the genes of siblings or parent/child pairs are identical. Pairs of individuals in

Table 6.5 Alleles shared in common by different classes of relatives.

Class of relative	Proportion of alleles shared (%)	Examples
First degree	50	Parent/child, siblings
Second degree	25	Grandparent/grandchild, aunt or uncle/niece or nephew
Third degree	12.5	Cousins, great grandparent/great grandchild

Table 6.6 Observed risks to relatives of probands suffering from congenital malformations. Pyloric stenosis affects newborn babies and infants. It results in projectile vomiting caused by a malformation of the pyloric sphincter at the junction between the oesophagus and stomach.

Category affected	Cleft lip	Pyloric stenosis
General population	0.001	0.001
First-degree relatives	x40	x10
Second-degree relatives	x7	x5
Third-degree relatives	x3	x1.5

an extended family may be classified as first-, second- or third-degree relatives according to the proportion of genes which are identical (Table 6.5). If genetic factors play a role in the occurrence of the disease, then the average risk that a relative of an affected individual will also suffer from the same disease should be correlated to the degree of relatedness to the affected person. Table 6.6 shows two examples of congenital disorders where this has been shown to be true.

The increased risk to the relative of an individual affected by a complex disease is quantified by the recurrence risk λ for a specified degree of relatedness. Thus the risk to a sibling is defined as:

$$\lambda_s = \frac{\text{Frequency of disease in the sibling of an affected individual}}{\text{Frequency in the general population}}$$

For example, the frequency of type 1 diabetes (T1D) (also known as insulin-dependent diabetes) in sibs of affected individuals is 6% compared with a population incidence in British populations of 0.4%. Thus λ_s for T1D = 15 (6/0.4). It is important to remember that familial risk includes risks due to a shared family environment as well as shared genes. In the case of T1D, it is thought that the shared environment contributes about 20% of the overall familial risk.

6.2.4 Adoption studies

One way to dissect the different contributions of nature and nurture is to measure the recurrence of a disease in biological relatives who have been reared apart through adoption. This may then be compared with the recurrence risk in a control group of non-adoptive families. For example, the risk that the offspring of a schizophrenic parent will be affected is 13%, compared with a population incidence of 1%. In one investigation, the frequency of schizophrenia in adopted offspring of schizophrenic parents was compared to the frequency of schizophrenia in a control group of adopted individuals of unaffected parents. In the 47 adopted children of schizophrenic mothers, five also suffered from the disorder. In contrast, none of the 50 individuals in the control group was affected. Apparently there is little change in the rate of recurrence, regardless of whether a child is reared by its biological or adopted parent. Thus the reason that schizophrenia is familial is that family members share the same genes, not that they share the same environment.

6.3 The genetic architecture of complex diseases

6.3.1 Polygenic variation and the threshold hypothesis

As we saw in the preceding section, genetic factors are likely to contribute to the risk of complex disease. However, these diseases do not segregate in a Mendelian fashion. In fact many, or even most, phenotypic characteristics such as height or weight show continuous variation, which is apparently inconsistent with Mendel's laws that predict discrete phenotypic classes. This inconsistency caused a great controversy at the beginning of the 20th century between those such as Francis Galton, who analysed traits showing continuous variation, and others such as William Bateson, who had rediscovered Mendel's laws of inheritance. The two schools were reconciled by Ronald Fisher in 1918, who showed that continuous variation could arise if a trait were controlled by a number of genes called polygenes. Each allele of the individual polygene segregates in a Mendelian fashion, but only makes a small contribution to the overall phenotype. The combined effect of many polygenes, modified by environmental factors, would produce the continuous variation observed. Figure 6.2 shows the possible genotypes and their relative frequency that arise from the simplest case where just two genes, each with two alleles, segregate in a population. In this example it is assumed that one allele of each gene shifts the phenotype by an equal amount towards one end of the phenotypic spectrum, that each allele has an equal population frequency and that there are no dominance effects. As we shall see in the next section, these are very simplistic assumptions, highly unlikely to apply in reality. Environmental factors also influence the phenotype, so individuals with the same genotype will not have identical phenotypes; rather, they will form a subpopulation that also shows a continuous distribution around a mean determined by the genotype. The observed

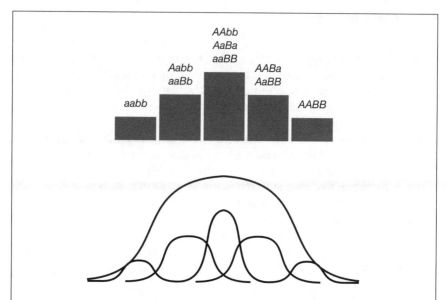

Figure 6.2 Two loci controlling a trait with environmental variation. A quantitative phenotype is controlled by two loci, each with two alleles, *A* or *a* and *B* or *b*. Each copy of *A* or *B* shifts the phenotype by an equal amount. There are thus five possible genotypic classes depending on the number of these alleles. The mean phenotype of each class is determined by the number of *A* and/or *B* alleles. However, environmental variation results in a continuous distribution about the mean of each class, shown by the black curves. The observed population variation will be the sum of these subgroups and approaches a continuous distribution, shown by the blue line.

overall population distribution of phenotypes will be the sum of these separate subpopulations.

Even with two alleles the distribution starts to resemble a continuous curve after environmental effects are considered. With more genes contributing, the match to a continuous curve becomes better. When two individuals with phenotypes towards one end of the phenotypic distribution mate, then the phenotypes of their progeny will form a distribution where the mean is closer to the population mean than the mean phenotype of the two parents, but still shifted relative to the population mean. Because the mean phenotypes of the progeny are shifted from the population mean, the chance that the phenotype of some of the children will be extreme is more likely than with members of the general population. So phenotypes will tend to run in families.

For medically important traits that show a continuous distribution, such as hypertension or obesity, this model is sufficient to account for the observed behaviour of such traits to run in families without showing Mendelian inheritance. In such a model we assume that one of the alleles at each locus increases the phenotypic value of the quantitative character being studied. If this increase is medically unfavourable it is called a risk allele or susceptibility allele. The overall phenotype is thus determined by the sum of the effects of all the risk alleles, modified by environmental variation. However, most diseases are discontinuous in that an individual either does or does not suffer from the disease. To account for this, it is supposed that the influence of polygenes and the environment combine to produce an underlying distribution of liability to the disease in the population. Only when this liability is greater than a threshold value is the disease triggered. This affects only a minority of the population towards one extreme of the liability distribution (Figure 6.3). For the reasons outlined above, the distribution of liability in relatives of an affected individual will be shifted from the population mean, thus they are more likely to be above the threshold and suffer from the disease themselves (Figure 6.3).

Direct evidence for the threshold hypothesis comes from congenital diseases that are more likely to affect one sex compared with the other. For example, pyloric stenosis is five times more common in boys compared with girls. The implication is that, for some reason, the liability threshold is higher in girls compared with boys. If this is the case, then affected girls must have more susceptibility alleles and so relatives of affected girls should be at greater risk than relatives of affected boys. As in the general population, male relatives are at a correspondingly greater risk than female relatives. However, as predicted, both male and female first-degree relatives of female probands are four to five times more likely to be affected than the corresponding sex in the first-degree relatives of male probands.

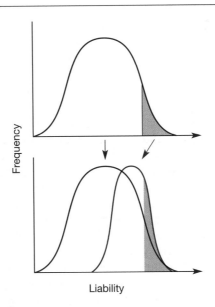

Figure 6.3 **The threshold model for discrete diseases.** There is a continuous distribution of liability in the population. When this exceeds a threshold the disease results (shaded blue). The relatives of individuals who have exceeded this threshold will have a distribution of liability that regresses towards the population mean but is still shifted (blue line). Thus a greater proportion of relatives will exceed the threshold.

6.3.2 The properties of polygenes

The two-gene example outlined in the preceding section makes a number of simplifying assumptions. In order to understand the contribution that each allele makes to the overall population variability, it is necessary to know properties of the genes such as the size of the risk conferred by each allele, the population frequency, whether it acts in a dominant or additive manner and whether there is epistasis with alleles at other loci.

Experimentally, the size of the effect it confers on the phenotype is determined in population-based association studies by the odds ratio (OR), defined as follows:

$$\frac{\text{Cases/controls genotype B}}{\text{Cases/controls genotype A}}$$

Genotype A and B refers to the diploid genotype of a biallelic SNP, which takes into account whether the allele is homozygous for one or other of the alleles or heterozygous. For example, genotype A could be the homozygote for the common allele, while genotype B is a heterozygote. Cases are the numbers of individuals with the disease and controls are the numbers of individuals without the disease matched for possible confounding effects such as age, sex, ethnicity, etc.

Most complex diseases appear to be caused by the combined effect of many alleles, each with OR values less than 1.2. There is a small number of cases where the OR values are significantly higher. For example, the risk allele at the HLA locus which increases the risk of type 1 diabetes results in an OR = 6 in the heterozygous state and OR = 18 in the homozygous state. Another well established example is the *ApoE*E4* allele that increases the risk of Alzheimer's disease, in the heterozygous state OR = 2.84 and in the homozygous state OR = 8. (These values, and other values of this type quoted in this chapter, are published figures from particular studies. The values should be treated as approximate as few studies are large or detailed enough to be precise, and other studies may arrive at slightly different values.) It is now clear that there are very few other alleles, if any, which result in such high OR values. Some alleles may be protective, in which case the OR will be less than 1.0. The risk due to a particular genotype is sometimes referred to as the **genotype relative risk (GRR)** which, as its name implies, is the ratio of the risk of the disease with the genotype compared with the risk without the genotype.

In family-based studies the effect of an allele can be defined in terms of a locus-specific recurrence risk λ that quantifies the effect of an allele in terms of the increased risk it confers to a sib of the affected person compared with the general population. This is defined as:

$$\frac{\text{Expected proportion of affected sib pairs sharing zero alleles identical by descent (25\%)}}{\text{Observed proportion of affected sib pairs sharing zero alleles identical by descent}}$$

In T1D the risk allele at the HLA locus has a locus-specific λ of about 3.0. By comparison, other known risk alleles for T1D, such as at the insulin locus, confer a locus-specific λ of 1.02.

The population frequency of the allele and the size of its effect influence the difficulty of its identification and mapping. An important concept in this context is **power**. In formal terms, power is defined as the percentage probability of rejecting the null hypothesis when it is false to a given threshold of statistical significance (α). In complex disease epidemiology the null hypothesis is that the allele being tested has no effect on the risk. In other words, power estimates the number of subjects needed to reliably detect a risk allele with a given set of properties. Figure 6.4 shows how power to detect a risk allele in an association study varies with the effect size, meausured by genotype relative risk (GRR), allele frequency and the numbers of cases and controls. Power can be seen to depend on all three of these variables. First, the power to detect the allele is greater with 3000 cases compared with 1000 cases. Second, power depends on allele frequency – power is lower with alleles that are either relatively rare (population frequency <0.2) or relatively common (population frequency >0.8). Third, power is proportional to the effect size of the allele – the power to detect an allele with a GRR of 2.0 is high with 1000 cases and controls, provided the allele is either not very rare or very common. However, an allele with a GRR of 1.2 cannot be reliably detected even with 3000 cases and controls. In fact, 6000 cases and controls would be needed to detect such an allele.

When a risk allele is identified it is clearly of interest to know what effect it has on the incidence of the disease in a population. This is measured by the **population-attributable risk (PAR)**, which is the fraction of the overall population incidence that would be prevented if the allele was

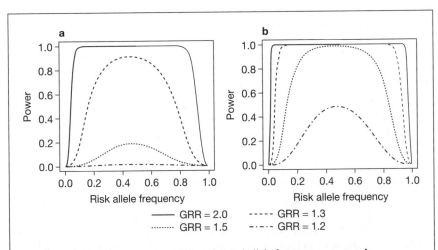

Figure 6.4 The effect of numbers, effect size and allele frequency on power in a population association study. Power to detect a risk allele with the indicated GRR values is plotted against allele frequency with 1000 (a) or 3000 (b) cases and controls.
(Reproduced from Iles, M.M. (2008) *PLoS Genetics*, **4**, e33.)

Table 6.7 Effect of allele frequency and genotype odds ratio on population-attributable risk (PAR) and familial risk measured by the sibling recurrence ratio (λ_s). (Data from Hemminski, K. *et al.* (2006) *Nature Reviews Genetics*, **7**, 958–964.

Allele frequency	Genotype odds ratio					
	1.5		5.0		10.0	
	PAR	λ_s	PAR	λ_s	PAR	λ_s
0.01	1.0	1.0	7.4	1.13	15.2	1.57
0.1	8.7	1.02	43.2	1.38	63.1	1.8
0.5	27.3	1.01	75.0	1.06	87.1	1.08

not present. Table 6.7 shows how PAR of an allele varies with allele frequency and effect size, meausure by its genotype odds ratio. Even though an allele may have a small effect size, it can account for a significant proportion of the population risk if it is common. Table 6.7 also shows the relationship between allele frequency, effect size and familial risk measured by the sibling recurrence risk λ_s. This reveals what may seem a surprising disparity between the fractions of familial and population risk attributable to an allele. An allele which confers a small risk may have a significant effect on population risk if it is relatively common, but may explain only a very small part of the familial risk. Consider for example the allele with a genotype odds ratio of 1.5 and allele frequency of 0.5. Such an allele accounts for 27% of the population risk (PAR) but increases the familial risk (λ_s) by only 1%. The reason for this is that the familial risk is closely related to the amount of population *variation* (heritability) attributable to the allele, and an allele that is common will not explain much of the variation. Consider the extreme situation where a risk allele has reached fixation in a population, i.e. everybody in a particular population carries the allele. It would explain none of the genetic variation in risk and thus none of the familial risk. However, if the allele was not present, say in a different population, then the incidence of the disease would be reduced according to the size of the risk the allele conferred. So it could explain a large proportion of the population risk.

Alleles at a risk locus may be **additive** or **dominant**. Where the alleles are additive the heterozygote has a phenotype midway between the two homozygotes; whereas if one allele is dominant, the heterozygote will have the same phenotype as one of the homozygotes. In population-based studies it is possible to measure the OR for all three genotypes at a diallelic SNP locus. So far the results indicate that most alleles act in an additive fashion.

A final property of risk alleles is the extent to which there is **epistasis**; that is, whether the contribution of the allele is modified by the presence of an allele at a separate locus. If there is no epistasis then the phenotype would be that expected from the independent contributions of the two alleles. This may be calculated on additive or multiplicative models. For example, if two alleles each conferred an OR = 3, then the resulting combined OR would be 6 on an additive model and 9 on a multiplicative model. If the observed OR = 30, then epistasis would be occurring. Epistasis could explain how large risks could arise from the combination of alleles that separately confer only a small risk. Epistasis is difficult to measure experimentally because the number of different combinations with which the alleles may combine means that numbers of individuals with each genotype will be small and the computation complex. Where it has been possible to examine the effect of epistasis, such as in T1D, the early indications are that it is not a big factor.

6.3.3 Possible genetic architectures of complex diseases

Common diseases can come about through different genetic architectures. A disease may be caused by a small number of dominant alleles which are each rare but confer a large increase in risk. This model may explain a minority of cases in diseases such as breast cancer, Parkinson's disease and Alzheimer's disease, where a small subset of cases shows Mendelian inheritance of a dominant allele. There may be many alleles, each of which is relatively common, but each exerts a small increase in risk. This is often referred to as the **common disease common variant (CDCV)** model. The number of polygenes may be of the order of tens to hundreds of genes and the odds ratio conferred by each risk allele may be lower than 1.2. Finally, there are intermediate situations such as T1D and the more common, sporadic forms of Alzheimer's disease, where one allele exerts a relatively large effect, which falls short of a highly penetrant Mendelian allele, and where the remainder of the risk is conferred by a number of polygenes.

6.4 Experimental approaches to identifying risk alleles

6.4.1 Family-based linkage studies

6.4.1.1 Parametric analysis

This refers to LOD score-based linkage analysis used so successfully to map and clone genes responsible for single-gene disorders. It requires a genetic model to be tested in which parameters for the mode of transmission, dominance, allele frequency, penetrance, genetic heterogeneity, etc. are specified. Because it requires the specification of parameters, this type of analysis is called **parametric**. It is the most powerful means of mapping a disease gene, but it works only when the genetic model is correct. In the context of complex disease, parametric analysis has proved successful for mapping the genes responsible for a subset of cases caused by alleles with a large effect segregating in a Mendelian fashion. For example, breast cancer is twice as common in the first-degree relatives of women with the disease as in the general population. In a small number of families, breast cancer segregates in an autosomal dominant pattern. Once the possibility had been discounted that this was due to chance clustering, it was possible to map two loci, BRCA1 and BRCA2, at which mutations led to a high risk of breast cancer, as described in Chapter 5. Alleles affecting DNA repair and checkpoint pathways have also been identified that increase the risk of breast cancer (see Section 6.6.4). However, altogether, the known alleles account for about 25% of the total familial risk of breast cancer. Exhaustive linkage studies have eliminated the possibility that further alleles exist with a similar effect to BRCA1 and BRCA2, so the remainder of the family risk is likely to be polygenic in nature – that is, many alleles, each contributing a small amount to the risk. Furthermore, there is evidence that the penetrance of BRCA1 and BRCA2 is modified by the genetic background of the carriers, which again is likely to be polygenic in nature. A similar situation arises in Alzheimer's disease where a small minority of cases, which are characterised by early onset, are caused by highly penetrant alleles that segregate in a Mendelian fashion as autosomal dominants. Table 6.8 lists some further examples where a small subset of complex diseases can be caused by single-gene defects.

Table 6.8 Examples of loci identified to be responsible for subgroups of complex disease which segregate in a Mendelian fashion. The restricting criteria column refers to indications that can be used to identify those cases where Mendelian segregation may be occurring.

Disease	Restricting criteria	Loci identified
Breast cancer	Age of onset Bilateral tumours Familial clustering	BRCA1, BRCA2
HNPCC (see Chapter 4)	Familial clustering Age of onset	MSH1, MLH1, PMS1, PMS2
FAP (see Chapter 4)	Extreme polyposis	APC
Alzheimer's disease	Age of onset Family clustering	Presenilin II, presenilin I, ApoE, APP
Premature heart disease	Hypercholesterolaemia Age of onset	LDL receptor
Non-insulin dependent diabetes	Onset in young adulthood Family clustering	MODY1 Hepatocyte nuclear factor 4α MODY2 Glucokinase MODY3 Hepatocyte nuclear factor 4α MODY4 Insulin promoting factor 1 MODY5 Hepatocyte nuclear factor 1β

6.4.1.2 Non-parametric linkage analysis

Realistically, parametric analysis can be used only where the disease is caused by a mutation at a single gene. It is not suitable for analysing diseases where more than one gene is involved because of the almost inevitable errors that will arise specifying all the parameters for two or more genes. However, family studies can still be used to search for risk alleles if a genetic model is not specified. The basic assumption of this approach is that if an allele is contributing towards the liability of a disease, then the region of the genome within which the affected gene is situated will be co-inherited from a common ancestor by affected members of a pedigree more frequently than would be expected by chance. Such a statement makes no assumptions about the way the gene is operating, nor any assumptions about how many other genes may also contribute to the risk. Because it does not require parameters to be specified, this type of analysis is called **non-parametric**.

Commonly inherited regions can be recognised by a genome scan following the inheritance of polymorphic microsatellite markers in affected members of a pedigree. If affected members co-inherit the same marker from a common ancestor more frequently than would be expected by chance, then the marker is linked to an allele that confers the risk. In non-parametric analysis genome scans are usually based on **affected sib pairs, (ASPs)**; that is, two sibs in the same family who both suffer from the disease. As well as affected sib pairs, other members of an affected pedigree can be examined (**affected pedigree member or APM**). A set of families containing two affected sibs is assembled. Typical datasets in early studies were composed of about 100–200 families but, as we shall see, this number is insufficient to reliably detect alleles that confer low risks. DNA samples were collected from the affected sibs along with their parents. The microsatellite loci were genotyped using PCR with fluorescent tagged primers. About 300 primers were used in the initial scan, which covered the genome at 10-cM intervals. At each locus there were two alleles in each subject (except for sex-linked loci in boys) distinguished by the length of the amplified PCR products. For each locus the genotype of the parents and both sibs can thus be specified. If the inheritance of chromosomes from each parent is analysed separately, the random expectation is that 50% of the sibs will both have received the same chromosome and so will be identical by descent (IBD) (Figure 6.5). Using the combined data from all the families, the IBD frequencies at each locus were determined and each tested to see whether there was deviation from the value expected from random segregation.

Table 6.9 illustrates how a genome scan was applied to search for alleles leading to susceptibility to type 1 diabetes (T1D). The different types of diabetes are described in Box 6.1. Table 6.9 shows that a clear deviation from a 1:1 random segregation ratio is evident for alleles linked to the HLA locus (see Box 6.2). The MLS score (see Chapter 5) exceeds the level of significance by a large margin. This locus had already been identified by population association studies (see Section 6.4.2) and shows that the method will detect alleles that have a significant effect, but which is less than that of a Mendelian locus.

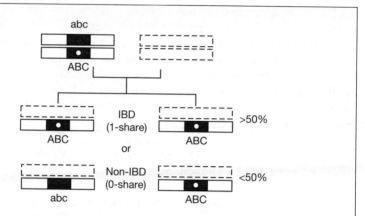

Figure 6.5 Identity by descent (IBD) in affected sib pairs. An allele contributing to the risk of the genome (white circle) is situated in an autosomal region of the genome linked to three mapping microsatellite markers (blue box). The diagram shows the transmission of the region to two affected sibs. Because the microsatellite markers are highly polymorphic, the two haplotypes from one parent (abc and ABC) can be distinguished from the other parent, so the transmission of the chromosomal region from that parent can be followed independently. The chromosomes from the other parent are represented as dashed lines. There are two possible outcomes: either the two sibs inherit the same chromosome (1-share) or they inherit different chromosomes (0-share). The random expectation is that each of these outcomes will occur with an equal (50%) frequency. However, if the allele contributes to the risk of the disease, the region will be inherited by both sibs more often than would be expected by chance. So, in a group of affected sibs the frequency of the 1-share outcome will be significantly higher than 50%.

Table 6.9 A genome scan for T1D. A set of 96 families from the UK where two sibs suffered from T1D were genotyped at 290 marker loci. The value of MLS for significance was set at 2.3 (this would now be considered to be too liberal, leading to the identification of false positive signals). A total of 20 chromosome regions showed some evidence of linkage, although only some are expected to be genuine. The table shows data for three of these where there is evidence of linkage to T1D. Two of the regions correspond to the HLA locus on chromosome 6p and the *INS* locus on chromosome 11q encoding insulin that had previously been implicated by other studies. The effect of the *HLA* genotype on other loci was analysed. However, association at 6q has not been replicated in further studies. (Data from Davies, J. *et al.* (1994) *Nature*, **371**, 130–135.)

	Marker	Locus	IBD		MLS
			1-share	*0-share*	
6p	HLA	*HLA*	97	35	7.3
6q	ESR	*IDDM5*	95	59	1.8
11p	INS	*IDDM2*	107	67	2.1

BOX 6.1: DIABETES MELLITUS

Diabetes mellitus is a disease characterised by the inability of body cells to take up glucose, leading to high levels of glucose in blood and urine, but intracellular starvation. In order to gain energy, cells break down fats. The end product of fat breakdown is acetyl CoA, which is metabolised to acetoacetate, β-hydroxybutyrate and acetone, collectively known as ketone bodies. These accumulate in the bloodstream (ketonaemia) and are excreted in the urine (ketonuria). The acetone contaminates the breath of diabetic people with a characteristic odour. The build-up of ketone bodies is known as ketosis. The uptake of glucose into cells is promoted by the hormone insulin produced in the islets of Langerhans in the pancreas.

There are two different diseases, which result in elevated blood glucose levels:

1. **Type 1 diabetes (T1D)**: this disease is characterised by lack of insulin, and can usually be treated by insulin injections. In most cases the lack of insulin is caused by the autoimmune destruction of the insulin-producing β cells in the islets of Langerhans in the pancreas. Typically the disease shows rapid juvenile onset – the median age of diagnosis is 12 years. However, a build-up of autoantibodies can be detected much earlier. Although the immediate symptoms can be treated with insulin, in the long term people with T1D may face complications which lead to blindness, kidney failure, amputation of lower limbs and heart attacks. It affects about 4 in 1000 Caucasian children of European descent. There are striking ethnic differences in its frequency, the condition being rare in Mongoloid and black populations.

2. **Type 2 diabetes (T2D)**: as its name implies, this disease is not caused by a lack of insulin but is characterised by high blood and urine glucose levels that are resistant to insulin. It typically affects people over age 40 and progresses more slowly than T1D. There are clearly predisposing genetic factors that are apparently unmasked by environmental factors. Insulin resistance is common in obese people and may result from prolonged exposure to high glucose levels resulting from sugar-rich diets or there may be genetic factors that predispose to both diabetes and obesity. T2D accounts for about 85% of diabetes cases and rates are much higher than T1D in developed societies, probably reflecting the effects of affluence on diet.

    A subtype of this disease is MODY (maturity onset diabetes of the young). This disease shows earlier onset than normal diabetes, affecting teenage children and young adults. It often shows a monogenic segregation (see Table 6.8).

Diabetes is a major source of ill health. It may lead to blindness, renal failure, loss of lower limbs, heart attacks and strokes. In children of European–Caucasian descent, T2D is the second most common cause of chronic ill health after asthma. In the United States, there are more than 14 million cases of diabetes. According to the National Diabetes Research Commission, the consequences each year of diabetes in the United States include:

- 15 000–39 000 new cases of blindness
- 13 000 cases of end-stage renal disease
- 54 000 amputations, mostly of lower limbs
- 162 000 deaths from heart attacks, strokes, etc.

Both types of diabetes show strong evidence of a genetic component, especially for T2D, as evidenced by the continued rise in its incidence.

BOX 6.2: THE HLA LOCUS

When organs are transplanted from one human to another, the transplanted organ is rejected because the host recognises the organ as foreign. The rejection is controlled by the major histocompatibility antigens (MHC) displayed on the cell surface. Many of these are encoded by a cluster of genes called the HLA locus (human leucocyte antigens). The HLA locus occupies 4 Mb of DNA on chromosome 6 and contains over 100 genes. The region is divided into three classes of genes: class I, class II and class III. Organs are accepted as self only if they are identical to the host at class I and class II genes which are highly polymorphic. The proteins encoded by these genes are displayed on the cell surface of all cells. During the neonatal period, they induce tolerance in cytotoxic T cells in the thymus. The extracellular domain of both class I and class II proteins is part of the immunoglobulin superfamily. Class III genes encode proteins and peptides with diverse functions of the immune system, including the production of cytokines such as tumour necrosis factor, and complement fixation.

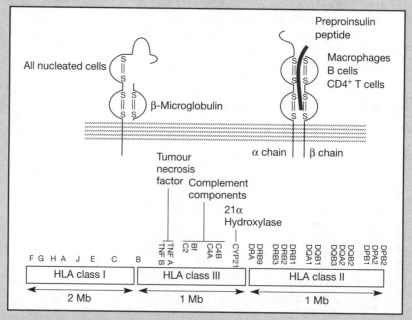

As well as marking self and non-self, class I and class II molecules are responsible for this presentation of foreign antigens to T helper cells. Class I molecules present intracellular antigens from all cells to T cells, which allows the recognition of intracellular infective agents such as viruses. This results in the stimulation of cytotoxic T cells, which kill the infected cells. Class II molecules present extracellular antigens in dendritic cells, B cells and macrophages, so triggering the humoral or antibody response. Class II molecules are heterodimers of α and β chains, encoded by pairs of closely linked genes. For example, DQA1 and DQB1 encode DQα1 and DQβ1 that form a heterodimer. The two chains come together so that a groove is formed between them in which the foreign antigen is located and presented to receptors on the T helper cell. In the figure, the DQα1 and DQβ2 chains are shown presenting a fragment of the insulin precursor chain to T helper cells.

This and other studies with T1D also suggested a number of other regions contributing to the overall genetic liability, but with a much smaller effect. The problem was that the record of replication from one study to another was poor. The apparent linkage at at loci such as the *ESR* locus at 6q shown in Table 6.9 did not replicate in other studies. A large number of genome scans were carried out for the common complex diseases in the late 1990s. Only a few produced robust evidence for linkage that was replicated independently. However, there were valuable examples where evidence of linkage subsequently led to the identification of risk alleles when the regions were examined in more detail by association studies. The identification of alleles at *NOD2/CARD15* involved in Crohn's disease (see Section 6.6.3) and *CFH1* gene in age-related macular degeneration (see Section 6.5.2) relied upon the previous detection of linkage in non-parametric genome scans with affected sib pairs. A further benefit of the negative results was that it eliminated the possibility that there were many more alleles with a large effect similar to the effect of alleles at the HLA locus in T1D.

6.4.2 Population-based association studies

Theoretical calculations (Table 6.10) showed that genome scans of affected sib pairs had a good chance of detecting rare alleles with a large effect (GRR > 4.0) but unless thousands, or even tens of thousands, of families were used there was only a low chance of identifying more common alleles, especially if they conferred a moderate or low risk (GRR < 2). The same theoretical calculations showed that population association studies could detect such alleles with realistically sized sets of cases and controls (Table 6.10).

In population association studies, an attempt is made to find an allele that is associated with the disease. That is, the frequency of the allele is higher in affected individuals (cases) than in unaffected members of a matched control group (see Figure 6.1). This can come about for two fundamentally different reasons. The most obvious reason is that the allele may actually affect a biologically relevant locus. However, association may also be due to linkage disequilibrium (LD), which can occur because cases in the present-day population with the disease have all inherited the same fragment of a chromosome from a common ancestor in whom the original mutation occurred. If the original configuration of closely linked alleles has not been disrupted by recombination in the intervening generations, then the haplotype surrounding the mutation will also appear associated with the disease (Figure 6.6).

Linkage disequilibrium is useful because it effectively increases the target size in association studies. To see the involvement of a gene it is not necessary to find an association with the causative allele itself; association with any of a number of alleles in LD with the causative allele will also identify the genomic region. However, the task then remains of identifying the causative allele. The degree of LD between two loci is measured by the

Table 6.10 Population association methods are more powerful than genome scans with affected sib pairs. For alleles of the indicated effect size and allele frequency, the table shows the numbers of subjects required for 80% power to detect their involvement in a complex disease. With GRR values of 1.5 or less, detection of risk alleles is no longer feasible by allele sharing but can still be achieved with association tests with realistically sized data sets. The calculations for association tests were based on the transmission disequilibrium test which measures the frequency with which an allele is transmitted from a heterozygous parent to affected offspring. So these calculations are based on genotyping two parents and either one or two affected offspring, as shown. Nevertheless, the general conclusion that association tests are inherently more powerful than linkage tests for alleles of modest or low effect size holds good. (Source: Risch, N. and Merikangas, K. (1996) *Science*, **273**, 1516–1517. Reprinted with permission from AAAS.)

GRR	Allele frequency	Allele sharing (number of families required)	Association (number of cases)	
			Single	Sib pairs
4	0.1	185	150	48
	0.5	297	103	61
	0.8	2013	222	161
2	0.1	5382	695	264
	0.5	2498	340	180
	0.8	11 917	640	394
1.5	0.1	67 816	2218	941
	0.5	17 997	949	484
	0.8	67 816	1663	941

parameter D' on a scale of 0 (no LD) to 1 (complete LD) (Box 6.3). A related parameter, r^2, also measured on a scale of 0 to 1, is useful because it is inversely proportional to the number of cases required to detect association if the true risk allele is in LD with the allele being tested.

6.4.2.1 Candidate gene studies

Historically, association studies have been based on examining candidate loci which may be plausibly involved in the aetiology of the disease. Examples of this are associations between alleles at the HLA locus and autoimmune diseases such as type 1 diabetes, ankylosing spondylitis and rheumatoid arthritis. This involvement of the alleles at the HLA locus in autoimmune diseases is the strongest association so far detected with complex disease. The HLA locus is responsible for encoding cell surface antigens that are used to distinguish self and non-self, so it not surprising that variation in these alleles could be a factor in risk of developing an autoimmune disease. In European populations, T1D is strongly associated with *DR3* and *DR4* alleles at the *DRBQ1* gene at the HLA locus (see Box 6.2). Because of LD, the other alleles in the haplotype will appear to be

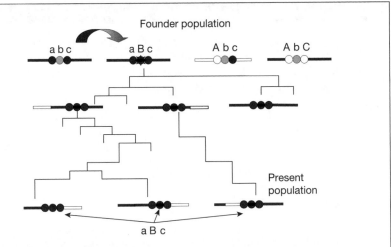

Figure 6.6 Linkage disequilibrium marks a chromosomal region carrying a risk allele.
In an ancestral population a mutation occurs to generate a new allele, *B* (black circle), that contributes to the risk of a complex disease. After the mutation occurs, there are four haplotypes in the ancestral population. Although recombination in subsequent generations gradually mixes the chromosomal content of the four ancestral chromosome types, the haplotype carrying the risk allele is preserved as the alleles are close together and so not shuffled by recombination. In the present-day population, the alleles *a* and *c* are associated with the disease as well as allele *B*; hence the haplotype carrying both *a* and *c* alleles will appear to be associated with the disease as well as the risk allele itself.

associated with the disease and so it is not always clear whether the allele that is observed to be associated with a disease is actually the true cause of the risk. In the case of T1D, extensive studies demonstrated that it is caused not by the *DR3* and *DR4* alleles but rather an allele in a nearby gene called *DQB1* that encodes a class II antigen called DQβ1 (see Box 6.2). Any residue apart from aspartate at this position increases the risk of diabetes. An allele which encodes aspartate at residue 57 is protective; that is, it lowers the risk of diabetes. DQβ1 is part of the dimer that presents intracellular antigens to T helper cells as part of the cell-mediated immune system. In the case of diabetes, it seems that the variant DQβ1 chain results in T-helper cells erroneously recognising the β cells in the pancreatic islets as foreign and initiating an immune attack. One theory is that during the neonatal period when immune tolerance to the body's own proteins is established, the presentation of insulin by the variant DQβ1 chain is inefficient, so immune tolerance to insulin is not developed.

A second clear case of association is the risk of Alzheimer's disease conferred by the *ApoE-ε4* allele. *ApoE* codes for apolipoprotein E, a plasma protein that mediates cholesterol transport and storage. It is synthesised in large amounts in the liver; however, it is also synthesised in the nervous system and its synthesis is increased after brain injury, a known environmental risk factor for Alzheimer's disease (AD). Moreover, it is present in

BOX 6.3: LINKAGE DISEQUILIBRIUM

Linkage disequilibrium is formally defined as a configuration of alleles at linked loci that occurs more often than would be expected by chance. The frequency that two alleles would occur on the same chromosome by chance is the product of their individual frequencies. Thus LD can be defined as:

$$D = f(AB) - f(A) f(B)$$

where

D = The magnitude of linkage disequilibrium
A and B are alleles at two separate biallelic loci
AB is the haplotype with both A and B alleles
f = allele frequency.

D cannot exceed the frequency of the less frequent allele, so on its own the absolute value of D does not give a meaningful measure of LD. It is necessary to normalise D to the maximum value possible, given the frequency of the two alleles.

$$D' = D/D_{max}$$

D' measures LD on a scale of 0 (no LD) to 1 (complete LD). D' provides a direct measure of the amount of historical recombination that has occurred between two alleles.

An alterative measure of LD is r^2, the correlation coefficient between two alleles:

$$r^2 = D^2/f(A) f(B) f(a) f(b)$$

r^2 is also measured on a scale of 0 (no LD) to 1 (complete LD), but $r^2 = D'$ only when LD is complete, i.e. when both r^2 and $D' = 1$. r^2 is important in association studies because it is inversely proportional to the degree to which the sample size must be increased because LD is less than complete. For example, suppose 6000 subjects were needed to detect an association between a disease and the risk allele itself. If an allele in LD with the risk allele is used to detect the association, where $r^2 = 0.8$, the number of subjects required is 6000/0.8 = 7500.

amyloid plaques and neurofibrillary tangles that characterise the disease. Finally, the *ApoE* locus lays within a chromosomal region implicated by linkage studies. All of these factors made *ApoE* a candidate gene. There are three *ApoE* alleles that differ by a single amino acid. *ApoE-ε4* has a population frequency of 16% and is thus a common allele. Association studies quickly showed that the individuals homozygous for the allele had a relative risk of 8.0 compared with individuals without the allele (Table 6.11).

6.4.2.2 The importance of numbers

The risk conferred by the HLA *DQB1* allele in T1D and *APOE-ε4* allele in AD is high, and the association has been repeated in many studies. However,

Table 6.11 Dosage-dependent risk of *ApoE-ε4* allele. The *ApoE* status of 95 AD patients from 42 families was ascertained. (Data from Corder, E.H. *et al.* (1993) *Science*, 261, 921–923.

ApoE-ε4 dose	% affected	Relative risk	Age of onset
0	20	1.00	84.3
1	46.6	2.84	75.5
2	91.3	8.07	68.4

there have been a very large number of reports of association of an allele with a disease which have not been reproduced. The major problem with many studies is that the numbers of cases and controls are too small, so that the study is statistically underpowered. In association studies, power is a function of the size of the risk conferred by an allele, its population frequency and the numbers of samples used in the study. The confusion caused by underpowered studies is illustrated by the involvement of the *pro12ala* polymorphism allele at the *PPARG* locus in the risk of type 2 diabetes (T2D). *PPARG* encodes a protein called the peroxisome activated-receptor γ (PPARγ), a transcription factor that stimulates the differentiation of adipocytes and the expression of adipocyte-specific genes in response to the presence of fatty acids. Moreover, it is known to be the receptor for thiazolidinedione insulin-sensitising drugs used to treat diabetes. Thus it is a logical candidate for involvement in T2D. The more common *pro* allele, which encodes proline at position 12, has a frequency of 0.8, while the less common *ala* allele encodes alanine. An initial report claimed that the *ala* allele conferred a substantial protection, with a GRR value of 0.35. Three subsequent studies failed to confirm this finding while three other studies replicated it. The matter was finally settled by a study that included a large number of subjects and a number of independent patient groups. This study confirmed the protective effect of the *ala* allele with a GRR value of 0.8.

The important point of this study was that the confidence limits overlapped with the confidence limits of all the previous studies. However, because the previous studies had fewer subjects, the confidence limits were so large that there appeared to be no statistically significant effect of the *ala* allele. Thus, the studies that were underpowered hindered the identification of a genuine risk allele. There are additional points to note from this example. First, the risk allele (*pro*) is carried by 80% of the population, a clear example of a common variant common disease (CVCD) model of complex disease. Second, although the risk associated with the *pro* allele is small (OR = 1.2), its contribution to the population risk is substantial (PAR = 25%) because it is so common. Third, while the initial report turned out to be generally correct – the *ala* allele was protective – it overestimated its effect. This is a common occurrence in these studies; the first report of an association often turns out to have inflated the effect of an allele (this issue is considered in more detail in Section 6.5.3.1).

6.5 Genome-wide association studies

Candidate gene studies are limited by our knowledge of biology and disease mechanisms. They are also very wasteful, because a group of patients and controls are collected at great trouble and expense and then tested for the involvement of only one or a small number of the many different variants that might contribute to the risk. It has been a longstanding goal of researchers in this field to be able to carry out so-called genome-wide association studies (GWAS); that is, to assess the possible association with a disease of every region of the genome in the same group of cases and controls. Two developments have allowed this to happen. First, following the completion of the Human Genome Project, technologies were developed that allowed the genotyping of very large numbers of SNPs. Second, when these technologies were applied to investigate patterns of linkage disequilibrium, it was found that the human genome is divided into blocks within which there was only a limited number of different haplotypes. This meant that determining the genotype of a small number of alleles in each block allowed the genotype of a large number of other alleles to be inferred. Together these developments made it feasible to genotype the large number of SNPs required to scan the genome in the large numbers of cases and controls required for statistical power.

6.5.1 SNPs and high-throughput genotyping technologies

As a result of the Human Genome Project, databases of single nucleotide polymorphisms or **SNPs** (pronounced 'snips') were quickly established by systematic resequencing of DNA clones from the different individuals. The ENCODE project resequenced ten 500 Mb regions in 48 individuals from the HapMap project to identify all common polymorphisms in those regions. By extrapolation from the results of the ENCODE project, it is estimated that there are about 11 million common SNPs in the human gene pool (common is generally taken to mean a minor allele frequency of >5%). In principle, any of these could be a risk allele. Once the SNPs had been identified and mapped, a variety of commercial companies developed platforms for the genotyping of up to 650 000 SNPs on a single chip. The technologies used were described in Chapter 4, Section 4.10.1.3.

6.5.2 Haplotype blocks and the HapMap project

Common SNPs are thought to have occurred in the ancestral human population before the expansion out of Africa (see Chapter 9). This means that they would have occurred about 10^4 generations ago. Because the rate of mutation is about 10^{-8} per generation, it is likely that each SNP arose as a single event and therefore arose in a unique haplotype context. Theoretical calculation shows that the amount of recombination that would have occurred would result in LD extending over only a few kilobases at most. However, when the actual SNP haplotypes of real people were genotyped using the high-throughput technologies, it was found that LD extended over much longer distances. In fact, it quickly became evident that the human genome was composed of blocks within which there is only a limited number of different haplotypes. These probably arose for a combination of reasons. First, the frequency recombination is not uniform across the genome, but is much higher in localised regions called recombination hotspots. Haplotypes between the recombination hotspots are preserved because of the low rate of recombination. Second, demographic events such as founder effects, population bottlenecks and chance events in small ancestral populations may have contributed to the extent of LD being longer than predicted on the basis of average recombination frequencies.

Haplotype blocks are often represented by a diagram which plots the average value of D' (see Box 6.3) between pairs of alleles across a chromosomal region. A block is defined as a contiguous region on a chromosome in which there is no evidence for historical recombination because the average D' value is greater then 95%. A criterion of 100% is not used to allow for genotyping errors and recurrent mutation. Figure 6.7 shows an example of such a diagram, showing LD in the region of duplicated genes encoding complement fixation factor H which acts as an inhibitor of complement fixation. An allele conferring a risk of developing age-related macular degeneration, the leading cause of blindness in older people, was mapped by linkage studies to this region. SNPs in block 1 all showed association to the disease which allowed the causative allele to be mapped to the *CFH*1 gene. It is likely that the causative SNP is a mis-sense mutation that lowers the affinity of CFH for complement; the resulting overactivated complement causes inflammation that damages the macula, the high acuity region in the centre of the retina. This *CFH* allele confers a relatively high risk (OR = 4.6) and accounts for about 50% of the total population risk. This is much higher than the risk associated with most risk alleles which, as we will see, typically have OR values of 1.2 or less.

The importance of haplotype blocks is that each haplotype can be distinguished by genotyping a small number of SNPs which therefore form **tags** for the haplotype within the block and they stand **proxy** for the other SNPs in the block in a GWAS (Figure 6.8). Thus the whole genome can be scanned for association to a disease using only a small proportion of the total SNPs present.

Haplotype blocks are not distributed evenly across the genome, nor are they identical in different populations. In order to exploit the haplotype blocks it was necessary to map where they occur and determine the allelic configuration in each of the blocks. This information can then be used to select a panel of SNPs which will efficiently stand proxy for all the other SNPs in the genome. This formed the basis of the international collaboration called the HapMap project. The pattern of haplotype blocks and allele composition was determined in 30–45 trios (two parents and one child) from each of four different populations: the Yoruba in Ibadan, Nigeria (YRI), Japanese (JPT), Chinese (CHB) and from the CEPH collection of families of Western European descent in Utah (CEU). Because these were

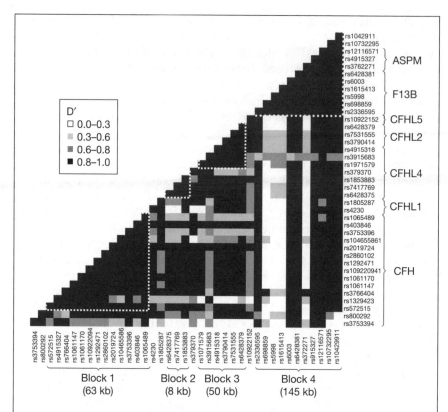

Figure 6.7 Haplotype blocks at the RCA locus on chromosome 1q31.3 regulating complement fixation. The locus consists of *CFH*, encoding complement fixation factor H, and five related genes derived from *CFH* through gene duplication. The diagram shows the extent of LD between all possible pairs of SNPs. Four haplotype blocks are evident. Each block consists of a series of adjacent SNPs where all the SNPs show high values of D' when compared with each other (white dashed lines). SNPs associated with age-related mocular degeneration (AMD) were found to be in block 1. One of these, rs1061170, causes a mis-sense mutation in *CFH* that is predicted to lower its affinity for complement, thus providing a plausible biological explanation for the association with AMD. Note that rs1061170 shows complete LD with rs1805287, the only SNP outside of block 1 that showed an association with AMD. (From Edwards, A.O. *et al.* (2005) *Science*, 308, 421–424. Reprinted with permission from AAAS.)

not collected to be representative of wider African, European or Asian populations, they are only referred to by their abbreviations. The allele frequencies were found to be very similar for the CHB and JPT populations so they are normally combined (CHB + JPT). The trio format allowed the phase in the child to be unambiguously determined. The project was carried out in two phases and the results of phase II were published in 2007. Altogether, 3.1 million SNPs were genotyped in 269 DNA samples so that one SNP was genotyped every 1 kb, which is estimated to account for 25% of all common SNPs.

About 85% of the human genome falls into haplotype blocks. These vary in length from 1 kb to over 100 kb. The blocks are shorter in the

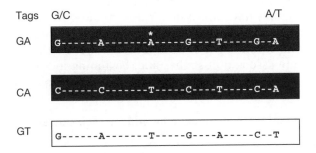

Figure 6.8 Limited chromosome diversity in haplotype blocks. A genomic region contains seven SNP sites; one of the alleles marked with an asterisk is a risk allele. There are $2^7 = 128$ different possible haplotype combinations of these SNPs, but only three different haplotypes are actually present in the population. These can be distinguished by genotyping the two SNPs shown, producing three different possible genotypes that identify which block is present. These tag SNPs act as proxies for the allele that is actually responsible for risk, and identify the haplotype which contains a risk allele. Note that data from the HapMap project allows the genotype of alleles not directly genotyped to be imputed and so tested whether they are associated. In this example, the allele *A* in the *GA*-tagged haplotype, second from left, is unlikely to be associated with the disease because it is also present in the *GT*-tagged haplotype which is not associated, whereas the risk allele is uniquely present in the *GA*-tagged haplotype.

African populations. This is probably because all non-African populations originate from a subset of the African population. The resulting bottleneck in the small group that founded all the non-African populations would have acted to decrease genetic diversity and increase LD (see Chapter 9). Using the 10 ENCODE regions as a reference where all the common SNPs have been catalogued, it was estimated that the mean maximum r^2 value of any SNP to an SNP typed by the HapMap project was estimated to be 0.96, 0.95 and 0.9 for CEU (CHB + JPT) and YRI populations respectively. However, 1% of the SNPs had no tags because they fell within recombination hotspots. Initially, there were some concerns that the HapMap data might only be useful in those populations from which it was derived, but additional studies soon allayed these fears. It was shown that using the HapMap data from the HapMap reference population with the closest ancestry to the population being studied resulted in only a modest loss of power, e.g. the CEU data could be used for Finnish populations. The exception was the use of the YRI for other African populations because of the greater genetic diversity of African populations. Although the HapMap data can be used in non-reference populations, nevertheless there is some loss of power, so mapping studies have been extended to seven more populations.

How well the different commercial platforms covered the total variation in the human genome was assessed by the HapMap project. Generally these performed well with the CEU and Asian (CHB + JPT) populations, so that 80–90% of all SNPs had a proxy where $r^2 > 0.8$. The figure was lower with the African population because of the smaller blocks. Some of the commercial platforms include an additional set of SNPs for use with African populations to compensate for this. Another feature of some commercial systems is tests to investigate the possible association of a disease with copy number and other structural variants. Thus, the means were to hand to initiate association studies and, beginning in 2007, a number of such studies were published with all of the common complex diseases.

6.5.3 Experimental issues in genome-wide association studies

6.5.3.1 Numbers

A major lesson that had been learnt from previous attempts to identify risk alleles, from both linkage and association studies, was that studies with small numbers of cases and controls not only did not have the power to detect alleles with a small effect, but would cause confusion. The *PPARG* example described above is a clear example of this problem. A further aspect of underpowered studies is that the first report often overestimates the true effect of an allele. This is due to a subtle statistical problem, known as the 'winner's curse'. The observed P values (see Section 6.5.3.2) of a particular allele will fluctuate from study to study simply due to random variation. For example, a computer simulation of a study with 1500 cases and 1500 controls investigating an allele with OR = 1.2 showed that the median P value was 10^{-4}, but in 5% of the simulations the P value was >0.025 and in 5% the P values were $<10^{-7}$. When the P value is low it will not be distinguished from the many other alleles being tested, so the effect of the allele will only be noticed and published when the effect in that particular test happens at the high end in the range of fluctuations. When the study is repeated or a meta-analysis combines the data from several independent association studies, the effect of the allele is likely to be less.

All of this emphasises the need for sufficient numbers so that the study has the power to detect alleles with small OR values reliably. Moreover, it is essential that any positive results are independently replicated. Many of the GWAS were carried out by consortia where a number of different centres collaborated to pool their resources of patient groups, genotyping technology and statistical expertise. This allowed numbers that were sufficiently large to have the power to identify effect sizes as small as OR = 1.2 or even less, and they were structured in such a way that the initial findings were replicated on independent sets of patients and controls. Because of this, more or less for the first time in the field, findings of associations with alleles with small effect sizes were consistently reproduced in independent studies. Moreover, the data from different consortia were combined to allow meta-analyses of tens of thousands of cases and controls and thus had the power to detect new associations and prove beyond any reasonable doubt the association of alleles identified in the individual studies.

6.5.3.2 Statistics

In principle, searching for an association is a simple statistical exercise. For each biallelic locus, the frequency of each allele in cases and controls is compared and the null hypothesis tested is that there is no significant differ-ence between these frequencies. To capture additive effects, association of the frequency of the three possible genotypes (two homozygotes and the heterozygote) can be compared in cases and controls. These statistical tests produce a *p value* or *p trend* statistic for each SNP. This is the probability of obtaining a deviation from random association at least as great as that observed if the null hypothesis is true that there is no association. In prac-tice, the statistical tests for significance are anything but simple. The problem is multiple testing, which makes it difficult to decide where to set the threshold of significance (α). In a GWAS, each of approximately 500 000 different alleles are independently tested for association to the disease. The traditional threshold of significance in a statistical test is that the observed result will occur less than one in 20 times by chance ($P \leq 0.05$). When 500 000 tests are carried out then we would expect 25 000 of them to exceed this threshold. Clearly the threshold for significance has to be modified to account for this. The problem is, how do we set the level of significance to be sufficiently stringent to avoid these false positives (type I errors) without setting it so high that we discard genuine associations (type II errors)? There is no single accepted way of solving this problem, and a variety of approaches have been adopted (Box 6.4). All of them reach a similar conclusion that the threshold P value needs to be set between 10^{-7} and 10^{-8}.

There are two commonly used ways of graphically representing the results to allow genome-wide significance to be appreciated. First is the quantile–quantile plot in which the negative logs of all the observed P values are plotted against the negative log of their expected value if the distribution of P values conforms to a normal distribution (Figure 6.9a). A normal distribution of P values will form a straight line. Deviations from a straight line at low P values indicate associations that may be significant. Deviations from the straight line at higher values indicate a systematic bias, such as population stratification (see below). Second, all p values can be plotted according to their genomic positions. Figure 6.9b shows such a plot for a GWAS on T2D. SNPs in three genes, *CDKAL1*, *TCF7L2* and *FTO*, show strong signals that clearly rise above the background of spurious signals (see Section 6.6.1). Note that in Figure 6.9b groups of adjacent SNPs all give a strong signal, providing confidence that the apparent association of the regions with T2D is genuine. The correlated association of adjacent SNPs can be more readily appreciated in a plot of SNPs around a locus showing strong association. Figure 6.9c shows such a plot demonstrating the associ-ation of SNPs in the *IL23R* gene with Crohn's disease (see Section 6.6.3).

6.5.3.3 Population stratification

Population association studies are subject to a particular error if there is a mixture of ethnic subgroups present in the test population, as is found in the USA or the UK because of the inward migration of a variety of different ethnic groups. If a disease is more common in one ethnic subgroup, then any allele that is more common in that subgroup will show a positive

BOX 6.4 ADJUSTING THE SIGNIFICANCE THRESHOLD TO COMPENSATE FOR MULTIPLE TESTING

Bonneferoni correction. This corrects for multiple testing simply by taking into account the number of independent tests carried out.

$$\alpha' = \alpha/n$$

Where α' = corrected significance threshold, α = uncorrected threshold and n = number of tests.

So, if there are 500 000 tests and $\alpha' = 0.05$, then $\alpha = 10^{-7}$.

This test is considered too conservative because the association observed for each SNP is not completely independent from that of adjacent SNPs as they can show correlated association (see Figure 6.9c for examples).

Bayesian approach. This approach is based on an estimation of the *a priori* odds that a particular SNP will be associated with a disease. According to Bayes' theorem the posterior odds of association = prior odds × conditional odds. To calculate the prior odds, suppose there are 10^6 different genomic regions and there are 10 genomic regions associated with a disease, then the prior odds of association would be 10^{-5}. The conditional odds are power/p value. Suppose the average power to detect association with each genuine allele is 0.5. Then a p value of 5×10^{-7} would result in posterior odds of 10:1. All of these values are very approximate estimates, but even if the true values were different by a factor of 10, the calculation gives an idea of the size of the significance threshold necessary.

Computer simulation. Repeated computer simulations are run in which the genotypes are kept intact to preserve LD structure but the phenotypes with which they are associated are randomised between cases and controls. This gives a baseline of the false discovery rate which can be compared with the actual experimental values.

association with the disease, whether or not it is actually involved in the aetiology of the disease. This problem is called population stratification. The analysis of SNPs in different populations in the HapMap project led to the identification of alleles which have a higher frequency in one population compared with another or in some cases are population-specific. These alleles can be used as internal controls because, if these alleles show association with the disease, it may indicate stratification. Population stratification is also revealed in quantile–quantile plots when there is substantial deviation from a straight line plot at moderate or low P values.

6.5.3.4 Tiered study designs

Although high-density arrays allow 500 000 SNPs to be genotyped, it is still a major undertaking to do this in each of the thousands of cases and controls necessary to detect an allele with a low OR (see Figure 6.4). Tiered study designs allow such a study to be carried out more efficiently with only a small loss of power. In the first stage only about a third of the cases and

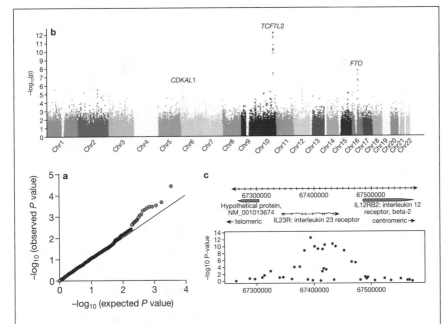

Figure 6.9 Graphical representation of *P* values. a Quantile-quantile plot: the observed
P value of each SNP is plotted against -log10 its expected value. SNPs that deviate from
the straight line at low *P values* (grey circles) indicate significant associations. **b** Observed
P value of each SNP is plotted against its chromosomal position. The plot shown is from
a GWAS of T2D. Strong signals rising above the background are clearly evident for the
indicated genes. **c** Association of Crohn's disease with IL23R. Top panel shows location
of genes on chromosome Ip31. Bottom panel shows -log10 *P* values for association of
individual SNPs with Crohn's disease. Note lack of association with the two genes either
side of IL23R. (a,b Reprinted by permission from Macmillan Publishers Ltd: McCarthy, M.I.
(2008) *Nature Reviews Genetics* **9**, 356–369; c Duerr, R.H. *et al.* (2006) *Science*, **314**,
1461–1463. Reprinted with permission from AAAS.)

controls are genotyped. Alleles are selected that show an association with a
liberal threshold of significance ($p = 0.05$). These alleles are then used to
test for association in the full data set. Although there may be 25 000 false
positives after the first stage, the procedure has reduced by 95% the number
of alleles that must be genotyped in the full set of cases and controls.
However, this cost benefit is reduced by the fact that this approach requires
a custom array to be constructed for the second stage. The cost per SNP in
a custom array is much higher than that in the standard array used to scan
the whole genome.

6.5.3.5 Quality control

It is very important that the experimental data is as perfect as possible
because small errors and biases can lead to the appearance of spurious
associations that appear to be highly statistically significant. This has led to
the incorporation of extensive quality control procedures to ensure that the
genotype calls have a high standard of accuracy – typically more than

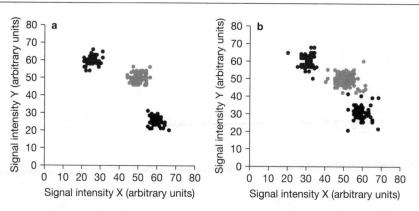

Figure 6.10 Cluster plots to check the accuracy of genotype calling. Each allele of an SNP (*X* and *Y*) has a separate cell to which DNA carrying that allele will hybridise. The signal intensity of DNA that is homozygous for that allele should be higher than the heterozygote, which in turn should be higher than the background signal from DNA that does not carry the allele. In a cluster plot, the signal intensities from the two cells are plotted against each other. So in the diagram, the DNA homozygous for the *Y* allele (blue) gives a signal of ~60 units against its correct cell on the y axis and ~20 units background on the *x* axis against the cell designed for the other allele. The *X* allele (black) gives the reverse pattern, 60 units on the *x* axis against 20 on the *y* axis. The heterozygote carrying one copy of each allele results in an intermediate signal on both axes (light blue). On the left, **a**, the three clusters of signals from a group of individuals are clearly separated into the two homozygotes and the heterozygote. However, on the right, **b**, the data points in the heterozygote cluster overlap points from both homozygote clusters. Thus, the genotype is uncertain and may be incorrectly called. (Based on McCarthy *et al.* (2005) *Nature Reviews Genetics*, **9**, 356–369.)

99.9%. An example of a quality control procedure is the cluster plot that checks that the signal intensities from the two homozygotes and the heterozygote of an SNP are clearly distinguished (Figure 6.10).

The quality control procedures can be too rigorous because discarding large amounts of data can itself distort the conclusions. For example, one allele may be more prone to technical error in genotyping and thus more likely to be rejected by the quality control procedures. This may result in data being removed from the sample in a non-random manner, which in itself may generate a spurious association. This type of error is called informative missingness. It is possible that the technical problems associated with the genotyping platforms may be different, so they each may have different types of biases. Therefore independent validation of positive associations is more convincing when the genotyping is carried out on a different platform. Often, the original investigation will genotype positive signals on a different platform to eliminate the possibility that the signal arises out of a technical artifact specific to that particular platform.

6.5.3.6 GWAS identify genomic regions not genes

The 500 000 tags used in most of the commercial genotyping platforms act as proxies for the rest of the genome. Thus, a positive association suggests

that the tag SNP is in LD with a susceptibility allele that is in the same
region of the genome. The region may contain several genes, so the observation of a positive association with a tag SNP does not directly identify the
gene which is involved in the disease. Often there is a strong indication of
which gene is involved because there is only one gene in the region or one
of the several genes present can be plausibly implicated in the aetiology of
the disease. However, it is important to remember that this is not direct
proof that alleles of that gene are actually the source of the increased susceptibility. Sometimes a strong signal comes from a tag SNP that appears
not to be close to a gene. Intergenic regions can be important either
because they exert long range influence on the expression of a gene which
is not in the immediate vicinity, or perhaps because a regulatory miRNA
originates from the region. These issues should be borne in mind in the
examples below where it is commonly assumed that a positive signal from
an SNP within or near a particular gene identifies that gene as being
involved in the disease.

6.5.3.7 Imputed SNP genotypes

The SNP associated with the disease in GWAS is unlikely to be the true
causative allele, but the region containing the causative allele has been
identified because the tag SNP is in LD with the true causative allele.
Because the LD is not perfect there will be a loss of power dependent on
the magnitude of LD between the tag and the causative allele measured by the
r^2 value. Just relying on the pairwise LD between the tag SNP and a putative
causative allele fails to make use of all the available information. It is possible
to impute the likely genotype of the other SNPs in the regions that were
genotyped in the HapMap project but missing from the panel of SNPs used in
the genotyping platform (see Figure 6.8). These can then be tested for association with the disease. Computer simulations show that this can increase
power to detect variants by decreasing the P value by about a factor of 10.

6.5.3.8 Structural variation

So far GWAS platforms have been primarily designed to detect association
between SNPs and complex diseases. However, since the human genome
sequencing was finished it has become apparent that structural variants are
important aspects of genetic variation in human populations. These take
the form of insertions, deletions, translocations and inversions. Of these, the
most common are insertions and deletions which change the copy number
of affected sequences. These are called copy number variants (CNVs).
These CNVs range in size from a few kilobases to several megabases,
although the majority are under 100 kb in size. Clearly, this variation may
affect susceptibility to disease and there are already a number of examples
where this has been demonstrated. For example, rare duplications of genes
have been shown to be responsible for some Parkinson's and Alzheimer's
disease families. These affect genes such as those encoding alpha-synuclein
in Parkinson's disease and amyloid precursor protein in Alzheimer's
disease, which were already known to be responsible for rare cases of these

diseases that segregate in families in a Mendelian fashion. Another example is susceptibility to HIV-1 infection, which has been shown to be associated with the copy number of the *CCL3L1*, encoding a chemokine involved in the regulation of the immune system.

CNVs and other structural variants are poorly tagged by SNPs defined by HapMap. There is actually a good reason for this: SNPs that deviate from the normal diploid copy number will show abnormal hybridisation intensities on microarrays and thus will be rejected on quality control grounds. We saw in Chapter 4 that a high-density microarray has been developed that is designed to genotype both SNPs and CNVs. When this microarray was used to examine the 274 individuals that were genotyped in the HapMap project, it was found that over 90% were common SNPs with a population frequency of over 1%. Thus, these variants probably arose once in a common ancestor. This means that they are likely to be taggable by SNPs in the same way as other SNPs. Examining association with CNVs is likely to be a feature of the design of future GWAS.

6.5.4 The Wellcome Trust Case Control Consortium

The Wellcome Trust Case Control Consortium (WTCCC) was a collaboration between 24 different groups to carry out a GWAS into different diseases using two common control groups. The study consisted of 2000 samples from each of seven case groups: type 1 and type 2 diabetes, cardiac disease, rheumatoid arthritis, Crohn's disease, hypertension and bipolar affected disorder. The controls came from two separate sources: 1500 came from blood donors in the UK national blood service and 1500 came from representatives from the 1958 Birth Cohort study, a longitudinal study of 17 500 UK babies born in one week in 1958. The GWAS used a commercial Affimetrix chip to genotype 500 000 tag SNPs in the control group and the seven case groups so that, altogether, 17 000 individuals were genotyped.

One of the aims of the study was to develop and test the GWA methodology. One issue investigated was whether it was necessary to genotype a separate group of controls for each disease matched for sociodemographic variables, or whether a common control group could be shared for all the different diseases. To investigate this issue the allele frequency of the SNPs used in the genotyping was compared between the two control groups and it was observed that very few differed significantly, suggesting that it would be safe to combine the two groups and use them as a common control for the seven different sets of cases. Another issue investigated was whether population stratification would distort the results. Even after excluding recent immigrants, the native British population derives from a mixture of different ancestries, one of which may be more prevalent to one or more of the diseases studied. To investigate whether heterogeneity existed that could bias the results, the control group was partitioned into 12 broad geographical areas and allele frequencies compared. Only 13 genomic regions showed regional variation, suggesting that population structure was not likely to bias the results. If any of the 13 genomic regions did show an association they would have to be treated with caution, but in practice this proved not to be an issue.

Using a high stringency threshold of $p \leq 10^{-7}$ for significance, the association tests identified 25 regions which showed a strong evidence of association with one of the diseases. These included 13 out of a list of 15 variants for which there was already robust evidence of association, thus at least the study appeared to be efficient in capturing known variants. One of the two regions missing from the list of known variants was not tagged efficiently in the commercial Affimetrix array used for the study and the other missing region did actually show evidence of association but had been removed from the results because it failed quality control tests. The largest number of strong associations identified in the WTCCC affected just two diseases, type 1 diabetes and Crohn's disease, which also showed the strongest familial aggregation. Most of these regions showing strong association had already been reported to show strong association or were replicated in parallel independent studies. The importance of this result is that it shows that the GWA methodology is one that at last provides reproducible results in the search for alleles associated with complex disease. As stressed before, one of the most important considerations is that adequate numbers should be used. To demonstrate the importance of numbers, WTCCC analysed how many of a list of 16 alleles which showed strong association would have been detected if they had used fewer numbers in the study. If only 1000 cases and 1000 controls had been used, they calculated that they would only have been certain of identifying two of them with an expectation of about 6.

Given the strength of the evidence and prior replication, most or all of these 25 strong associations are likely to be real. A further 58 regions showed weaker evidence of association, with p values between 10^{-5} and 10^{-7}. Among these were regions that included genes such as $PPAR\gamma$ that had already been proven to be associated with T2D. In fact, although the p value was greater than 10^{-7}, the risk associated with alleles in this gene was consistent with previous studies. This suggested that many of these 58 regions that show weaker evidence of association are nevertheless likely to prove real when further extensive replication studies are carried out. Indeed, this proved the case when the T2D data from the WTCCC study were combined with those of other studies in a meta-analysis (see below).

One of the most important conclusions to come out of the WTCCC study, which has become a recurrent theme in studies of this type, is that even the strongest associations confer only a modest risk and explain only a small fraction of the overall familial risk. Most of the OR ratios were less than 1.2. It is likely that the familial risk is made up from the combined effect of many, possibly hundreds, of alleles, each conferring a small risk.

6.6 Case studies of complex diseases

6.6.1 Type 2 diabetes

Candidate gene association studies established three genes involved in T2D. *PPARG* was described above as an example of the confusion caused by underpowered studies. *KCNJ*11 encodes a subunit of an ATP-dependent potassium channel required for β cell function. It is already known to be the receptor for sulfonylurea drugs that stimulate insulin production and are widely used to treat T2D. So, although the *KCNJ11* allele confers a relatively low risk, the example illustrates how finding such alleles could lead to the identification of new drugs. *TCF7L2* encodes a transcription factor required for pancreatic development. These associations were confirmed in the WTCCC study and, in addition, a number of other regions showed suggestive signals. A number of international consortia independently carried out GWAS on T2D as well as UK groups, extending and replicating the WTCCC findings. There was impressive consistency in the genes implicated by the positive signals in all of these studies. Three groups collaborated to pool their data so that altogether an analysis of 32 554 samples could be carried out consisting of 14 586 cases and 17 968 controls providing ample power to uncover variants with a small OR and to confirm unambiguously the association of six new regions with T2D. Table 6.12 lists the genes implicated by the variants, briefly describes their function and gives the OR associated with the variant.

The studies provided an unexpected conclusion concerning the biology of T2D. First, one of the associations was located in an intergenic region, two hundred kilobases from *CDKN2A/CDCKN2B*, which were the nearest genes. These genes can be plausibly implicated in the aetiology of T2D, but it is surprising that the variants are located so far from the coding sequences and the association would probably have been missed in candidate gene studies. Second, previous candidate studies have focused on the signal transduction pathways that mediate reception of the insulin and the cellular response. Most of the functions of the genes implicated in the causation are concerned with the development and function of the insulin-producing pancreatic β cells. The unexpected focus of the results on β cell function demonstrates the value of carrying out unbiased genome-wide scans instead of selecting candidates which are now revealed to be based on incomplete prior knowledge. Research in T2D can now be redirected to understanding the roles of these genes. Moreover, the revelation that the development and function of pancreatic β cells is important in the aetiology of T2D opens up new avenues for therapeutic intervention.

In genetic terms the results provide interesting insights into the genetic architecture of complex diseases. The region with the largest effect is *TCF7L2*, with an OR of 1.37. These studies were more than adequately powered to identify all variants with ORs of this size so it is unlikely that there are any common variants that remain to be discovered that confer a similar or greater risk than *TCF7L2*. The other eight regions have ORs of between 1.12 and 1.2. Together, the sibling recurrence risk attributable to all these loci combined is 1.07. Since λ_s for T2D is estimated to be 3.01, together the variants account for only 2.3% of the total familial risk. Clearly there are many more undiscovered variants that contribute to the risk of T2D. Each of these variants is likely to confer only a small risk. An important corollary to this conclusion is that so far these results do not provide the means to identify individuals at risk of developing T2D.

Table 6.12 Risk alleles for type 2 diabetes.

Region	Function	OR[a]
TCF7L2	Transcription factor required for pancreatic development. Fine structure mapping confirmed that causative variants affect this gene	1.37
SLC30A8	β cell-specific zinc transporter required for insulin biosynthesis, maturation and storage. Variant allele associated with decreased insulin secretion	1.12
IDE/HHEX/KIF1	Variant associated with reduced β cell function. Three candidate genes: – HHEX homeobox protein target of Wnt signal transduction pathway; essential for pancreatic development – IDE (insulin degrading enzyme) – implicated from studies with mouse model – KIF1 (kinesin interacting factor) – kinesin is a plus-end directed motor protein transporting cargo along microtubules	1.13
CDKAL1	Variant associated with reduced β cell function. CDKAL1 is similar to CDK5 regulatory subunit 1. CDK5 is a cyclin-dependent kinase implicated in glucotoxic loss of β cell function	1.12
CDKN2A/2B (intergenic)	Associated variant is several hundred kilobases from these two genes encoding proteins that inhibit p16^{ink4A} which accumulates in β cells and induces an age-dependent decline in proliferation and regeneration	1.12/1.20 (2 alleles)
IGF2BP2	Binds to IGF2 mRNA and regulates its translation. IGF2 stimulates cell responses to insulin	1.14
FTO1	Involved in body weight regulation. Association with diabetes disappears when cases and controls are matched for body mass index, so the primary effect is through body mass	1.17
PPARG	Nuclear hormone receptor required for adipogenesis. Target for thiazolidinedione drugs used to treat T2D	1.14
KCNJ11	ATP potassium channel subunit required for β cell function. Target for sulfonylurea drugs used to treat diabetes	1.14

[a] Per allele risk from meta-analysis of four independent studies (Source: Zeggini, E. *et al.* (2007) *Science*, **316**, 1336–1341. Reprinted with permission from AAAS.)

6.6.2 Type 1 diabetes

The $DQ\beta1$ allele at the HLA locus confers by far the greatest risk, and has long been identified by association studies. The OR = 15 for homozygotes of the $DQ\beta1$ risk allele is an order of magnitude greater than the risk conferred by the other alleles in which the OR values are all less than 2. Five other genomic regions have been shown to be associated with T1D, including a minisatellite polymorphism in the promoter of the *INS* gene encoding insulin. Five of these six regions were readily detected in the WTCCC study as well as other regions which were suggestive but fell short of the strict P value $\leq 10^{-7}$ threshold of significance. The follow-up replication studies of suggestive associations in the WTCCC confirmed four regions that had previously not been associated with T1D. Together, the known alleles are calculated to account for about 48% of the familial clustering of T1D; variants at the HLA locus alone account for 40% of the familial risk. The environmental component contribution from twin studies is estimated to be 20%. So, about one-third of the familial clustering remains unexplained, apparently arising from contributions from many unidentified genes, each contributing a small amount to the risk.

In terms of biological mechanism, the HLA antigens determine the recognition of self and non-self. Other genes implicated are *INS*, encoding insulin, *CTLA4* that encodes a negative regulator of the T lymphocyte immune response, *CD226* that stimulates TH1 lymphocytes, and *PTPN2* which encodes a T cell-specific tyrosine phosphatase. From this a picture emerges where T1D appears to arise from the inappropriate immune recognition of β cell antigens, including insulin, together with an alteration in the regulation of cell-mediated immunity.

6.6.3 Crohn's disease

Crohn's disease is inflammatory disease of the bowel. Part of its aetiology is thought to be an abnormal inflammatory response against the resident bacterial microflora in the bowel. A locus on chromosome 16q12 was mapped by a non-parametric genome scan with affected sib pairs. The mapped region was large (20 Mb) and thus contains a large number of genes. Within this region a gene variously called *NOD2* or *CARD15* had been mapped. This gene was an excellent candidate because it encoded an intracellular version of Toll-like receptors. Toll-like receptors recognise PAMPs (pathogen-associated molecular patterns), which allow the innate immune system to recognise classes of potential pathogens without the delay involved in developing adaptive immunity to specific antigens. One such PAMP is the lipopolysaccharide (LPS) of bacterial cell walls. *NOD2* activates NFκB, a transcription factor that regulates the immune response to infection. Incorrect regulation of NFκB has been implicated in inflammatory diseases, so *NOD2* was clearly an excellent candidate gene for Crohn's disease. Two groups independently showed that the frequency of a variant of this gene, 3020*insC*, was higher in patients compared with controls.

3020*insC* is a frameshift mutation that truncates a leucine-rich repeat region (LRR) in the C terminus of the protein where LPS binds. Cells expressing this truncated protein show reduced stimulation NFκB in response to LPS.

Separate WGAS identified the involvement of two further genes, *IL23R* and *ATG16L1*, in Crohn's disease. *ATG16L1* encodes a protein in the autophagosome pathway that controls intracellular bacterial parasites. *IL23R* encodes a receptor for the proinflammatory cyctokine, interleukin-23. Strong signals at *NOD2/CARD15*, *IL23R* and *ATG16L* were readily identified in the WTCCC GWAS. In addition, six other signals passed the stringent p value $\leq 10^{-7}$ threshold. Replication studies confirmed four of these. One of these signals was *IRGM*, which is also a gene in the autophagy pathway. Another signal in the WTCCC implicated a gene called *NKX2* that encodes a transcription factor which, when deleted in mice, results in abnormalities in the intestine and a failure of T and B cells to segregate to their normal position in the spleen and lymphoid organs. Taken together, these results suggest that Crohn's disease pathology arises from defects in the early immune response to bacteria in the bowel and in recognising and eliminating intracellular bacteria. There were two further points of interest from the GWAS. First, there were two strong signals on chromosomes 1 and 5 that arise from regions of the genome that seem to be devoid of genes – so-called gene deserts. Second, there was evidence of association in the *PTPN2* gene that, as we have seen, is also implicated in T1D.

6.6.4 Breast cancer

Breast cancer is one of the most common cancers in western society, with a lifetime risk in women of 12%. Breast cancer is twice as common in first-degree relatives of women with breast cancer compared with the general population, and the risk increases further when more than two relatives are affected. Twin studies also suggest a strong genetic component in the incidence of breast cancer. Mutations in two genes, *BRCA1* and *BRCA2*, segregate as autosomal dominant alleles and result in a high (10–20-fold higher) risk of breast cancer by age 60 but are responsible for only a minority of familial cases, especially when there is only one affected relative. Mutations in the *PT53* gene that encodes the p53 checkpoint protein, result in Li–Fraumeni syndrome, which is associated with a high risk of breast cancer. Mutations in genes involved in DNA repair, *CHEK2*, *ATM*, *BRIP* and *PALB2*, lead to a risk that is moderately elevated – 2–4-fold compared with the general population – but individually the population incidence of these alleles is low (<0.6%). Together, the known mutations account for 20–25% of the familial risk. The remainder is likely to be polygenic in nature. Until the advent of GWAS, little was known of the nature of the alleles that make up this polygenic risk.

GWAS have now been carried out to identify these polygenic alleles. A three-stage GWAS identified five SNPs that were strongly associated with breast cancer at P values $<10^{-7}$. The third stage examined 30 SNPs with a suggestive association in 21 800 cases and 22 500 controls. Additional WGAS replicated these associations and identified a sixth associated allele. The strongest association was with a SNP, rs2981582, located in the *FGR2* gene, a tyrosine kinase receptor for fibroblast growth factor. The identification of *FGR2* was significant because there is a clear functional link to breast cancer. Furthermore, this gene is overexpressed in 5–10% of breast cancer cases, and somatic mutations in this gene have been detected in many common cancers. Because of the significance of this gene, an effort was made to identify the causal variant, which serves as an illustration of the additional work needed once an association has been found with a tag SNP. The *FGR2* region was resequenced in 45 individuals, revealing 29 previously unrecognised SNPs. These SNPs were then tested for association to breast cancer in the second stage group and three Asian groups where the LD was weaker, providing greater power to detect the causal variant. Six of these novel SNPs proved 100 times more likely to be the causative variant than the original rs2981582 tag SNP. It was not possible to decide which of these six SNPs was more likely to be the causal variant, but there were some interesting clues. All six were located in intron 2, which shows a high level of sequence conservation among the *FGR2* gene from other mammals, indicating a functional role. One of the SNPs, rs10736303, generates a putative binding site for the oestrogen receptor. Another, rs35054928, is adjacent to a binding site for a transcription factor called the POU domain protein octamer (Oct). The implication is that one or more of these SNP variants may affect expression of *FGR2*.

The *FGR2* rs2981582 allele conferred the highest risk of the five SNPs that passed the $P < 10^{-7}$ significance threshold. The OR was 1.23 for heterozygotes and 1.67 for homozygotes of the rs2981582 allele. In the UK, 39% of women are heterozygous and 19% homozygous for this allele. Because this allele is so common, it has a PAR value (16%), which means that a significant number of breast cancer cases would not occur in the absence of this allele. However, population screening for this allele would not be appropriate. The elevated cancer risk by age 70 is estimated to be 10.5% for rs2981582 homozygotes and 6.7 for heterozygotes, compared with 5.5% for women who are not carriers. To put it another way, 94.3% of heterozygotes and 90% of homozygotes will not get breast cancer.

The power to detect the association at the *FGR2* locus was estimated to be 93%, and, indeed, the same association was detected in two independent GWAS. Thus, it is unlikely that there are any other alleles conferring a risk similar to the *FGR2* variant. Four other SNPs for significance were associated with breast cancer. Interestingly, the power to detect the two of these other SNPs that passed the $P < 10^{-7}$ threshold was as low as 1% and 3%, where the per allele OR = 1.08 and 1.10 respectively. This implies that a large number of SNPs conferring a similar risk were not detected in the study. It seems likely that there is very large number of alleles, each conferring a small risk of breast cancer. Together the five alleles account for only 3.6% of the familial risk. There is clearly a long way to go in accounting for all of the genetic factors elevating the risk of breast cancer.

So far we have focused on alleles which affect the susceptibility to common diseases. However, as we saw in Section 6.3.1, susceptibility to common diseases shows a similar mode of inheritance as common phenotypic traits, and so GWAS can also be used to identify alleles that control these phenotypes. Such studies have now been carried out on traits such as body mass index, height, skin pigmentation, hair colour and eye colour.

The study on body mass index identified an allele in a gene called *FTO* (fat mass and obesity-associated). Adults who were homozygous for the risk allele were 3 kilograms heavier and had a 1.67-fold increased risk of obesity compared to individuals who did not carry the risk allele, which affects children from the age of 7 years. The risk allele is relatively common – 16% of the UK population are homozygotes. This allele has also been found to be associated with T2D. The risk of diabetes was specifically associated with the increase in fat mass. When patients and controls were matched for weight, the association of diabetes with *FTO* disappeared.

Pale skin pigmentation, freckling, blue or green eyes and red hair are already known to be risk factors associated with skin cancers such as cutaneous melanoma and basal cell carcinoma. GWAS identified nine regions where alleles have a simultaneous effect on skin pigmentation, eye and hair colour as well as susceptibility to skin cancers. Although the causative allele has not been identified, several of the regions contain genes in the melanin pathway or have homologs in mice which affect coat colour, thus providing plausible biological candidates for the association.

Height is a classic phenotype showing polygenic inheritance, with an estimated heritability of 0.8. Indeed, it was an example used by Fisher to show how quantitative traits could be explained by the Mendelian inheritance of alleles at many polygenes. Although a few rare mutations cause a dramatic effect on height, until the advent of GWAS no polygenes had been reliably identified, despite many attempts to detect association using candidate genes. GWAS have now identified two loci. Alleles in the region containing the *HMGA2* gene are associated with about 0.4 cm increase in height. *HMGA2* encodes a high mobility group protein that probably binds DNA to alter chromatin structure. Rare truncation mutations in humans cause a severe overgrowth syndrome in children (stature of 167 cm in children aged 8 years) and rearrangements are common in certain cancers. A separate GWAS found a region associated with two genes, *GDF5* and *UQCC*, both of which could plausibly be involved in height control. *UQCC* is part of a network that responds to fibroblast growth factor 2 (FGF2) to promote bone growth during development. GDF5 is a growth factor that is expressed in the primordial cartilage of long bones, and mutations in *GDF5* have been associated with severe disorders of skeletal development. An allele near *GDF5* has also been associated with osteoarthritis. This allele, which decreased *GDF5* expression, was also associated with a decrease in height. Thus the evidence suggests that the most likely effect of the associated allele was to decrease *GDF5* expression. The combined effect of the *HGMA2* and

GDF5 alleles accounts for only 1% of the population variation in height, so there are likely to be many more alleles to be discovered that affect height.

6.6.6 Conclusions from genome-wide association studies

6.6.6.1 The genetic dark matter

While GWAS are now reproducibly identifying regions of the genome that contain risk alleles, the OR of the alleles identified is small – the mean is about 1.2 – and together account for only a small fraction of the overall familial risk. Clearly, so far GWAS in general are finding only a small fraction of the overall risk. There are three possible reasons for this:

1. Structural variation is important but has so far been poorly captured by the genotyping arrays.
2. The genetic risk may consist of hundreds of alleles, each conferring only a very small risk. In the breast cancer study it was calculated that the power to detect the confirmed allele with the smallest risk (rs3817198 in the LSP1 gene, OR = 1.07) was only 1%. The evidence for the association was strong ($P = 2 \times 10^{-9}$), but the implication of the power calculation is that there are many more alleles that confer a similar effect. According to this scenario, identification of all, or at least the majority of, risk alleles has only just started. GWAS with very large numbers of subjects will be required to identify all the risk alleles.
3. The majority of the risk is carried by risk alleles that confer a moderate effect (OR 2–5) and are moderately rare (MAF 0.01 or less). These alleles would not have been recognised in the GWA scans with chips based on the HapMap data because only common alleles (MAF > 0.05) were mapped. It is interesting to note that MAF = 0.15 for the least frequent allele, showing strong association in the WTCCC GWAS (rs9272346 at the MHC locus conferring a risk of T1D). Consider an allele with a population frequency of 1% conferring a moderately high risk (OR = 3). Thirty alleles of this type would explain the familial clustering of T2D, yet this allele would not have been identified in the GWA screens carried out so far because they would not be sufficiently common to have been included in the GWAS. Nor would they have been discovered in the linkage-based studies because their effect is too low. So such alleles fall in a gap between common alleles with a small effect identified in GWAS and rarer, highly penetrant alleles that can be identified by linkage studies. Low-frequency alleles with moderate effect may constitute the 'dark matter' of genetic risk – responsible for a large part of the risk but invisible to present techniques. Ultimately, the only way these alleles may be identified is complete resequencing of genomes allied with population association. Large-scale, but relatively rare, structural variation may also be a factor that needs to be investigated.

Ultimately, it may be that GWAS based on tag SNPs is only an interim measure until whole genome sequencing on a population scale becomes possible.

6.6.6.2 Risk alleles are not only located in coding sequences

It is natural to assume that risk alleles will change the coding sequence of proteins or affected regulatory sequences located in immediately adjacent coding sequences. In fact, many of the alleles so far identified are located in introns or in genomic locations a long way from the nearest ORF. In the case of Crohn's disease, one of the alleles with the largest effect is located in a so-called gene desert on chromosome 5. Clearly, it is not just coding regions and associated regulatory regions that contribute to the phenotype in a sequence-dependent fashion. Identifying the true causative allele in many cases will involve resequencing the region in multiple cases and controls. The fine structure mapping of the *FGR2* region to track down the causative allele in breast cancer susceptibility is an example of the sort of analysis necessary.

6.6.6.3 Identification of risk alleles opens up new avenues of investigation

In many cases the alleles identified in the GWAS were not involved in pathways or processes which were previously the focus of attention. In T2D this is illustrated by the fact that GWAS highlighted β cell function and differentiation rather than insulin signalling. In the case of breast cancer, *FGR2* functions in growth factor signalling rather than DNA repair. These novel biological insights justify the GWA screens. Even if the alleles confer a small risk, the genes they affect are part of pathways that will repay further investigation and may provide new avenues for therapeutic intervention. For example, two alleles result in a modest risk of T2D, yet the genes they affect, *PPARG* and *KCNJ*11, are already known to be targets of drugs used to treat the disease. Thus, many of the other risk alleles may identify novel drug targets directly or at least draw attention to pathways where novel targets may be found.

6.6.6.4 Predictive genetic testing would be premature

Although there is now a growing list of alleles contributing to the risk of common complex diseases, with some of these alleles being responsible for a significant fraction of the population risk, it would be premature to use the information for predictive genetic testing. As pointed out in the case of breast cancer, 94% of women with one risk allele at the *FGR2* locus will not get breast cancer; and even in the case of women homozygous for this allele, 90% will not get breast cancer. Even when the effects of all the known risk alleles are combined, they account for only a small proportion of the overall familial risk. Although risks cannot be accurately predicted, it may be possible to use such risk profiles, imperfect as they are, to prioritise groups of the population for routine screening. For example, it has been argued that such tests could be used to identify women who should be screened for breast cancer from age 40 rather than age 50, the present policy in the UK. The increased risk in such women is still small, and extensive counselling would be required to explain this to women identified to be in the risk group. A major current concern is that genetic tests for common diseases are already being offered commercially, where the proportion of predictable individual risk remains small.

6.7 Summary

- Common diseases tend to run in families. This could be either because families share the same environment or because they share the same copies of genes.

- Evidence for genetic involvement in common diseases comes from twin concordance studies, the increased risk to relatives of affected individuals and adoption studies.

- Identifying these factors would help to identify those at risk, define environmental risk factors, illuminate biological pathways involved in the origin of the diseases, and lead to novel and more rational methods of treatment.

- The inheritance of these diseases is similar to quantitative traits that show a normal distribution in the population. This inheritance is explained by postulating that the trait is controlled by the action of a number of polygenes and influenced by environmental factors. Since most diseases are not quantitative in nature it is proposed that the disease occurs when the combined genetic and environmental liability exceeds a certain threshold.

- To understand the contribution of each risk allele to overall susceptibility it is necessary to measure the magnitude of the increased risk it confers and its population frequency. Experimentally, the magnitude of the risk can be measured by the odds ratio (OR) of the risk to a person carrying the allele compared with the risk of a person who does not. Other relevant factors include the interaction with alleles at other loci (epistasis) and whether the allele acts in a dominant or additive fashion.

- In some complex diseases, a minority of cases are caused by the Mendelian segregation of highly penetrant alleles. These can be mapped using the same LOD score-based linkage studies in families that were used to identify the gene affected in monogenic disorders.

- Genome scans of affected sib pairs have been used in attempts to map polygenic risk alleles by linkage studies. The underlying principle is that if a genomic region contains a risk allele it will be inherited from a heterozygous parent by both affected sibs more often than would be expected by chance. The results were generally disappointing, although in some cases genomic regions were identified that subsequently turned out to contain genuine risk alleles.

- Population association studies search for alleles that are more frequent in groups of cases compared with matched controls. Theoretical analysis showed that this approach is inherently more powerful than linkage-based studies.

- Candidate gene studies look for associations in genes chosen because of the biological properties of the encoded proteins. Although genuine associations have been identified by this method, many reports failed to be replicated in independent studies. The main problem was that too few subjects were used. In statistical terms, the studies lacked power.

- Alleles may be associated with a disease because they affect a biologically relevant gene. However, they may also be associated because they are in linkage disequilibrium with the causative allele. That is, the original mutation leading to the risk occurred on a founder chromosome and the configuration of closely linked alleles on that chromosome has not been disrupted by recombination. So the linked alleles, as well as the risk allele, are associated with the disease.

- The human genome consists of blocks in which the diversity of haplotypes is limited because of linkage disequilibrium. The HapMap project mapped the extent of these blocks in four representative human populations. This allowed the selection of a panel of tag SNPs that efficiently stand proxy for most of the 11 million common SNPs that are estimated to be present in the human genome pool. Commercial microarrays have been developed that allow these tag SNPs to be genotyped in numbers of subjects.

- Genome-wide association studies (GWAS) carry out unbiased investigations of the possible association of every region of the genome in a population-based study.

- GWAS need to be based on thousands or even tens of thousands of subjects to detect the small increase in risk conferred by most risk alleles.

- A particular statistical problem arises in GWAS because each of the 500 000 individual SNPs tested for association to the disease represents an independent test. By chance, a large number of SNPs will appear associated and in statistical tests will exceed the traditional threshold of significance ($P \leq 0.05$). To correct for this problem of multiple testing, a very stringent threshold of significance needs to be set ($P \leq 10^{-7}$–10^{-8}).

- Properly designed and powered GWAS have now been carried out on a number of common complex diseases. Examples include type 1 and type 2 diabetes, Crohn's disease and breast cancer. These studies involve large-scale collaborations and have been designed to allow meta-analysis of the combined data from different studies, permitting analysis of tens of thousands of cases and controls.

- GWAS have also been carried out on phenotypic characteristics such as height, weight and skin pigmentation. For the first time, alleles in polygenes have been identified that influence these phenotypes. Some of these alleles also influence disease susceptibility. For example, alleles at the *FTO* locus increase both body mass and susceptibility to T2D; alleles in multiple genes affect eye and hair colour, skin pigmentation and freckling and also increase susceptibility to melanoma.

- These studies have confirmed previous associations and made new discoveries of genomic regions associated with diseases. Importantly, the new findings of association are now being reliably replicated in different studies.

- The GWAS identify regions, not alleles. But in many cases the region contains a plausible candidate gene, providing new insights into the molecular pathogenesis of the diseases.

- The alleles being discovered have a small effect on the risk; most OR values are less than 1.2. So far, they account for only a small fraction of the total familial risk, yet the studies were sufficiently powered to discover any alleles with larger effects that were tagged by the micro-arrays. So either the familial risk of complex diseases arises from the combined effect of very large numbers of alleles, possibly hundreds, each conferring a very small risk, or the methodology is systematically failing to identify a class of alleles that confer a bigger risk.

- The alleles mapped in the HapMap project are common alleles with a population frequency >5%. So the risk alleles are common; in some cases they are the major allele in a polymorphism. It is possible that alleles that confer a higher risk are present at lower population frequencies and so are not present on the genotyping arrays. These alleles may be responsible for the bulk of the familial risk. They may be discovered by resequencing whole genomes if newly developed high-throughput sequencing technologies reduce the unit cost of sequencing whole genomes.

Further reading

BALDING, J. (2006) A tutorial on statistical methods for population association studies. *Nature Reviews Genetics*, **7**, 781–791.

Provides a detailed account of the statistical tests used to determine association in a GWAS.

HEMMINKI, K., LORENZO, B.J. and FORSTI, A. (2006) The balance between heritable and environmental aetiology of human disease. *Nature Reviews Genetics*, **7**, 958–965.

As well as a general account of heritability, this review emphasises the environmental contribution to genetic diseases.

HIRSCHHORN, J.N. and DALY, M.J. (2005) Genome-wide association studies for common diseases and complex traits. *Nature Reviews Genetics*, **6**, 95–108.

A general review of GWAS.

ILES, M.M. (2008) What can genome-wide association studies tell us about the genetics of common disease? *PLoS Genetics*, **4**, e33.

Presents computer simulations to show that GWAS are likely to detect common alleles with a small effect, even if most of the risk is due to rare or moderately rare alleles with a large effect. Contains a discussion of why common alleles with a small effect may account for a large proportion of population attributable risk (PAR) but only a small proportion of the familial risk.

KRUGLYAK, L. (2008) The road to genome-wide association studies. *Nature Reviews Genetics*, **9**, 314–318.

A personalised review of the history of the search for risk alleles for common diseases, pointing out the key milestones in the journey; also speculates on future directions.

MAHER, B. (2008) The case of the missing heritability. *Nature*, **456**, 18–21.

An accessible discussion of why GWAS has identified only a small fraction of the heritability of complex diseases.

MANOLIO, T.A., BROOKS, L.D. and COLLINS, F.S. (2008) A HapMap harvest of insights into the genetics of common disease. *Journal of Clinical Investigation*, **118**, 1590–1605.

Presents a detailed list of the GWAS that had been published at the time of writing (June 2008). Contains references to the research papers presenting the results of GWAS described in this chapter.

McCARTHY, M.I., ABECASIS, G.R., CARDON, L. *et al.* (2008) Genome-wide association studies for complex traits: consensus, uncertainty and challenges. *Nature Reviews Genetics*, **9**, 356–369.

A general and comprehensive review of GWAS along with the experimental issues that must be addressed.

PEARSON, T.A. and MANOLIO, T.A. (2008) How to interpret a genome-wide association study. *Journal of the American Medical Association*, **299**, 1335–1344.

General review of GWAS.

PHAROAH, P.D.P., ANTONIOU, A.C., EASTON, D.F. and PONDER, B.A.J. (2008) Polygenes, risk prediction, and targeted prevention of breast cancer. *New England Journal of Medicine*, **358**, 2796–2803.

Argues that although the current knowledge of breast cancer risk alleles is not sufficient to predict individual breast cancer risks, it may be possible to identify groups who would benefit from breast cancer screening at age 40 rather than at the current age of 50.

RISCH, N.J. (2000) Searching for genetic determinants in the new millennium. *Nature*, **405**, 847–856.

Explains the basic polygenic model of inheritance and threshold model of complex diseases. Shows how different genetic architectures, such as rare alleles with large effect or common alleles with a small effect, could explain the observed population incidence of common diseases.

The International HapMap Consortium (2005) A haplotype map of the human genome. *Nature*, **437**, 1299–1320.

The International HapMap Consortium (2007) A second generation human haplotype map of over 3.1 million SNPs. *Nature*, **449**, 851–861.

Two papers describing phases I and II of the HapMap project. The 2005 paper contains a fuller account of the general implications of the findings.

The Wellcome Trust Case Control Consortium (2007) Genome-wide association study of 14 000 cases of seven common diseases and 3000 shared controls. *Nature*, **447**, 661–678.

The WTCCC GWAS that acted as the pilot study for GWAS generally and addressed issues such as whether a common control group could be used for different diseases and whether population stratification was a serious issue in a population with mixed ancestry such as the UK population.

VISSCHER, P.M., HILL, W.G. and WRAY, N.R. (2008) Heritability in the genomics era – concepts and misconceptions. *Nature Reviews Genetics*, **9**, 255–266.

A general review of the theoretical background to the concept of heritability.

WANG, W.Y.S., BARRATT, B.J., CLAYTON, D.G. and TODD, J.A. (2005) Genome-wide association studies: theoretical and practical concerns. *Nature Reviews Genetics*, **6**, 109–118.

Contains a detailed analysis of issues such as the relationship between power, allele frequency and sample numbers, the different possible architectures of common disease and the theoretical basis of LD.

CHAPTER SEVEN

Small RNAs in human genetics

Key topics

- The structure of miRNAs
- The miRNA pathway
- How do miRNAs reduce gene function: translational repression versus mRNA degradation?
- Identifying miRNAs
- Identifying the targets of miRNAs
- MiRNA dysfunction in disease
- The use of miRNA profiles for diagnostics
- Use of RNA interference to reduce gene function
  - ○ The RNAi pathway
  - ○ Methods for inducing RNAi
  - ○ Designing potent RNAi inducers
  - ○ Specificity of RNAi
- RNAi for studying gene function and for drug target identification
- RNAi therapeutics
- The diversity of small RNAs

7.1 Introduction

The central dogma of molecular biology states that genetic information flows from DNA to RNA to protein. Until recently RNA was, with a few notable exceptions (such as tRNA and ribosomal RNA), thought to be simply a messenger. However, several discoveries have changed that view. First, scientists performing experiments using anti-sense RNA in an attempt to block the expression of a gene in the worm *Caenorhabditis elegans* noted that the introduction of sense RNA molecules was as effective as the introduction of the anti-sense RNA oligonucleotides. The most potent molecule at blocking the expression of the gene was in fact double-stranded RNA. It was shown that the introduction of double-stranded RNA led to the degradation of complementary mRNA. A similar process was also noted in plants, where it was referred to as co-suppression. This so-called **RNA interference** or RNAi (Table 7.1) mechanism has been shown to be conserved across a very wide variety of both animal and plant species, including humans. RNAi has been shown to be triggered by 21-bp long double-stranded RNAs known as **short interfering RNAs** or siRNAs, which may be derived from a longer double-stranded RNA. The second discovery was also made by scientists working in *C. elegans*. While studying genes involved in developmental timing, they identified a 21-nucleotide RNA which was complementary to the **3′ untranslated region (3′ UTR)** of several genes and seemed to regulate them. This was not the first time a small RNA had been shown to regulate genes negatively in *C. elegans*. However, in contrast to the previously identified loci, this new **microRNA** (miRNA) was conserved across a wide variety of species, including vertebrates, showing that gene regulation by small RNAs was not a quirk of *C. elegans*, but in fact was a novel layer of gene regulation found in all plant and animal species. Thousands of miRNAs, including hundreds in humans, have now been identified and have been shown to be involved in a large variety of processes. Meanwhile, experimentally introduced double-stranded RNA has emerged as an important new tool in investigating gene function, allowing the sort of genetics experiments previously available only in model organisms to be performed in human cells. Further, it holds great promise as a possible therapeutic agent, allowing misregulated or mutant genes to be switched off in patients.

7.2 Structure and function of microRNAs

MicroRNAs are short single-stranded RNAs that trigger a reduction in the expression of genes to which they are partially complementary. They are transcribed as an often imperfect **hairpin** structure in a larger transcript (the **primary miRNA**), the mature, single-stranded miRNA being contained in the stem of this hairpin (Figure 7.1). The processing of primary miRNA to the mature miRNA takes place in two stages. In the first stage, the tip of the hairpin (known as the precursor miRNA or pre-miRNA; see Figure 7.1) is liberated from the primary transcript. This stage takes place in the nucleus and is carried out by two proteins (Drosha and DGCR8 – known as Pasha in other organisms), which form a complex known as the microprocessor. These **endonucleases** leave a characteristic overhang on the 3′ end of the hairpin (Figure 7.2). The pre-miRNA is then exported from the nucleus into the cytoplasm. Here the second stage of processing takes place, where the endonuclease dicer cleaves the pre-miRNA approximately 22 nucleotides (nt) or two helical turns from the site of the previous cleavage, again leaving a 3′ overhang, removing the loop region and leaving a 22-nt

Table 7.1 Useful abbreviations and jargon.

Abbreviation	Full title	Description
RNAi	RNA interference	The process by which double-stranded RNA leads to the degradation of complementary mRNA
siRNA	Short interfering RNA	A 19–21-bp double-stranded RNA which causes an RNAi effect
miRNA	**(mature) microRNA**	A single-stranded short RNA, derived from the processing of RNAs with a hairpin structure, which causes translational repression and destabilisation of mRNAs with target sites in their 3′ UTRs
3′ UTR	3′ Untranslated region	The region at the end of an mRNA which is after the stop codon and therefore is not translated into protein
Pri-miRNA	Primary microRNA	The RNA transcript that is transcribed from the genome which contains the piece of RNA that will eventually form the miRNA
Pre-miRNA	**Precursor microRNA**	A hairpin structure, released from the pri-miRNA, that will be processed to release the mature miRNA
(mi)RISC	(micro)RNA-induced silencing complex	A protein complex, targeted by small RNAs, that silences mRNAs either by cleavage of the mRNA, translational repression or mRNA destabilization
—	miRNA gene	The genomic locus from which the primary miRNA originates
shRNA	Short hairpin RNAs	Artificially constructed RNA hairpin structures that are expressed from DNA sequences and processed to form siRNAs
piRNAs	Piwi-interacting RNAs	Small, single-stranded RNAs (29–30 nucleotides) of unknown function found complexed with a protein known as piwi
snRNAs	Small nuclear RNAs	Small RNAs that are found in protein complexes localized to the nucleus
snoRNAs	Small nucleolar RNAs	Small RNAs which guide the modification of other RNAs, particularly snRNAs and ribosomal RNAs. Found localized in the nucleolus

double-stranded RNA duplex (see Figure 7.2). One strand of this duplex is the mature miRNA. The duplex is unwound by an unknown helicase protein and the mature miRNA is loaded into an **argonaute family** protein, which, along with dicer and several other proteins, forms the **microRNA**

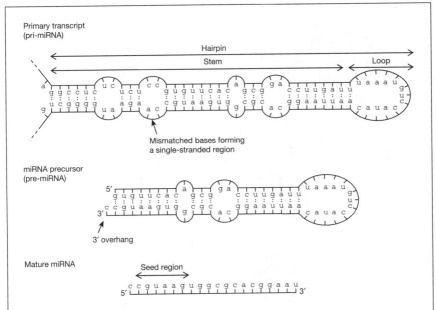

Figure 7.1 Anatomy of a miRNA. Part of the primary, the precursor and the mature forms of a miRNA. The primary transcript (or pri-miRNA) contains a double-stranded region, known as the stem, formed by the single-stranded RNA folding back and base pairing with itself. This double-stranded region forms a helix, similar to that of DNA. The two strands of the double-stranded regions are joined by a single-stranded region, known as the loop. Together these make up a 'hairpin' structure. The stem is not made up exclusively of complementary base pairs and occasionally contains short stretches of bases which do not pair. The pri-miRNA may be very long and the hairpin generally forms only part of it. The primary form is cleaved to form the precursor (or pre-miRNA), which is around 70 nucleotides long, and is entirely a hairpin structure, with an overhang on the 3' end. The cleavage usually takes place in a non-paired region, approximately two turns in the helix from the loop. The mature miRNA is a short (around 19–22 nucleotides) single-stranded RNA formed from one strand of the precursor stem. The strand of the precursor which does not form the mature miRNA is known as the star (or *) strand. The sequence in the primary transcript and precursor that will form the mature miRNA is highlighted in blue. Nucleotides 2–7 (sometimes 2–8) are referred to as the seed region, and are important in determining the targets of the miRNA.

induced silencing complex (miRISC). The other strand is degraded. Which strand is loaded into the miRISC is determined by the relative stability of the two ends of the duplex; the strand with its 5' end at the less stable end of the duplex is preferentially loaded into miRISC. In some cases it is possible that either strand can become the mature miRNA.

The miRNA-containing miRISC causes a reduction in the expression of a set of genes. Which genes are affected (the 'targets' of the miRNA) is determined by the sequence of the miRNA. The mechanism by which miRISC silences the expression of its targets is controversial and not fully understood. Where the miRNA is perfectly complementary to a portion of an mRNA, then the mRNA is cleaved at a specific position. However, this

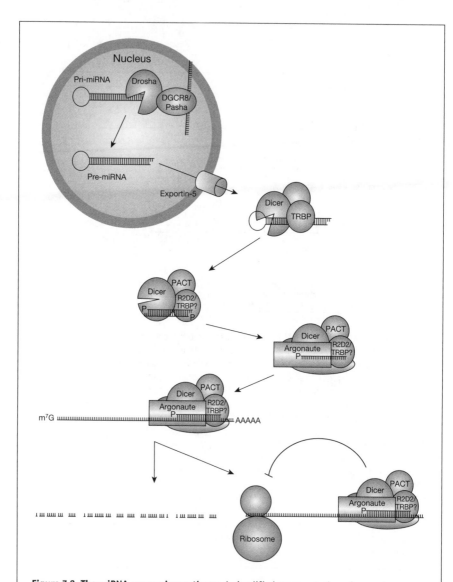

Figure 7.2 The miRNA processing pathway. A simplified representation of current knowledge of the miRNA pathway. The hairpin in the primary transcript is cleaved in the nucleus by the microprocessor complex, consisting of dicer and DGCR8, to form the pre-miRNA. This is then exported from the nucleus by exportin-5. Once in the cytoplasm the pre-miRNA hairpin is further cleaved by dicer, with the double-stranded RNA binding protein TRBP acting as a co-factor. The resulting double-stranded RNA duplex is unwound and loaded into one of four argonaute family proteins which forms the miRNA-induced silencing complex (miRISC). miRISC is targeted to mRNAs using the mature miRNA, represses translation of the mRNA, and may also promote the degradation of the mRNA.

mechanism is rare in animals, since generally targets are not perfectly complementary to the sequence of the miRNA. Instead, it appears that most targets contain, in their 3′ UTR, sequences complementary to the so-called 'seed sequence' of the miRNA, a 6- or 7-nucleotide sequence at the 5′ end of the miRNA. It was initially believed that the most important mechanism by which miRNAs repress the expression of proteins was by inhibiting translation by disrupting either translation initiation or elongation. However, there is now increasing evidence that, while translational inhibition is an important effect, miRNAs imperfectly matched to their targets can reduce target mRNA levels. The mechanism by which this might happen is again not completely understood, although in some cases it may involve the destabilisation of the mRNA by promoting the removal of the **methylguanosine cap** (m^7G cap) which is added to mRNAs to promote their nuclear export and prevent nuclease degradation and also the removal of the polyadenine tail. miRNAs, their target mRNAs, as well as Argonaute proteins and enzymes involved in the removal of the m^7G cap, are all found localised together at places in the cell known as processing- or P-bodies. The formation of such bodies is important for miRNA function. It is possible that destabilisation of the mRNA by removal of the m^7G cap and removal of the **polyadenine tail** is due to the localisation of miRNA target mRNAs in P-bodies where the required enzymes are found.

The relative importance of these two mechanisms – translational repression and mRNA destabilisation – appears to be different for different miRNAs and different targets.

7.3 MicroRNA genomics

There are 678 human microRNAs listed in miRBase, which is a database of miRNAs whose existence is supported by experimental evidence. These may be transcribed from three different types of primary transcript: single miRNAs, clustered or **polycistronic miRNAs** and **intronic miRNAs** (Figure 7.3). Of the miRNAs listed in miRBase, 36% are found as clusters in the genome, and are therefore predicted to be expressed from a single transcript (a polycistronic miRNA; see Figure 7.3). Fifty per cent of miRBase miRNAs overlap with a previously annotated transcript, such as that of a protein-coding gene, and are therefore expected to be expressed as part of this transcript and then excised by the splicing machinery (intronic miRNAs; see Figure 7.3). One consequence of this is that the miRNA will be expressed at the same times and in the same places as the protein-coding gene. miRNA genes are found on all human chromosomes with the exception of the Y chromosome.

miRNAs can be grouped together on several levels. The same mature miRNA may be expressed from several different precursors or the same precursors from different places in the genome. Such miRNAs are denoted by a number following the name; for example, the microRNA genes *hsa-mir-128-1* and *hsa-mir-128-2* produce microRNAs of exactly the same

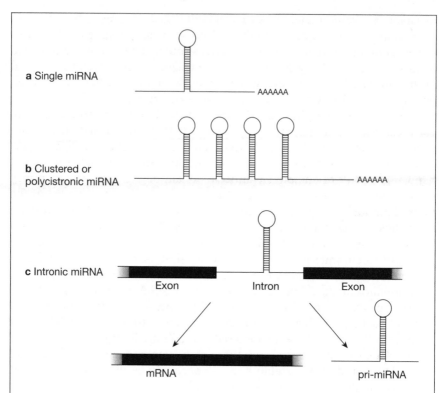

a Single miRNA

AAAAAA

b Clustered or polycistronic miRNA

AAAAAA

c Intronic miRNA

Exon | Intron | Exon

mRNA

pri-miRNA

Figure 7.3 Genomic sources of miRNAs. miRNAs can come from one of three different sources. **a** Transcripts, transcribed from a polymerase II promoter ending with a poly-A tail and containing a single miRNA hairpin structure. **b** Clustered miRNA, where multiple hairpins are expressed on the same transcript. **c** Intronic miRNAs, where the miRNA is found in the intron of a previously defined non-coding or protein-coding gene. When the transcript is spliced to produce the mature mRNA, the intron containing the miRNA is released and can be processed by the miRNA processing machinery.

sequence (mir-128), but are on chromosomes 2 and 3, respectively. If two mature sequences are very closely related, but not identical, they are denoted by a letter after the name, e.g. *hsa-mir-26a* and *hsa-mir-26b*, whose mature sequences differ at only two bases (*mir-26a* is also one base longer). Finally, miRNAs can be grouped together into microRNA families based on the seven bases from base 2 to base 8, known as the seed sequence, which are important in targeting miRNAs. For example, *mir-103* and *mir-107* share the seed sequence GCAGCA. miRNAs in the same family are expected to share many of the same targets.

7.3.1 Identifying miRNAs

There are in principle three ways in which miRNAs can be identified. The first miRNAs were identified through traditional **forward genetics**, whereby model organisms are mutated at random, individuals displaying a particular characteristic or phenotype are selected, and the mutated gene is identified. This is not possible in humans because we cannot randomly mutate human subjects, and isolated cells, which may be mutated, cannot be mated to produce homozygotes. However, if miRNAs are identified in model organisms in this way and the same sequence is also found in the human genome, then it can be marked as a miRNA. A total of four miRNAs have been identified in this way, only one of which is conserved in humans. The small number of miRNAs identified this way is probably due to the size of miRNAs, making them a small mutational target.

The second way in which miRNAs may be identified is through isolating RNA from cells or tissue, purifying the small RNA, and cloning and sequencing this fraction. Sequences corresponding to the breakdown products of other RNAs such as tRNAs or ribosomal RNA are discarded. The remaining RNA must then be mapped back onto the genome, which is not easy owing to their small size. Once mapped back, the sequences are examined to determine which could have originated from genomic sequences that could form precursors-like hairpins. In some cases it is not possible to map the sequenced small RNAs to a single locus in the genome, and so the actual position in the genome from which a particular miRNA is transcribed could be one or more of several potential sites. The new, massively parallel sequencing technologies (see Chapter 4) have massively accelerated the process of gathering small RNA sequences, allowing an almost exhaustive cataloguing of all small RNAs in a particular sample. However, determining which of these RNAs are genuine miRNAs remains a problem.

The third possibility takes the opposite approach. Instead of detecting small RNA sequences and using the genome sequences to decide if these are real miRNAs, genomes are scanned for sequences that look like potential miRNA sequences. Several features can be used to decide if a sequence could be a miRNA, the most obvious being that the sequence must be capable of forming a hairpin structure when transcribed into RNA. However, there are over 11 million potential hairpins in the human genome; clearly these are not all miRNAs, so several other criteria are used, the two most important being the detailed structure of the hairpin and conservation between organisms. miRNA hairpins often show several particular features, such as an absence of large sections of unpaired sequence in the stem of the hairpin (see Figure 7.1). Such structural features can be used to narrow down the number of hairpins that correspond to real miRNAs. Many of the miRNAs identified thus far have highly conserved sequences which are found in many organisms. Therefore conserved hairpins are more likely to represent real miRNAs. In some cases, more complicated patterns of conservation are used. For example, the mature miRNA sequence is often more highly conserved than the loop sequences or the non-functional strand of the hairpin. These features are combined into artificial intelligence learning programs which learn what miRNAs look like and are able to find new candidates. However, no prediction can be completely trusted until the miRNA has been shown to be expressed experimentally. Often large numbers of predictions are tested by looking for their expression on a microarray, or by using techniques that aim to clone predicted miRNAs.

7.3.2 Identifying miRNA targets

Identifying the targets of miRNAs is far more difficult than identifying the miRNAs themselves. This is due to the fact that miRNAs are short and are rarely fully complementary to their targets. In fact, the full rules determining which genes are the targets of miRNAs are not understood. Target sites are generally in the 3′ UTR of targeted genes. Complete complementarity between the target site and bases 2–7 or bases 2–8 of the mature miRNA (known as the seed sequence; see Figure 7.1) is important, sometimes sufficient on its own, but target sites do exist that have mismatches in this region. The small size of this seed sequence means that such sites occur very frequently in the 3′ UTRs of human genes, which are on average 1 kb long. The presence of more than one target site in the 3′ UTR of a gene increases the chance that it is a target.

Currently only 570 genes have been shown experimentally to be miRNA targets in any organism. These 570 are targets of just 128 miRNAs out of the thousands known in any species. Experimental confirmation of a miRNA target usually involves fusing the 3′ UTR of the gene in question to a sequence which encodes a reporter gene (Box 7.1).

In the absence of experimental methods that are suitable for high-throughput attempts to find miRNA targets, computational predictions are important. Algorithms for finding targets generally look for regions of complementarity between the miRNA and 3′ UTRs or regions where the miRNA binds the 3′ UTR strongly. Often a perfect match to the seed region is required or rewarded. These matches are then filtered and ranked using a number of other criteria. These often include the thermodynamic properties of the binding (that is, how strong the chemical bond between the miRNA and the site would be), the conservation of the site in one or more different organisms, and the number of sites in the 3′ UTR. It is expected that the highest scoring of these predicted targets are real targets, while it is accepted that a number of the lower scoring targets will be false positive results. Thus predictions from such programs are not automatically accepted, but must be verified experimentally.

Target prediction programs often predict hundreds or, occasionally, even thousands of targets for each miRNA. If even a proportion of these targets are real, it means that a large proportion of genes in the human genome are regulated by one or more miRNA.

7.4 MicroRNAs in human disease and diagnostics

The roles of microRNAs in biology are as diverse as the protein-coding genes which they regulate. Given the potentially very large number of genes that are regulated by miRNAs, it is not surprising that roles for miRNAs

BOX 7.1: EXPERIMENTAL METHODS FOR CONFIRMING miRNA TARGETS

There are several methods by which the targets of miRNAs can be confirmed. Most of these methods involve fusing the open reading frame of a reporter gene to the putative target 3′ UTR, or a fragment of the 3′ UTR that contains the putative miRNA binding site(s). Commonly, **luciferase** is used as the reporter gene. Luciferase is a protein, isolated from fireflies (among other organisms), which produces light when its substrate, luciferin, is present. The amount of light produced can be used to determine the amount of luciferase protein present.

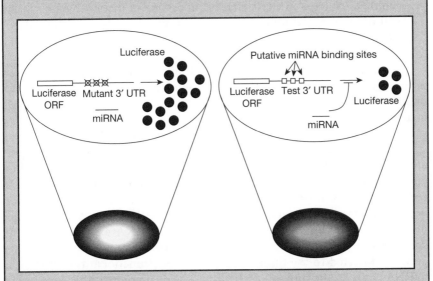

In the design shown, luciferase is fused to the test 3′ UTR in one sample and to the same 3′ UTR with the putative miRNA binding sites mutated or removed. These constructs are introduced into cells which produce the miRNA in question. If the miRNA interacts with the 3′ UTR in question, via the putative binding sites, the expression of luciferase will be suppressed in the test sample but not in the sample with the mutant 3′ UTR.

A different design involves introducing a luciferase–3′ UTR fusion into cells along with artificially made miRNAs with the same sequence as the miRNA thought to target the 3′ UTR (the test miRNA) or a miRNA made with a scrambled sequence (the control miRNA). The test miRNA should suppress luciferase, while the control miRNA should not.

Since some miRNAs function to destabilise the target mRNA, microarrays have also been used to investigate miRNA targets.

have been described in many processes, including development cell type specification, the immune system, apoptosis and regulation of the cell cycle. It is clear, then, that dysfunction of miRNAs will be involved in a range of human diseases.

MicroRNAs in human disease and diagnostics

There are, in principle, three different ways in which a miRNA might be involved in disease:

- Misregulation of the miRNA, leading to an increase or decrease in miRNA levels, which in turn will affect the levels of its many targets.

- Mutation of the miRNA's targets, leading to the creation or destruction of miRNA targets, and thus changes of expression for the mutant gene.

- Mutation of miRNAs themselves. This might be in the primary transcript, leading to changes in expression or processing of the miRNAs, or in the mature sequence, leading to a change in which transcripts the miRNA will target.

In practice, most studies of the role of miRNAs in disease have focused on changes in the expression level. A common approach is to take samples of RNAs from normal and diseased cells or tissues and measure the levels of a large number of miRNAs from each sample. The levels of miRNAs in a sample can be determined using a number of different techniques, including specially modified oligonucleotide arrays and sequencing using highly parallel next-generation sequencing methods (Box 7.2). miRNAs are then identified that show a different pattern of regulation in the disease sample compared to the normal sample. Such studies may reveal a correlation between a disease state and expression of particular miRNAs, but cannot distinguish between cases where changes in miRNA expression are the causal agent in the disease, rather than simply the consequence of other events connected to the disease. To address any functional role that these miRNAs have in the disease mechanism, it is common to either increase or decrease the level of the miRNA experimentally in an animal or cellular model of the disease.

There follow some examples of diseases where miRNAs have been implicated, starting with cancer.

7.4.1 MicroRNAs in cancer

The processes involved in the formation of cancer cells, such as the cell cycle and apoptosis (cell suicide), are tightly regulated processes that depend on the correct control of the expression level of many different genes. Since miRNAs are thought to be involved in fine tuning the expression of large numbers of genes, this makes miRNAs attractive targets for studying the molecular events that lead to cancer. A large fraction of miRNAs are located in regions of the genome known to be involved in cancer, such as commonly deleted regions or sites of common rearrangements. Many miRNAs have been identified that are involved in a large range of cancers (Table 7.2); indeed, changes in miRNAs have been observed in every cancer sample studied.

Table 7.2 shows miRNAs that are commonly associated with cancer, some examples of cancers with which they have been connected and

BOX 7.2: METHODS FOR DETERMINING miRNA EXPRESSION LEVELS

miRNA arrays

Just as oligonucleotide arrays are the most common method for determining the expression levels of many mRNAs at once, such arrays are also commonly used for quantifying the levels of miRNAs. However, owing to the short length of miRNAs, and the fact that they can differ at only a single base, modified bases which increase the specificity of the oligonucleotide probes on the array are often used.

Bead-based miRNA profiling

In this method oligonucleotide probes corresponding to each of the miRNAs to be measured are attached to beads of different colours, with one colour for each miRNA. miRNAs in the sample being measured are reverse transcribed and amplified by PCR using primers that have a small molecule called biotin attached. The amplified miRNAs are hybridised to the bead-attached oligonucleotides and are then stained with a fluorescent molecule called **phycoerythrin** attached to a molecule called **streptavidin**. Streptavidin binds **biotin** very strongly. Thus the amount of the phycoerythrin attached to each bead shows how much miRNA is attached to it. The sample is then passed through a machine called a **flow cytometer**, which uses lasers to measure the identity of miRNAs attached to each bead (by bead colour) and the amount of miRNA (by the amount of the phycoerythrin). The advantage of this method is that it is cheap and is more specific than standard microarrays because hybridisation takes place in solution.

Sequencing

New sequencing technologies allow millions of molecules to be sequenced simultaneously. If small RNAs are isolated from a sample and sequenced using these new technologies, then the number of times a particular sequence is seen is correlated with the level of expression in the sample.

Table 7.2 A selection of miRNAs commonly implicated in cancer.

miRNA	Cancer	Effect
miR-15, 16	Chronic lymphocytic leukaemia (CLL), gastric	Deleted, mutated and downregulated
miR-17-92	Colon, lung, thyroid	Upregulated
miR-145	Breast, colorectal, cervical, ovary	Downregulated
miR-21	Breast, cervical, gastric, glioblastoma multiforme, hepatic, head and neck, ovary, pancreatic	Upregulated
miR-155	Breast, B cell lymphomas, cervical, lung, pancreatic, prostate	Upregulated

whether the miRNAs in question were upregulated or downregulated, deleted (which results in a downregulation), or mutated.

miRNAs can act both as oncogenes, promoting cancerous growth, and as tumour suppressors, protecting against cancerous growth. When upregulation of a miRNA is associated with a cancer, the miRNA is expected to target protein-coding genes that function as tumour suppressors. An example of such a miRNA is miR-155 which acts as an oncogene in **Hodgkin's** and **Burkitt's lymphomas**. **Lymphomas** are tumours of the lymphatic system, and Hodgkin's and Burkitt's lymphomas are both tumours of the B cells of the immune system. Examination of RNA expression in Hodgkin's lymphoma cells revealed that a gene called *BIC* was expressed at a higher level than in the equivalent non-tumour cells. This was also found to be the case in children with Burkitt's lymphoma. This upregulation was not seen in other lymphomas studied. The *BIC* transcript includes the microRNA miR-155, and mature miR-155 is also upregulated in these cancers. miRNAs that act as oncogenes have been named **oncomiRNAs**.

MicroRNAs that are downregulated in cancer cells potentially act as tumour suppressors by repressing the expression of oncogenes in normal cells. It is important to remember that increased or decreased accumulation of a particular miRNA in a cancer tissue does not necessarily imply that such miRNAs are directly involved in carcinogenesis. However, in some cases there is better evidence for direct involvement.

Chronic lymphocytic leukaemia (CLL) is the most common form of leukaemia in adults in the western world, and more than half of all cases have either a homozygous or heterozygous deletion on chromosome 13 at 13q14. This deletion is also present in other cancers. Several genes in this region were identified but no evidence for their involvement was found. A mapping of the smallest deleted region that is contained in all the rearrangements seen in patients identified a 30-kb region which contains two miRNAs, miR-15a and miR-16, which are transcribed as one cluster. Measurements of miRNA expression in a collection of CLL samples as well as samples from other cancers revealed that the miRNAs were downregulated in the majority of CLL cases as well as most of the prostate cancer cells tested. A mutation has also been located in miR-16. This mutant was shown to reduce expression of the miRNA. The individuals carrying this mutation had CLL and had a family history of the disease. Thus miR-15 and miR-16 fit the classic pattern for inherited predisposition to cancer (see Chapter 5, Section 5.7). Individuals inherit a mutant copy of the tumour suppressor gene and so need to lose only one more copy to lose expression of the gene, whereas those that do not carry a mutant copy of the gene must lose both to develop the cancer (Figure 7.4). The target of miR-15 and miR-16 is *BCL2*, an oncogene which inhibits **apoptosis**, allowing the tumour cell to proliferate in an uncontrolled manner while escaping signals to kill itself.

7.4.2 MicroRNAs in cardiovascular diseases

Development of the mammalian body is a complex process which requires the coordinated expression of many different genes at different times. At the

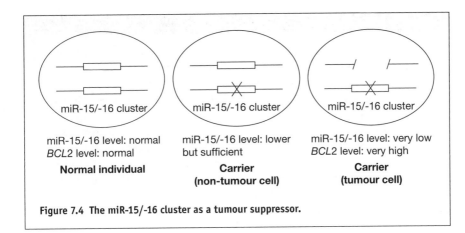

Figure 7.4 The miR-15/-16 cluster as a tumour suppressor.

end of the process, each different cell type must express a different set of genes from the same genome. One of the major roles of miRNAs is thought to be reinforcing and fine tuning tissue-specific patterns of gene expression. In support of this idea, many miRNAs show tissue-specific expression. For example, the miRNAs miR-1 and miR-133 show expression specifically in adult muscle tissue.

Developing cardiac muscle cells express different genes from their fully developed and differentiated adult counterparts. However, adult cardiac muscle cells respond to stress by reactivating these fetal genes. This leads to **hypertrophy**, an increase in the length and volume of cells without an increase in the number of cells. This helps the heart to sustain its output in the face of increased stress. However, prolonged hypertrophy leads to a condition called **hypertrophic cardiomyopathy**, where the wall of the heart is thickened. Hypertrophic cardiomyopathy is a leading cause of sudden death in young athletes and is also a cause of sudden death in individuals of any age. miRNA profiles of cardiac tissue from animal models of hypertrophy revealed that miR-1 and miR-133 were downregulated compared with normal cardiac tissue. It was also shown experimentally that increasing the expression of miR-133 could reduce the effects of hypertrophy. Measurements of miR-1 and miR-133 levels in individuals undergoing surgery for hypertrophic cardiomyopathy revealed that these individuals also had lower levels of these miRNAs than normal individuals. While hypertrophic cardiomyopathy does have an inherited genetic component, there is no evidence yet that alterations to miR-1 and miR-133 are involved in predisposition in humans; however, there is evidence of a genetic link between miR-1 and hypertrophy in sheep (Box 7.3).

In contrast to the reduced levels in hypertrophy, levels of miR-1 are increased in individuals with coronary artery disease. It was found that blocking miR-1 in the hearts of rats with an experimentally induced heart attack suppressed **arrhythmia** in the period following the heart attack. The rhythm of the beating heart is controlled by depolarisation of the cell membrane of cardiac cells caused by the flow of ions across it. It was found

BOX 7.3: TEXEL SHEEP

The Texel sheep is a breed which has an exceptionally high muscle mass. Mapping of the genetic locus responsible for this commercially valuable phenotype identified the region of chromosome 2 which contains the myostatin gene. No polymorphisms were found in the open reading frame, but a single base change in the 3′ UTR was identified which created a target site for miR-1 and its family member miR-206. MiR-1 and miR-206 are both muscle-specific microRNAs, suggesting that this polymorphism might lead to a downregulation of myostatin. Texel sheep do indeed have a decreased expression of myostatin. As loss of myostatin in mice, cattle and humans leads to a muscle doubling, this explains the large muscle mass of Texel sheep.

that in hearts with elevated levels of miR-1 there was a defect in the flow of positively charged potassium ions following a depolarisation. The gene *KCNJ2* codes for the main potassium channel responsible for this flow of ions back into the cell and is a target of miR-1. Thus increased levels of miR-1 lead to decreased levels of *KCNJ2*, and thus an impairment of the mechanism controlling the rhythm of the heart.

Not all associations between miRNAs and disease are due to changes in the level of miRNAs. As outlined above, other mechanisms include mutation of the miRNA and mutation of target sites. Mapping known single nucleotide polymorphisms (SNPs) to known miRNA target sites revealed a SNP in the 3′ UTR of a gene called *AGTR1*. The minor allele of this SNP destroys a target site for miR-155 and causes an increase in the level of the protein product of this gene. This SNP has been associated with hypertension (high blood pressure) and so implicates miR-155 in the control of blood pressure. Interestingly, since miR-155 is located on chromosome 21, it is triplicated in **trisomy 21** (Down's syndrome). People with Down's syndrome often have reduced blood pressure, which could be a consequence of an increase in miR-155 expression.

7.4.3 The use of miRNAs as a diagnostic tool

One of the aims of human genetics is to identify **biomarkers** which may be used to develop diagnostic tests. Given that miRNAs have been found to be misregulated in many diseases, particularly cancer, they provide good candidates for use as diagnostic markers in cancer, and there has been much interest in investigating them for this.

There are several tests for which miRNAs could be useful in this setting:

- **Distinguishing normal tissue from tumour tissue**. Early diagnosis of a tumour before it becomes symptomatic can greatly aid the chances of successful treatment. The use of miRNA diagnostics may be able to aid in the identification of tumorous or pretumorous cells in patient samples.

- **Determining tissue of origin**. Between 3% and 5% of **metastatic** cancer is diagnosed as cancer of unknown primary site (CUP). This makes correct diagnosis of the type of cancer and the appropriate treatment difficult.

- **Determing tumour subtype**. Simply knowing the tissue of origin may not be enough to diagnose the type of cancer definitively. Several cancers are known to have distinct molecular subtypes despite coming from the same original tissue and showing the same gross morphological phenotype.

- **Prognosis**. Patients with later-stage or more aggressive tumours may have a poor prognosis. In these cases it may be suitable to use more aggressive treatments. Conversely, in patients with a relatively good prognosis it may be more suitable to avoid harsh treatments with serious side effects.

All these methods rely on knowing the expression levels of miRNAs in particular cancer cell types, known as the miRNA profile of that cell type. Such profiles can be generated from patient samples of known tumour type, origin or prognosis and compared with profiles from tumour samples of other types, prognosis or origin, or compared with normal, non-cancerous samples. There are different ways in which knowledge of miRNA expression might be used in such tests, and the suitability of each method relies on the question asked and the number of miRNAs that can distinguish between different answers.

If a test to distinguish between cancerous and non-cancerous samples is required, for instance to aid in a screening programme, then it is likely that large numbers of samples will have to be processed cheaply and quickly. Here a test that uses a single miRNA or a very small number is appropriate. Such a test might work by using *in situ* **hybridisation**, where a labelled **oligonucleotide probe** corresponding to the miRNA in question is added to a sample of cells from the patient. The oligonucleotide probe will bind to the miRNA and therefore mark cells where the miRNA is expressed. Many such samples can be included on a **tissue microarray** (a block of paraffin, with multiple tissue biopsy samples, arranged in a grid, embedded in it) to allow for rapid screening (Figure 7.5). For example, in normal breast tissue, miR-145 is expressed in a particular subgroup of cells called myoepithelial cells, but its expression is greatly reduced in tumour samples.

Once a tumour is identified, more care and time may be taken to assign the tumour to a subtype and to give an indication of the prognosis. This allows more accurate quantitation of a larger number of miRNAs to find a more specific pattern of dysregulation. For example, chronic lymphoid leukaemia can be divided into two distinct subtypes. One subtype, indolent CLL, is fairly slow growing and patients can live many years without treatment. The second subtype is more aggressive and has a very high mortality rate. Comparison of miRNA expression in these two types of CLL identified 13 miRNAs which could accurately distinguish between the two subtypes. Similarly, the expression levels of the miRNAs miR-155 and Let-7 can be used to predict prognosis in lung cancer samples (see Figure 7.5).

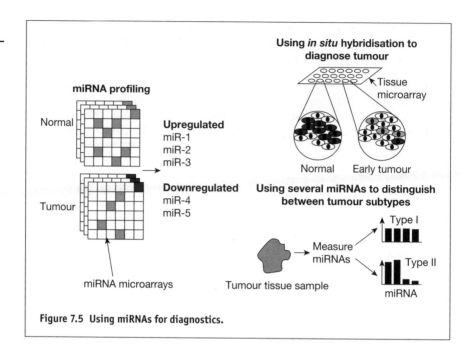

Figure 7.5 Using miRNAs for diagnostics.

In both these cases, miRNA profiling is used to identify miRNAs that are misregulated and can then be used for testing. A different approach involves using the whole profile to identify patterns that can predict characteristics of the sample (known as the **expression signature** of that characteristic). Such approaches generally involve using a number of examples of each type of sample to train a computer program, known as a classifier, and then using this program to classify new samples of unknown type.

While certain miRNAs are deregulated between normal and tumour samples, and between tumours of different subtypes, the underlying pattern of miRNA expression is determined by the history of the tumour, particularly the cell type from which it originates. It has been demonstrated that it is possible to distinguish the tissue of origin of a tumour (where this is not possible by looking at the cells) by using miRNA expression patterns (Figure 7.6). Another example of the use of complete miRNA profiles is in distinguishing different types of breast cancer, which can be divided into four different categories. miRNA profiles can be used to distinguish between these different subtypes.

7.5 RNA interference

While miRNAs reveal a new layer of endogenous gene regulation in the human cell, providing new insights on disease and new diagnostic tools and drug targets, the other half of the small RNA revolution, RNA interference (RNAi), provides new tools for both the study of function of human genes

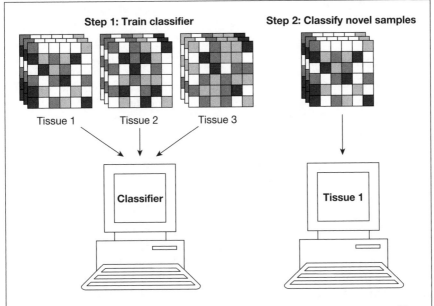

Step 1: Train classifier   **Step 2: Classify novel samples**

Tissue 1   Tissue 2   Tissue 3

Classifier   Tissue 1

Figure 7.6 Using miRNA signatures to define tissue of origin. Examples of miRNA profiles from each tissue type of interest are used to train a classifier. This program can then be used to define the likely tissue of origin of a profile from a tissue of unknown origin.

and for developing novel therapeutic strategies for treating genetic disease. The term RNA interference refers to the effect whereby the introduction of double-stranded RNA causes the cleavage of complementary mRNAs, leading to a reduction in the expression of the protein for which they code. Although originally observed in *C. elegans*, RNAi mechanisms are active in all animals studied, including the fruit fly, the zebrafish, mice and cultured human cells.

The natural function of the RNAi pathway is variously hypothesised to be an antiviral mechanism to suppress double-stranded RNA viruses, a mechanism to suppress repetitive elements, or a simple hijacking of the miRNA pathway.

While researchers have been able to create gene knockouts in model organisms for many years by either random mutagenesis or more directed methods, such methods have not been available to those studying humans until now. For the first time, RNAi allows scientists to reduce gene function in human cells. There are three principal drawbacks to RNAi. The first is that, unlike gene knockout techniques in model organisms, RNAi only induces a knockdown; that is, the expression of a gene is reduced but not totally eliminated. The level of knockdown required to elicit a phenotype differs greatly from one gene to another and from one cell type to another. Therefore it is impossible to say whether an RNAi experiment has ever reduced the level of a transcript 'enough'. The second is that, unless the double-stranded RNAi is continually delivered to the cell, the effects are transient. The third is that there are increasing questions about the specificity of the knockdown triggered by RNAi (see Section 7.5.4).

7.5.1 Triggers of RNAi

The triggers of the RNAi effect within the cell are small double-stranded RNAs of approximately 21 bp known as short interfering RNAs or siRNAs. siRNAs have 2-bp 3′ overhangs and phosphate groups on the 5′ end of each strand (Figure 7.7).

In many model organisms, RNAi can be triggered by the introduction of long double-stranded RNAs, representing all or part of the mRNA of the target gene. These are normally introduced into the animal by microinjection, soaking of the animal or cells in a solution containing the RNA or, in the case of worms, feeding of bacteria engineered to express the RNA. These long double-stranded RNAs are digested within the cell into siRNAs by dicer (see below). However, this is not possible in mammalian cells as the introduction of long double-stranded RNA triggers an antiviral response known as the **interferon response**, which leads to a general transcription shut-down and eventual death of the cell. This was originally seen as a major impediment to using RNAi in mammals. However, it was found that the interferon response can be avoided by directly introducing the 21-bp siRNAs, which are too small to trigger the response.

There are three major ways in which siRNAs can be introduced into mammalian cells. The most common way is to use chemically synthesised siRNAs (Figure 7.7), which are synthesised as two separate single-stranded RNAs of 19 bp plus two DNA bases at the 3′ end. These single-stranded RNA oligonucleotides are annealed to form a double-stranded RNA, with the two DNA bases forming the 2-nt overhang. These are most commonly introduced into cultured cells by enclosing them in a coat of cationic lipids or polymers which allow them to pass through the cell membrane – a process known as transfection (see Chapter 9 and below for more details). Chemically synthesised siRNAs are easy to use as the synthesis is usually conducted by specialist companies. The RNAi triggering molecule is well defined, and it is easy to control the amount of siRNA that enters the cell. Modifications to the siRNA can also be introduced into the molecule to make the siRNA more effective, more specific, more stable or fluorescent. The disadvantages are that chemical synthesis is currently expensive and produces only a limited amount of siRNA, which must be re-synthesised

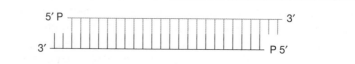

Figure 7.7 A chemically synthesised siRNA.

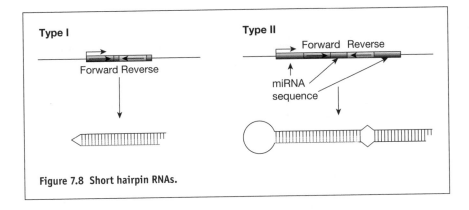

Figure 7.8 Short hairpin RNAs.

once used up. Further, the effect triggered by the introduction of such siRNAs is transient.

The second method is to introduce vectors containing inverted repeat sequences into the cell. When transcribed in the cell these sequences fold back to form hairpin structures, known as **short-hairpin RNAs (shRNAs)**. shRNAs mimic miRNAs and are processed to siRNAs in the cell. There are two types of shRNAs, type I and type II. **Type I shRNAs** are simply forward and reverse sequences separated by a short loop sequence. **Type II shRNAs** are modelled more closely on miRNAs, and the forward and reverse sequences are flanked by sequence from a real miRNA and connected by the loop sequences from the same miRNA (Figure 7.8). Type II shRNAs are thought to enter the small RNA processing pathway more easily and therefore more efficiently silence their target.

Such constructs can be introduced into cells either by transfection or by including the hairpin sequence in a vector containing the necessary sequences to allow it to be included in a virus (see below). However, since the DNA plasmid vector is much bigger than the chemically synthesised siRNA, most methods for introducing the constructs into cells are more difficult and less efficient. Almost any laboratory can make shRNA vectors, although this takes more time than chemically synthesised siRNA. The plasmid vectors containing the shRNA sequence can be propagated in a bacterial host, giving an unlimited source of the construct. Further, the sequence can also be propagated in the host cell, either by including a drug resistance marker or by using a viral vector that integrates into the genome, meaning that the hairpin is constantly produced, and so making the knockdown of the target gene permanent. It is also possible to use a range of inducible or tissue-specific promoters to control the knockdown.

The final method, which is currently the least commonly used, is to use *in vitro* transcription techniques to produce long double-stranded RNAs corresponding to part or all of the sequence of the target gene and then use a purified **endonuclease** such as dicer to digest the long RNA into a pool of short RNAs (Figure 7.9). The advantages of these endonuclease-prepared siRNAs (esiRNAs, pronounced easy-RNAs) are that they can be produced

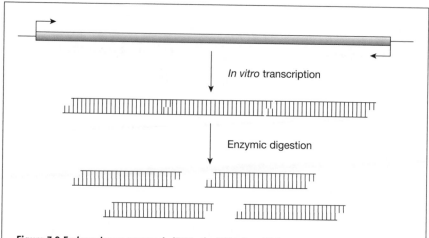

Figure 7.9 Endonuclease prepared siRNAs (esiRNAs). esiRNAs are produced by bidirectional *in vitro* transcription of a sequence from the gene to be targeted. The products of transcription from each direction are annealed to form a long double-stranded RNA. This RNA is digested into a pool of siRNAs using a purified endonuclease enzyme, such as dicer.

in any laboratory, and introduced into cells using the same methods as the chemically synthesised siRNAs. The wide range of different siRNA molecules in the pool overcomes problems involved in designing the sequence for siRNAs (see below). However, this very same property means that the actual agent introduced into cells is ill-defined. Like chemically synthesised siRNAs, the knockdown induced by esiRNAs is transient. Unlike siRNAs, esiRNAs cannot be chemically modified.

7.5.2 The RNAi pathway

The RNAi pathway, by which double-stranded RNA triggers the cleavage of mRNA, overlaps a great deal with the miRNA pathway, although the machinery is better understood. Indeed, there is much discussion as to whether the miRNA processing machinery and the RNAi machinery should be regarded as two separate pathways or simply separate outcomes from the same pathway.

As mentioned above, there are three separate ways in which siRNAs can enter the RNAi pathway: the cleavage of long double-stranded RNAs (although this is not used experimentally in mammals), the cleavage of short-hairpin RNAs, or the direct introduction of siRNAs. One strand of the siRNA, known as the guide strand, is incorporated into a RISC complex and guides this complex to mRNAs which are complementary. Which strand is specified as the guide strand is determined by differences in the stabilities of the two ends of the siRNA. The argonaute-2 protein in the RISC complex cleaves the target mRNA between the bases which are complementary to bases 10 and 11 of the siRNA. Figure 7.10 gives more

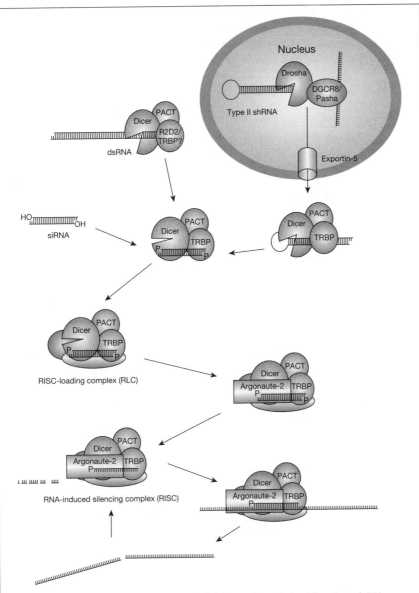

Figure 7.10 Possible mechanism of RNAi. siRNAs are either derived from long dsRNAs, shRNAs or directly introduced. The 5' end of each strand is phosphorylated. The 5' phosphate at the most stable of the two ends of the siRNA is bound by a double-stranded binding protein, thought to be TRBP in humans. This event defines which strand of the siRNA will be used to guide RISC to its target; the strand not bound will be loaded into the RISC complex and become the guide strand. This loading event is catalysed by the RISC-loading complex, which contains at least the double-stranded RNA binding protein and dicer. The RISC-loading complex loads the guide strand into the argonaute-2 protein, the catalytic part of the RISC complex. The other strand, known as the passenger strand, is degraded. RISC then uses the siRNA guide strand to guide it to and bind target mRNAs. The target mRNA is cleaved and then degraded. RISC is recycled and repeats the mRNA binding and cleavage cycle.

details. In some cases it has been shown that siRNAs targeted at the promoters of genes can cause the silencing of transcription.

While the mechanism by which miRNAs and siRNAs work is very similar, there are differences. The most obvious difference is that siRNAs cleave their target at a particular base, while miRNAs cause a translational repression or a destabilisation of the mRNA. It has been suggested that this is due to the level of complementarity between the small RNA and its target. If the RNA is perfectly complementary to its target, then it is cleaved, whereas if the complementarity is less than perfect then the target is translationally repressed/destabilised.

7.5.3 Potency of siRNAs

Not all siRNAs are equally potent. That is, two siRNAs chosen at random from all the 21-bp sequences that could target a gene will not cause the same reduction in the level of the mRNA. Thus, it is necessary to choose which of the possible sequences will be the most effective at knocking down the target gene. There are two possible approaches to this.

The first is a rule-based approach. A set of rules which define effective siRNAs is established, and then the siRNAs that adhere to all these rules (or most of them) are selected. Such rules may come from examination of the biochemistry of the RNAi mechanism; for example, that the siRNA duplex should be less stable at the 5′ end of the guide strand. Others have been deduced from systematically comparing different sequences. For example, by comparing siRNAs with every possible combination of bases in the 3′ overhang it was found that siRNAs with two uridines or a uridine and a guanine (or the equivalent DNA bases) are more effective than siRNAs with other bases in the overhang. Finally, some have been found by comparing a large number of effective and ineffective siRNAs.

The second approach does not bother deducing such rules. Instead, large sets of effective and ineffective siRNAs are used to train machine-learning programmes. These are then used to predict new effective siRNAs. Such programmes get better as more and more examples of effective and ineffective siRNAs are available to train them.

7.5.4 Specificity

It was originally believed that RNAi was exquisitely specific, with a single base change drastically reducing the amount by which a siRNA would reduce the levels of a transcript. However, it is now known that it is possible for an siRNA to have an effect on a large number of transcripts. When this occurs it is called an **off-target effect**, which may be either **sequence-specific** or **non-sequence-specific**:

- **Sequence-specific off-target effects**. These were first observed in microarrays measuring the levels of mRNAs following knockdown of a particular gene. Not only did the siRNA cause a reduction in the levels of many mRNAs unrelated to the intended target, but different siRNAs targeting the same gene produced different patterns of expression changes. In some cases this occurs because the RNAi machinery is more tolerant to base changes in particular locations in the siRNA than was initially thought. However, it is increasingly clear that a siRNA may act as a miRNA and regulate any transcript which has matches in its 3′ UTR to the 6–7-bp seed sequence at the 5′ end of the siRNA (see Figure 7.7). This may affect a very large number of transcripts, which makes predicting which mRNAs will be affected by an siRNA difficult.

- **Non-sequence specific off-target effects**. While siRNAs are thought to be too short to activate the normal interferon response, in some cell types a general response to the addition of siRNAs to the cell has been observed that partially overlaps with the interferon response. In some cells it has also been shown that small double-stranded RNAs with particular GU-rich motifs can bind and activate two immune receptors belonging to the **toll-like receptor** family and trigger an immune response to the siRNAs. The determinants of such responses are not fully understood, but it seems to be different in different cell types and also dependent on the reagent used to introduce the siRNA into the cell.

While off-target effects do pose a significant problem for RNAi, there are several approaches available for dealing with them. To avoid sequence-specific off-target effects, the genome is scanned for sequences that are very similar to that of the siRNA. If such sequences are found in genes other than the intended target, then the siRNA can be discarded at the design stage and another siRNA can be designed. In addition, it is usual to conduct experiments using more than one siRNA that targets the same gene. The assumption is that while both siRNAs will target the gene of interest, any other transcripts which they might regulate will be different. Action to avoid non-sequence-specific effects generally involves checking for activation of several components of the immune responses that might be stimulated by the introduction of the siRNAs. If such effects are found, then a different method for introducing the siRNA can be tried, as well as siRNAs with different sequences, chemically modified siRNAs, or, if possible, different cell types can be used for the experiment.

In the end, the only conclusive evidence that the phenotype produced by a siRNA is not caused by off-target effects is to attempt to rescue the phenotype by introducing a copy of the targeted gene into the cells where a silent mutation (that is, a mutation which does not change the encoded amino acid sequence) is introduced in the siRNA target site. mRNA expressed from this gene should be resistant to the effects of the siRNA and therefore such cells should not show a phenotype when the siRNA is introduced. However, this is a technically difficult experiment and is therefore rarely used in practice.

7.6 RNAi screening to identify gene function and identifying drug targets

Geneticists working in model organisms have long been able to study gene function by deliberately causing mutations to an organism and studying the resulting phenotype. However, human geneticists have had to rely on naturally occurring variation, such as disease-causing mutations, for material to work on. This is because it is virtually impossible to generate homozygous mutants in humans. Humans are diploid – they have two copies of every gene. To inactivate a gene, both copies must be mutated. In other diploid organisms this is achieved by generating a mutation in one allele and then using breeding programmes to isolate homozygous offspring. This is obviously not desirable in humans and not possible in cell culture. However, RNAi allows this problem to be circumvented in cell culture, allowing the effects of a gene knockdown to be studied. Even though RNAi is a relatively new technology, it has become routine to use knockdown of single genes or small numbers of genes to confirm a particular hypothesis about the function of genes: RNAi allows the knockdown of all the candidate genes in a genomic region associated with a particular disease to identify the causative gene. It can also be used to confirm that a statistical correlation between gene expression and a phenotype, such as those observed in a microarray experiment, are causal, at least at the cellular level.

The complete sequence of the human genome, and the knowledge of the location of most of the genes contained within it, allows the construction of sets of RNAi-inducing reagents targeting every gene in the genome. Collections of such reagents exist for siRNAs, shRNAs and esiRNAs. This allows experiments to be conducted where every gene in the genome is knocked down and the effect of this on a particular phenotype observed. This process is known as **RNAi screening**. If knockdown of a gene alters the phenotype of interest then that gene can be associated with the process under study. RNAi screening can be carried out in several ways (Figure 7.11). Perhaps the most common way to perform an RNAi screen is to perform a large number of individual experiments, usually in a multi-well plate, where one gene is knocked down in each well and the effect on the phenotype of interest (perhaps the activity of a fluorescent compound) recorded, either by specialised plate-reading equipment or by observation with a microscope. Such methods gather a very large amount of information about the effects of each gene knockdown but are very slow, expensive and labour-intensive. An alternative is to perform a pooled selection. Here a pool of RNAi-inducing reagents is introduced into a population of cells. A selection is then applied to separate cells showing the phenotype of interest from those that do not. The phenotype causing RNAi reagent can then be identified. Here, shRNAs are particularly useful because the shRNA causing the phenotype can be identified by sequencing hairpins present in selected cells.

The phenotype of interest can be connected to a basic biological function, such as the cell cycle, or could be related to a particular disease. In the case of disease phenotypes, any gene identified in an RNAi screen is a potential target for drugs to treat the disease (for an example, see Box 7.4).

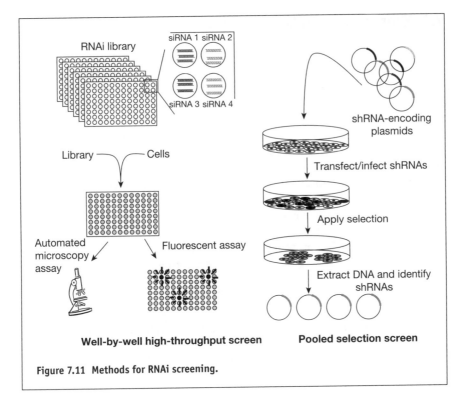

Well-by-well high-throughput screen   **Pooled selection screen**

Figure 7.11 Methods for RNAi screening.

BOX 7.4: FROM SCREEN TO CLINIC

One potential use for RNAi screens is to identify new drug targets for the treatment of disease. Familial cylindromatosis is a rare autosomal dominant predisposition to the formation of multiple tumours in the skin, often on the scalp. Sometimes large numbers of tumours merge to form large multi-tumour masses which are sometimes referred to as turban tumours. Such tumours are usually benign, but cause considerable disfigurement and discomfort. It is known that this disease is caused by mutations in the *CYLD* gene. A group in the Netherlands identified the *CYLD* gene when they screened for genes which modulate the NF-κB protein, a transcription factor that is involved in regulation of apoptosis. They found that when *CYLD* was knocked down, this led to an increase in the activity of an activator of NF-κB known as IKKβ. They hypothesised that in individuals who had a mutation in *CYLD*, NF-κB was highly activated, which led to protection from apoptosis, and therefore an increase in tumours. Further, it was already known that aspirin is an inhibitor of IKKβ. They decided that application of aspirin to the tumours of familial cylindromatosis might be an effective treatment for this disease. In a pilot study, application of aspirin cream caused two out of 12 tumours tested to go into complete remission whilst a further eight showed some response.

7.7 Therapeutic RNAi

As well as allowing researchers to probe the effects of reducing gene function to study basic biology or to find new drug targets, RNAi also has exciting clinical uses as a therapeutic. Traditional gene therapy aims to introduce a wild type copy of a malfunctioning gene to compensate for its loss. This means that gene therapy is restricted to treating conditions that result from the loss of function of a protein, such as many recessive genetic disorders, or using gene expression as a toxic agent; for instance, in cancer treatment. As RNAi triggers knockdown of gene function, it can be used in a far wider range of applications, including the treatment of dominant conditions or in gene replacement, where the mutated gene is de-activated and a copy of the wild type is expressed in its place. RNAi also allows the treatment of disorders not necessarily caused by the mutation in the coding region of a patient's genome, but where the expression level of a gene is increased or where alien DNA has been added to a genome, such as in viral infection. Many drug targets identified in RNAi screens (see above) may be directly targeted using therapeutic RNAi.

Both siRNA- and shRNA-induced RNAi can be used in a clinical setting and the choice of which is most appropriate depends on several factors, including the required duration of knockdown. siRNAs trigger a knockdown that lasts for only a limited time. This time could possibly be extended by reapplication. However, shRNAs can provide a long-lasting reduction in gene activity and may be more suited to the treatment of chronic conditions with a single dose. Treatments for a wide range of diseases are currently being developed, using both siRNAs and shRNAs, and several of these are now in clinical trials, including treatments for: **age-related macular degeneration**, diabetic macular oedema, acute renal failure, pachyonychia congenita, infection with hepatitis B virus (HBV), human immunodeficiency virus (HIV) and **respiratory syncytial virus** (RSV).

The challenges of RNAi-based therapy are similar to those for other forms of gene therapy, the two biggest hurdles being delivery of the RNAi-inducing agent to diseased cells and various safety issues that surround the effects of introducing these agents. In the following section various delivery methods will be briefly discussed for both siRNA and shRNA, along with a discussion of some of the safety issues facing the use of RNAi as a therapeutic agent. Three examples of the potential use of RNAi therapeutics will then be described. This is not intended to be an exhaustive list of all possible uses of RNAi therapeutics, as such a list would require a book of its own; rather, it gives an idea of different strategies being explored.

7.7.1 Delivery

Delivery of RNAi-inducing agents to the correct cells poses similar problems to those for traditional gene therapy. For shRNA-based approaches, the solutions to this are usually the use of the same viral vehicles as are used for other types of gene therapy (see Chapter 9). Delivery of siRNAs to cells often involves the delivery of modified or unmodified siRNAs to their target by enclosing the siRNA in a layer of lipids, complexing it to cationic polymers or peptides, or conjugating it with a variety of small molecules.

7.7.1.1 Delivery of naked, modified and conjugated siRNAs

For some easily accessible tissues such as the eye, the respiratory tract and the central nervous system, it is possible to deliver naked siRNA to the required site. siRNAs have been delivered to the eye and to the brain by injection of saline solutions containing the siRNA. Delivery via the nose of siRNAs in either a saline solution or 5% dextrose has been demonstrated to be effective for inducing RNAi in the lung. It is not known why a few cell types will take up naked siRNA like this while others will not. However, such delivery strategies are not suitable for delivery to the bloodstream as siRNAs are rapidly degraded by nucleases in the blood and cleared by the filtering action of the kidney.

The stability of siRNAs can be improved by altering their chemical structure. Changes to the sugars on the phosphate backbone of the sense or passenger strand of the siRNA can prevent their degradation by nucleases. Another possibility is to replace one of the strands of the siRNA with a DNA oligonucleotide. Such RNA/DNA hybrids are not degraded as readily by blood nucleases.

Targeting of siRNAs to particular tissues can be achieved by conjugating the passenger strand of the siRNA to a carrier molecule. Conjugation of siRNAs to cholesterol has been shown to improve the uptake of siRNAs into the liver. The addition of an **aptamer** molecule to the end of an siRNA has been used to target siRNAs to prostate cancer cells. Aptamers are RNA molecules that bind specifically to a particular target, in this case a protein called prostate-specific membrane antigen.

7.7.1.2 Lipid-based delivery

Perhaps the most common method for transferring siRNAs into cells is through the use of **liposomes** and **lipoplexes**. Liposomes are small vesicles surrounded by a lipid bilayer which contain an aqueous interior and can enclose charged molecules such as siRNAs. Lipoplexes are complexes of positively charged lipids with negatively charged nucleic acids. The lipids surround the siRNA and mask its negative charge from the negatively charged cell membrane. Lipoplexes and liposomes are routinely used for the introduction of siRNAs into cells *in vitro* and are also probably the most common method for the *in vivo* delivery of siRNAs. These methods protect the siRNA from nucleases, reduce clearing by the kidney, reduce activation of immune responses to naked siRNA and improve the uptake of siRNA across the cell membrane. The introduction of nucleic acids using lipids is known as **lipofection**.

A particular type of liposome known as a stable nucleic acid-lipid particle (SNALP; Figure 7.12) has been used to deliver an siRNA targeted

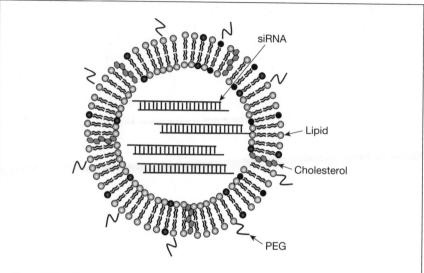

Figure 7.12 A liposome. The liposome shown is based on a formulation known as a SNALP. It is made up of a lipid bilayer containing different kinds of lipids, cholesterol and polyethylene glycol (PEG). PEG and cholesterol help to stabilise the liposome and shield the positive charges on the lipids. (After de Fougerolles, A. *et al.* (2007) *Nature Reviews Drug Discovery*, **6**, 443–453. Reprinted by permission of Macmillan Publishers Ltd.)

against the *apoB* gene in monkeys, leading to a reduction of the expression levels in the liver and an 11-day reduction in blood cholesterol.

Lipoplexes can be used to target mucosal membranes such as those in the vagina and intestine. Mucosal surfaces are generally fairly easily accessible from the outside and so are often used by viruses to enter the body, and so targeting these tissues might prove a useful way of combating these viruses.

It is possible to target liposomes to particular tissues by including targeting molecules in the liposome. These targeting molecules might be ligands for cell-specific receptors or antibodies to proteins expressed on the surface of certain cells.

7.7.1.3 Nanoparticle delivery
Some cell types are resistant to lipofection. An alternative to the use of lipid-based delivery vehicles is the use of tiny particles made out of positively charged polymers and peptides known as **nanoparticles**. Many different such particles have been created, but they usually involve the complexing of peptides or polymers to the siRNAs. siRNAs contained within such nanoparticles can be protected from nuclease degradation and clearance by the kidney. The positive charge of the compounds used and the addition of targeting molecules, such as ligands for cell receptors or antibodies to cell surface proteins, promote the uptake of nanoparticles into the cell.

7.7.2 RNAi for macular degeneration

Age-related macular degeneration (AMD) is the most common cause of vision impairment in the over-65s. The most common form, 'dry' AMD, is caused by the slow degeneration of the macular tissues over time. Dry AMD accounts for about 85–90% of AMD cases. In the remaining 10%, the disease progresses to 'wet' AMD. In these cases new blood vessels form below the retina and leak fluids, damaging the light-sensitive cells. The growth of these new blood vessels is known as **neovascularisation** and is stimulated by the **angiogenesis-promoting** vascular endothelial growth factor (VEGF). Suppressing this gene leads to an inhibition of neovascularisation and halts the progress of the condition.

At the time of writing there are currently two different RNAi-based treatments in clinical trials. A siRNA targeting the *VEGFA* gene is currently entering phase III trials. This is the first RNAi-based drug to enter phase III clinical trails, which are designed to test the efficacy of a drug in a large sample of patients. The trials are scheduled for completion in 2011. Additionally, a siRNA targeting the VEGF receptor *VEGFR1* is in phase II clinical trials. In both cases the siRNA is introduced via injection into the eye.

While this is very promising, a recent finding has cast doubt on these treatments. It has been found that in a mouse model of AMD, injection of any siRNA into the eye, irrespective of the sequence, has the same effect. This is due to the activation of the toll-like receptor *TLR3* by double-stranded RNAs of 21 bp. Mice lacking this receptor show no improvement when treated with siRNAs. Thus the anti-AMD effect observed with the injection of siRNAs targeting *VEGF* or its receptor is due not to the knockdown of these genes, but to an immune response to short double-stranded RNAs. While this does not mean that siRNA-based treatment of AMD will not work, it is a serious setback for RNAi-based therapies in general, particularly those involving the delivery of naked siRNA, as in the majority of situations the desired outcome cannot be achieved by activating the immune system.

7.7.3 RNAi for viral infections

Viruses are attractive targets for RNAi-based drugs because viral infections are essentially caused by the expression of foreign viral genes within the cell. Silencing of these genes should prevent viral replication without affecting the levels of endogenous genes. Another good target for antiviral treatment in the large number of viruses encoded by an RNA genome would seem to be the genome itself. However, the genomes of RNA viruses have proved hard to target. Many of these viruses appear to have mechanisms designed to avoid the RNAi machinery, which is not surprising if such machinery evolved as an antiviral system. A final important set of targets for RNAi-based antiviral drugs are host co-factors. These are proteins produced by the host cell that are required for the virus to replicate. However, care must be taken that knockdown of these endogenous genes does not harm the host cell or other non-infected cells which might be affected.

For acute viral infections, such as those of the respiratory tract, application of siRNAs may be a good way to reduce activity of viruses, alleviating symptoms and allowing the immune system time to clear the infection. The respiratory system is easily accessed and cells can be treated with either naked siRNA, or siRNAs packaged into liposomes and administered via the use of a device similar to those used for asthma treatments. Treatments for the **severe acute respiratory syndrome coronavirus** (SARS-CoV), influenza and **respiratory syncytial virus** (RSV) have been demonstrated in animal models and an siRNA-based treatment for RSV is currently in clinical trials. Respiratory syncytial virus is an RNA virus that infects the respiratory tracts of a very large number of people. It is estimated that nearly 60% of all infants in the US will be infected in their first year of life. Usually the symptoms are mild (not unlike the common cold) but are occasionally more severe and can lead to death. This is especially the case for premature babies and immunocompromised individuals. The RNAi-based treatment targets the **nucleocapsid** or 'N' protein of the virus which coats the viral genome and prevents it from being degraded by nucleases. The siRNA is delivered unmodified via the nose to the lung. In contrast to the siRNA-based treatment for AMD covered above, care has been taken to ensure that the clinical effect is not due to an elevated immune response and that the siRNA does reduce the levels of mRNA for the N gene.

There are also hopes that RNAi-based treatments may be applied to more chronic viral infections. Here RNAi must be induced by shRNAs which will continue to exert their effect long after treatment. Much work has focused on the development of treatments for HIV. One of the problems with any treatment for chronic virus infections, such as HIV, is that viruses can mutate very quickly and therefore evolve to escape silencing by shRNAs. This has been observed in models of HIV infection. One possible way around this is combinatorial treatment, where more than one target in the virus genome is hit at a time; thus if one target mutates, the virus is still hit by RNAi targeted at other genes. This has been shown to reduce so-called 'escape' in models of HIV infection. However, there are concerns that hitting cells with multiple shRNA constructs could lead to competition between them for the RNAi machinery, and even prevent the normal functioning of the miRNA pathway by saturating it. One way to avoid this is to use several different gene therapy approaches, only one of which is RNAi. This is the approach taken by one treatment which is currently in the early stages of clinical trials. In this treatment, **haematopoietic stem cells** (progenitor cells for all blood cells) from the patient are infected *ex vivo* with a lentiviral vector containing sequences which code for three RNAs:

- A shRNA targeting the essential HIV genes *tat* and *rev*; these code for a transcriptional activator and the reverse transcriptase respectively.

- An RNA which serves as a decoy for the TAR element of the HIV genome (TAR is the binding site for *tat*). Here the idea is that any *tat* protein that escapes RNAi will bind this decoy rather than the genuine TAR element.

- A small catalytic RNA, known as a **ribozyme**, which targets and cleaves the mRNA for the human **CCR5** gene, which acts as a co-receptor for the HIV virus.

The modified stem cells are then reintroduced into the patient in the hope that over time their resistance to HIV will mean that they replace the unmodified stem cells.

7.7.4 Anti-cancer RNAi

Much attention has been paid to developing RNAi-based treatments for cancer. In many ways cancer presents a more difficult target than the conditions discussed above. Treatments for cancer must be effective at killing cancer cells while minimising harm to normal cells. Cancer cells can also be distributed all over the body, arguing for systemic delivery. Despite these difficulties, several RNAi-based approaches to combating cancer have been demonstrated in animal models.

There are many possible targets for anti-cancer RNAi; perhaps the most obvious targets are the oncogenes promoting the unregulated proliferation of the cancer in question. Some of the most attractive targets are forms of oncogenes that are not present in normal cells. Fusion proteins are one example of such oncogenes; here the oncogene is formed by a rearrangement event that fuses two genes together. Since these genes exist only in tumour cells, siRNAs targeting these genes will have an effect only in tumour cells. Such an approach has been successfully used to target the *EWS-FLI* fusion gene found in the tumour Ewing's sarcoma in a mouse model. Another alternative is to create siRNAs that target a mutant allele of an oncogene while not targeting the wild type. Despite problems with **off-target effects**, it is possible to design siRNAs that are far more active against a mutant allele of the EGF receptor, often found in lung cancer, than the wild type allele *in vitro*. Finally, where tumours are promoted by the expression of viral genes, these genes provide attractive targets that are not present in normal cells. The human papillomavirus (HPV) protein E7 represses the tumour suppressor Rb. Knockdown of E7 by siRNAs *in vitro* leads to massive apoptosis in cancer cell lines infected with HPV.

Cancer cells are often resistant to apoptosis signals, which would usually trigger cell death. This can be caused by the overexpression of various anti-apoptotic genes. Targeting these genes would sensitise cells to apoptotic signals and lead to cell death. This would affect cancer cells more than normal cells because the pro-apoptotic signals which the genes are protecting tumour cells from will be weaker or absent in normal cells.

Genes which mediate the interaction between cancer cells and the host may also be good targets for anti-cancer RNAi treatment. siRNAs targeting the *VEGF*, which promotes the formation of new blood vessels in tumours, has been shown to reduce tumour growth in mice. Many tumours evade detection by the immune system which would usually attack such cells. One factor which is thought to be involved in this avoidance is the growth factor

EGF-β, which suppresses the immune system and is secreted by many tumours. Knockdown of genes involved in secretion or detection of EGF-β in mice has been shown to have an anti-tumour effect.

A different approach is to use RNAi to sensitise cancer cells to other drugs. Many tumour cells are resistant to anti-cancer drugs because they overexpress a protein known as **multi-drug resistance (MDR) protein**, which pumps drugs out of the cell. Targeting of this protein by RNAi *in vitro* has been shown to sensitise cells to treatment with other anti-cancer drugs. Alternatively, targeting DNA repair enzymes sensitises cancer cells to DNA-damaging chemotherapy drugs or radiotherapy.

We have seen that there is no shortage of targets for RNAi-based therapy. However, a major problem in RNAi-based cancer therapy is targeted delivery of the RNAi-inducing agent to cancer cells. In several experiments in animal models, successful anti-tumour activity has been shown after injection of siRNAs enclosed in liposomes or nanoparticles directly into the tumours. However, this would not allow the targeting of metastases in other parts of the body, and also it does not avoid the targeting of non-cancer cells in the same area. Various targeting approaches have been used to target siRNAs to tumour cells. The inclusion of peptide ligands for receptors expressed preferentially on the surface of tumour cells has been used to target liposomes or nanoparticles to tumour cells. The transferrin receptor is involved in iron metabolism and is overexpressed in tumour cells. Liposomes or nanoparticles conjugated to transferrin are taken into the cell by internalisation of the receptor which happens at a higher rate in tumour cells. Thus more siRNA is taken up into tumour cells than normal cells. siRNAs targeting the *EWS-FLI* oncogene have been successfully targeted to cancer cells by enclosing them in a nanoparticle containing transferrin. An alternative is to use antibodies or antibody fragments targeting tumour cell-specific proteins. A liposome coated in the antigen binding part of an anti-transferrin antibody has been successfully used to deliver a siRNA targeting the *HER*-2 oncogene in a mouse model of pancreatic cancer, including delivery to metastatic tumours. This led to an inhibition of tumour growth in these well established tumours.

One problem with using siRNA for anti-cancer RNAi treatments is that the effects of siRNAs are particularly short-lived in rapidly dividing cells. However, the use of viral delivery methods suffers from a lack of specificity. One solution may be to use viral vectors which infect only rapidly dividing cells, or to express the shRNA from a promoter which is active only in tumour cells. This would, however, require very high doses of the virus to guarantee altering all the tumour cells in the body.

7.7.5 Safety

With any new treatment, safety is of the utmost concern. Apart from immune responses to the vehicle used to deliver the RNAi-inducing agent to cells, there are several safety concerns with using RNAi for gene therapy.

As outlined in the section on specificity, immune responses can be stimulated directly by small RNAs themselves either by a response to double-stranded RNA within the cell or by the recognition of siRNAs by receptors on some immune cells which are designed to detect invading pathogens. This has already posed issues for some RNAi-based drugs (see above). There are several possible ways to avoid such responses. First, the receptor-mediated responses for at least one of the receptors involved seems to be restricted to siRNAs that contain certain GU-rich sequences. Second, chemical modification of the siRNA or the replacement of one of the two strands with a DNA oligonucleotide may reduce some of the effects. Such responses appear to be concentration-dependent, so using only a very small amount of siRNA or shRNA may help. Also, the use of some vehicles may shield the siRNAs from the immune system. In some cases, activation of the immune system might be advantageous, such as when targeting infections.

Another concern is that flooding cells with a large amount of RNAi-inducing agents may saturate the RNAi and miRNA machinery and prevent the natural functioning of miRNAs. This is particularly a problem for shRNA-based therapies since shRNAs share more of the machinery with miRNAs. This effect has already been observed in a mouse model where viral delivery of a shRNA to the mouse liver led to serious liver damage and death. Again, this effect can be avoided by using smaller amounts of shRNA.

Apart from these general responses to RNAi-inducing agents there is also the problem of the specificity of particular siRNAs or shRNAs. As mentioned above, siRNAs and shRNAs can act as miRNA to silence the expression of genes which are complementary to only six or seven bases in the siRNA/shRNA. It may be impossible to completely avoid such short matches to genes other than the intended target. Therefore all siRNAs/shRNAs must be very carefully checked to ensure that any off-target effects they trigger are not detrimental.

7.8 The diversity of small RNAs

In this chapter we have mostly discussed miRNAs and siRNAs. However, these are just two of a large family of small non-coding RNAs in the human genome. As well as the regulation of gene expression, these small RNAs are involved in suppression of transposons, splicing, RNA modification, maintenance of the telomere, translation and endonuclease activity. In this next section we will quickly review the functions of the other large groups of small non-coding RNAs.

7.8.1 Small RNAs related to miRNA and siRNA

The discovery of miRNAs and RNAi triggered the identification of several other related classes of RNA in many organisms. Endogenous siRNAs are siRNAs that are produced by transcription from the genome. While they are most prominently found in the worm *C. elegans*, there is some evidence that transcription from an anti-sense promoter in LINE1 elements (see Section 2.2.2.1) leads to the formation of a double-stranded RNA which is processed by dicer into siRNAs which act to suppress the transposition of LINE1 elements.

Piwi-interacting RNAs (piRNAs) are single-stranded RNAs of 28–33 nucleotides, found associated with the Argonaute homolog Piwi in mammals. The expression of piRNAs is restricted to the testes. piRNAs are similar to a family of RNAs found in the fly known as repeat-associated siRNAs (rasiRNAs), many of which originate from repeat sequences and are thought to be involved in the suppression of transposons. However, fewer piRNAs than would be expected are found in repeat sequences, and the function of piRNAs remains to be determined.

7.8.2 Small nuclear RNAs

Small nuclear RNAs (snRNAs) are a group of RNAs that are found predominantly in complexes with proteins (ribonuclear proteins or RNPs) which localise to the nucleus. They can be divided into several groups.

7.8.2.1 Spliceosomal RNAs

The spliceosome is a ribonuclear protein complex which catalyses the removal of introns from genes. The most common form, the major spliceosome, contains the small RNAs U1, U2, U4, U5 and U6. Another spliceosome, the minor spliceosome, catalyses the removal of introns from a rare class of introns known as U12 introns and contains the snRNAs U11, U12, U4atac, and U6atac and U5. With the exception of U6 (which is transcribed by DNA polymerase III), spliceosomal RNAs are transcribed by DNA polymerase II as long precursor forms and then processed to form the final snRNA.

7.8.2.2 Small nucleolar RNAs/small cajal body RNAs

Small nucleolar RNAs (snoRNAs) and small cajal body RNAs (scaRNAs) guide RNPs which catalyse the modification of other RNAs, particularly snRNAs and rRNAs by the addition of methyl groups to the 2′ position of the sugar ring of a base, or the conversion of uridine bases to its isomer **pseudouridine**. Most snoRNAs/scaRNAs are found in the introns of other genes and are processed from the excised introns.

There are two major groups of snoRNAs: C/D box and H/ACA box snoRNAs. C/D box RNAs contain two motifs, the C and the D box near either end of the RNA, as well as sequences complementary to their targets. C/D box snoRNAs usually direct 2′ O-methylation. H/ACA box snoRNAs contain two hairpin structures and two single-stranded regions, containing H and ACA motifs and also sequences complementary to the target RNA. H/ACA box snoRNA containing RNPs usually catalyse pseudouridylation. scaRNAs are similar to snoRNAs, although some contain both C/D box and H/ACA box motifs. Orphan snoRNAs are snoRNAs that are not complementary to snRNAs or rRNAs. The orphan snoRNA HBII-52 is involved in the alternative splicing of serotonin receptor 2C, and its loss has been linked to the genetic disorder Prader–Willi syndrome.

7.8.3 Ribozymes

Ribozyme is the name given to an RNA that acts as an enzyme. Examples of ribozymes are the ribosomal RNAs and RNaseP. The ribosome is made of the small ribosomal subunit (made up of the 18S RNA) and a large subunit (containing the 5S, 5.8S and 28S RNAs). These RNA components form the active sites of the ribosome, hold the tRNA in place, and catalyse the reaction which ligates amino acids to the growing polypeptide during translation. Ribosomal RNAs are transcribed from hundreds of repeats of rRNA genes found in five clusters on five chromosomes and are transcribed by RNA polymerase I, except the 5S subunit which is transcribed by RNA polymerase III.

The catalytic subunit of RNaseP is an RNA which catalyses the removal of nucleotides from the precursor of tRNAs. It is now possible in some cases to design artificial ribozymes to cleave specific targets, such as the ribozyme which cleaves the CCR5 transcript in the combinatorial HIV treatment described above.

7.9 Summary

- miRNAs are small single-stranded RNAs that repress the expression of genes to which they are partially complementary.

- miRNAs are transcribed from the genome as long primary miRNAs containing a hairpin structure. The primary transcript is processed to release the hairpin as the pre-miRNA. The hairpin is exported into the cytoplasm where it is further processed to make the single-stranded mature miRNA.

- The mature miRNA is incorporated into a protein complex known as miRISC which can direct cleavage of target mRNAs if they are perfectly complementary, or translational repression and destabilisation of the mRNA if the match is not perfect.

- miRNAs can be transcribed from transcripts containing one or more pre-miRNA hairpins or from the introns of other transcripts.

- miRNAs can be identified by directional cloning or by computational methods. In order to be classed as a miRNA it must fulfil several criteria, including:
  - It must be 18–22 nt long
  - It must come from a hairpin structure
  - It must be expressed

- Bases 2 to 7 or 8 of the miRNA are known as the seed and are important for targeting.

- Targets can be predicted computationally, but must be confirmed experimentally.

- MicroRNAs are involved in diseases through:
  - Being misregulated
  - Mutations in the miRNA
  - Mutations in miRNA targets

- MicroRNAs can act as both oncogenes and tumour suppressors in cancer.

- MicroRNAs 1 and 133 are involved in the development of muscle tissue, and their deregulation can lead to several cardiovascular disorders.

- MicroRNA expression patterns can be used as diagnostic markers for diseases. In cancer they can be used both to diagnose cancer and to identify the subtype and prognosis.

- RNA interference is the process by which double-stranded RNA complementary to mRNA leads to cleavage and degradation of the mRNA.

- RNAi can be triggered in mammals by three different types of trigger:
  - Chemically synthesised siRNAs
  - shRNAs transcribed from constructs introduced into cells. shRNAs are processed into siRNAs by the cellular miRNA machinery
  - *In vitro* transcription of long double-stranded RNA which is digested by endonucleases before introduction into cells

- RNAi acts through a pathway very similar to the miRNA pathway, and in fact these may represent two parts of one small regulatory RNA pathway.

- Not all siRNAs are of equal effectiveness, and effective siRNAs must be designed using sets of rules or complex computer algorithms.

- siRNAs may affect transcripts other than those to which they are targeted. These effects may be:
  - **Sequence-specific**. This happens when the siRNA is complementary to other transcripts with one or two mismatches. Alternatively, siRNAs may act as miRNAs and target transcripts by virtue of matches to their seed sequence.
  - **Non-sequence-specific**. siRNAs may trigger several general responses, including the interferon response and toll-like receptor pathways.

- Sequence-specific off-target effects can be controlled for by using multiple siRNAs targeting the same transcripts.

- RNAi can be used to find genes involved in a process by testing the effects of knocking down large numbers of genes. If this process is important for a disease phenotype, then the genes discovered may serve as drug targets.

- RNAi may also be used to treat medical conditions. In order to do this, relevant targets must be identified and the RNAi trigger delivered to the cells in question.

- shRNAs can be delivered using the same techniques used for gene transfer in traditional gene therapy.

- siRNAs may be delivered by direct application in a few cases, by conjugating various molecules to the siRNA or enclosing it in a liposome or nanoparticle. The siRNAs may be targeted by including molecules such as receptor ligands, antibodies or aptamers.

- Treatments for macular degeneration targeting the angiogenesis-promoting *VEGF* are in the late stages of clinical trails. However, it has been shown that these treatments may work by activating immune responses in the eye.

- siRNAs can be used to treat acute viral infections such as RSV; chronic viral infections may require the long-term production of shRNAs.

- Using several shRNAs or shRNAs in combination with other treatments can help to avoid viral escape by mutation of RNAi targets.

- Much effort is being expended trying to develop RNAi-based cancer therapies but, while there are many targets, systemic delivery to tumour cells while leaving non-tumour cells intact is difficult.

- miRNAs and siRNAs are just two examples of small non-coding RNAs in the human genome. Other small non-coding RNAs include piRNAs, spliceosomal RNA, snoRNAs and ribozymes.

- Small RNA biology is a rapidly growing and changing field, with many new discoveries and advances made every year.

Further reading

microRNA structure and function
VALENCIA-SANCHEZ, M.A., LIU, J., HANNON, G.J. and PARKER, R. (2006) Control of translation and mRNA degradation by miRNAs and siRNAs. *Genes and Development*, **20**, 515–524.

More detail on the modes of action of microRNAs.

microRNA genomics
KIM, V.N. and NAM, J. (2006) Genomics of microRNA. *Trends in Genetics*, **22**, 165–173.

A round-up of many aspects of miRNA genomics, including genomic locations and methods for identifying both miRNAs and their targets.

LANDGRAF, P., RUSU, M., SHERIDAN, R. *et al.* (2007) A mammalian microRNA expression atlas based on small RNA library sequencing. *Cell*, **129**, 1401–1414.

A massively parallel sequencing experiment to catalogue the miRNAs expressed in animal tissues.

microRNA and disease
CALIN, G.A. and CROCE, C.M. (2006) MicroRNA signatures in human cancers. *Nature Reviews Cancer*, **6**, 857–866.

MATTES, J., COLLISON, A. and FOSTER, P.S. (2008) Emerging role of microRNAs in disease pathogenesis and strategies for therapeutic modulation. *Current Opinion in Molecular Therapeutics*, **10**, 150–157.

Two reviews of the function of miRNAs in various diseases and the uses to which this information can be put.

LU, J., GETZ, G., MISKA, E.A. *et al.* (2005) MicroRNA expression profiles classify human cancers. *Nature*, **435**, 834–838.

An example of using miRNA expression signatures for diagnostics.

RNA interference
JAKSON, A.L. and LINSLEY, P.S. (2004) Noise amidst the silence: off-target effects of siRNAs? *Trends in Genetics*, **20**, 521–524.

A review of the different possible types of off-target effects. The detail is a little out of date, but it has clear explanations of the concepts.

JACKSON, A.L., BURCHARD, J., SCHELTER, J. *et al.* (2006) Widespread siRNA 'off-target' transcript silencing mediated by seed region sequence complementarity. *RNA*, **12**, 1179–1187.

A microarray study of the off-target effects of a group of siRNAs which identifies matches to seed sequences and the cause of the effects.

RNAi for studying gene function and for drug target identification
BOUTROS, M. and AHRINGER, J. (2008) The art and design of genetic screens: RNA interference. *Nature Reviews Genetics*, **9**, 554–566.

ECHEVERRI, C.J. and PERRIMON, N. (2006) High-throughput RNAi screening in cultured cells: a user's guide. *Nature Reviews Genetics*, **7**, 373–384.

OOSTERKAMP, H.M., NEERING, H., NIJMAN, S.M.B., DIRAC, A.M.G., MOOI, W.J., BERNARDS, R. and BRUMMELKAMP, T.R. (2006) An evaluation of the efficiency of topical application of salicylic acid for the treatment of familial cylindramatosis. *British Journal of Dermatology*, **155**, 182–185.

A very short report on the trial for using aspirin for cylindromatosis which came out of an RNAi screen.

RNAi therapeutics
de FOUGEROLLES, A., VORNLOCHER, H., MARAGANORE, J. and LIEBERMAN, J. (2007) Interfering with disease: a progress report on siRNA-based therapeutics. *Nature Reviews Drug Discovery*, **6**, 443–453.

KIM, D.H. and ROSSI, J.J. (2007) Strategies for silencing human disease using RNA interference. *Nature Reviews Genetics*, **8**, 173–184.

The diversity of small RNAs
KAWAJI, H. and HAYASHIZAKI, Y. (2008) Exploration of small RNAs. *PLoS Genetics*, **4**, 3–8.

CHAPTER EIGHT

Genetic testing

Key topics

- When gene testing is used
- Types of mutation
- General principles of testing
- Testing for mutations in the *CFTR* gene
- Testing for mutations in the dystrophin gene
- Testing for uncharacterised mutations
  - ○ Single-stranded conformational polymorphisms
  - ○ Denaturing gradient gel electrophoresis
  - ○ Mismatch cleavage detection
  - ○ Protein truncation test
  - ○ DNA chips
  - ○ DNA sequencing
  - ○ Chromosome tracking

8.1 Introduction

The advances in gene isolation and characterisation described in this book enhance enormously the ability to diagnose genetic disease by the more traditional methods based on a physician's skill supported by biochemical tests. This chapter is concerned with the techniques that can be used routinely to detect mutations in genes of interest. The application of these techniques often raises significant ethical problems; these are considered in Chapter 12.

There are a number of different situations in which gene testing might be carried out:

- Where there is a known risk of a particular disease, embryos may be tested with a view to termination; alternatively, prospective parents may wish to know their carrier status in order to make reproductive decisions. The optimum situation in embryo testing is where the nature of the mutation has previously been defined, thus allowing its presence to be specifically tested in the embryo. In an autosomal recessive disorder this will require two mutations to be identified. In the absence of such information, the chromosome(s) harbouring the mutation may be tracked by linked polymorphic markers.

- Neonatal population screening for genetic diseases where there is a clear benefit derived from early diagnosis and there is a simple and reliable test available. Neonatal screening for phenylketonuria is common using the Guthrie 6-day blood-spot test. A small sample of blood is taken from the baby and preserved by drying on to a card. Phenylalanine levels in the blood spot are then assayed. However, the same blood spot can provide material for other tests, such as screening for the ΔF508 *CFTR* mutation in European populations.

- To confirm or rule out a diagnosis of genetic disease made on the basis of medical symptoms.

- To diagnose the status of an individual, normally an adult, at risk from a late-onset genetic disease such as HD or one of the inherited cancer predispositions.

- To determine the status of genes predisposing to complex diseases. This area is likely to grow in importance as tests are developed for susceptibility to diseases such as type 1 diabetes mellitus.

- Where a candidate gene involved in a genetic disease has been identified, genetic testing is used to search for a correlation between the presence of a mutation and the onset of disease symptoms. This is often the final step in the hunt for a disease gene.

8.1.1 DNA testing has been revolutionised in the last few years

A few years ago work in a DNA diagnostic laboratory would mainly consist of tracking disease chromosomes through a family using RFLPs analysed by Southern blotting. Southern blotting is laborious, involves the use of radioisotopes, with the attendant safety problems, and requires significant amounts of biological material. RFLP analysis required extensive work to establish the phase of mutations and linked RFLP polymorphisms segregating in a family. Many meioses were uninformative because the RFLPs were not heterozygous, and if the linked markers were not sufficiently close to the mutation the diagnosis could be falsified by recombination.

Techniques used to screen for mutations have been revolutionised in the past few years by two developments. First, cloning of disease genes has allowed tests to be directed towards the gene sequences themselves rather than tracking the inheritance of mutant chromosomes through families. Second, PCR is used to amplify sequences. This generates enough product to be analysed by conventional agarose or polyacrylamide electrophoresis. However, increasingly the technology used for automated DNA sequencing is applied to analysing PCR products. Apart from increasing throughput and accuracy, a big advantage of this is that it allows the quantification necessary to detect heterozygous deletions.

8.1.2 Three different challenges to the gene tester

In some cases the nature of the mutation likely to affect a gene is known. First, where only one or a small number of different mutations in a population are responsible for a disease; typically, this results from founder mutations. Second, where the disease commonly occurs because of a particular type of mutation. The obvious example of this would be trinucleotide repeat expansion (TRE) mutations, but we may also consider cases such as DMD, where deletions are responsible for 60% of cases.

In other cases such information will not be available. This is inevitable when a gene is first cloned, during the attempt to correlate the presence of mutations with disease symptoms. There are also many genes where a large number of different mutations have been catalogued. Sometimes these mutations are unique to a particular family or individual and are known as **private mutations**. Testing for the presence of mutations in these genes is a formidable technical challenge. Often the genes may be large, and it is possible that the aetiological mutation may be a single base change.

We may thus summarise three different problems that a gene tester will be called upon to solve:

1. The disease is caused by one or a small number of well characterised mutations. This will be approached by using tests designed to detect specific mutations. These tests are described in Section 8.2.
2. The gene is cloned but there are many possible mutations. This requires tests that can detect differences between normal and mutant genes without necessarily defining the nature of the mutation. This is a more difficult challenge, requiring a different approach that is described in Section 8.2.
3. The mutation cannot be characterised, but the chromosome harbouring the mutation can be tracked through a family. This would have been the normal situation 10 years ago; it is less common now but the need still arises on occasion. An example is given in Section 8.2.

8.1.3 Nature of mutations

Before we consider techniques for detecting mutations, it is necessary to consider the different types of mutation that may occur (the nomenclature used to describe mutations is shown in Box 8.1):

BOX 8.1: NOMENCLATURE USED TO DESCRIBE MUTATIONS

A formal description of the recommended nomenclature can be obtained from http://www.interscience.wiley.com/jpages/1059-7794/nomenclature.html.

Amino acid substitution

Amino acid	Code	Amino acid	Code	Amino acid	Code
Alanine	A	Glycine	G	Proline	P
Arginine	R	Histidine	H	Serine	S
Asparagine	N	Isoleucine	I	Threonine	T
Aspartic acid	D	Leucine	L	Tryptophan	W
Cysteine	C	Lysine	K	Tyrosine	Y
Glutamine	Q	Methionine	M	Valine	V
Glutamic acid	E	Phenylalanine	F	Nonsense	X

Each amino acid is represented by a single-letter code, shown in the table above. The amino acid substitution is shown by a number representing the residue in the polypeptide flanked by the single letter code of the original and replacing amino acid, e.g.

R553X   A truncating mutation in which arginine at position 553 in the polypeptide is replaced by a stop codon.

N1303K   Asparagine at 1303 is replaced by lysine.

Nucleotide substitutions

These are described by a number representing the position of the nucleotide in the cDNA or genomic DNA (prefaced by c or g if it is not clear which is being used), followed by a letter (or letters) representing the original nucleotide (G, A, T or C) and a chevron before the letter (or letters) representing the replacing nucleotide. The A of the ATG codon is designated by +1. The nucleotide immediately preceding the initiator codon is designated by −1. Zero is not used. If the change occurs in an intron, the change is specified by indicating the number of nucleotides to the nearest nucleotide in the cDNA. Changes in introns may also be specified by reference to intron number using the abbreviation IVS. Nucleotide changes are usually specified only when they occur in introns, because changes in exons, which result in mutations rather than polymorphisms, will cause a change in the amino acid sequence, the nomenclature for which is described above.

621 + 1G>T   G is replaced by T at the first nucleotide in the intron following nucleotide 621 in the cDNA

1717 − 1G>A   G is replaced by A at the last nucleotide in the intron preceding nucleotide 1717 in the cDNA

IVS4 + 1G>T   G is replaced by T at the first nucleotide in intron 4

Continued ▶

Deletions and insertions

Deletions are represented by 'del'; insertions are represented by 'ins'. This is preceded by the number indicating the first nucleotide affected by the deletion, or the nucleotide before the insertion, and followed by the nucleotide(s) that are inserted or deleted, e.g.

1609delCA	The C and the A nucleotides starting at position 1609 are deleted
394 delT	The nucleotide T at position 394 in the cDNA is deleted
3905 insT	T is inserted after nucleotide 3905

Formerly, the greek letter Δ was used to indicate a deletion. This is no longer used because it cannot be represented in HTML files. However, the practice is continued with mutations such as the CFΔF508 (the phenylalanine at position 508 is deleted) and CFΔ1507 (the nucleotide at position 1507 is deleted), because they have become so well known by their original names. Often these mutations are represented by spelling out the Greek letter, e.g. deltaF508.

- Large-scale chromosomal changes such as deletions and inversions, which may be detected by cytogenetic examination.

- Small-scale deletions.

- TREs: this type of mutation is easy to identify and proved very useful in the hunt for genes such as the HD gene.

- Mutations affecting RNA production and processing, such as promoter and splice site mutations.

- Protein-truncating mutations, such as nonsense and frameshift mutations.

- Base substitutions that result in a change in the amino acid sequence of the encoded protein.

Some of these mutations, such as protein-truncating mutations and deletions, will quite obviously affect gene function. Others may be more problematic. For example, as discussed in Chapter 10, apparently neutral polymorphisms are common, so it is sometimes difficult to know whether a base change has functional consequences. Sometimes the nature of the change may be significant; for example, it may affect a highly conserved amino acid or a splice acceptor site.

8.1.4 Material used for tests

Material used for tests may be recovered from a number of different sources. An important advantage of PCR-based tests is that they can be used with very small amounts of material:

- A single cell from an eight-cell embryo produced by *in vitro* fertilisation; this allows preimplantation testing. Parents known to be at risk of having an affected child produce an embryo by *in vitro* fertilisation. This is then tested for the presence of the genetic disease and implanted only if it is shown not to be affected.

- **Chorionic villous sampling.** The chorion is derived from the zygote but is not part of the developing embryo. It is a membrane with projections or villi that surround the embryo. Chorionic villous sampling is carried out by introducing a catheter through the vagina until it touches the chorion. The test can be safely carried out in the tenth to eleventh weeks of pregnancy. It thus has an advantage over amniocentesis because, combined with PCR amplification of the sample, it allows an earlier termination of pregnancy should this be indicated by the results of the genetic test.

- **Amniocentesis.** Sampling of cells for genetic testing from the amniotic fluid surrounding the developing fetus. The procedure involves inserting a large tube through the mother's abdomen. These cells are cultured *in vitro* for about 3 weeks to increase numbers. The procedure cannot be attempted before the sixteenth week of pregnancy.

- Blood from the Guthrie card. DNA is stable so the card can be readily posted to different locations or retrieved for tests later.

- Buccal cells recovered from a mouth wash.

- Blood sample, most commonly used for adult tests.

Genetic tests are usually carried out on material amplified by PCR. Both mRNA and DNA can be used as templates. Genomic DNA can be recovered from any convenient source, such as a chorionic villous sample, a blood sample or mouth wash. It contains potentially relevant information in sequences such as promoters that are not expressed in mRNA. The main disadvantage of genomic DNA is that many genes are large with a complex organisation of introns and exons. If the structure of the gene is known, the exons can be amplified by specific primers. DNA samples are usually used to characterise the presence of specific mutations.

Because the introns have been removed by processing, cDNA produced using mRNA as a template is a much simpler target to analyse when the structure of the gene is not known. This is always the case when a gene is first isolated by positional cloning. In this situation it is necessary to prove that the gene is responsible for the condition by showing a correlation between the presence of a mutation and the occurrence of the disease. This requires that the gene from affected and normal individuals is scanned for differences, using techniques described in Section 8.3. The expression of a particular mRNA will often be restricted to a particular tissue and may be present in very low quantities. Thus in many cases it is not practicable to use mRNA.

8.2 Testing for known mutations

This section considers the general types of test that can be used to detect specific mutations and then illustrates how these are used, with examples drawn from the diagnosis of CF, DMD and a TRE mutation.

8.2.1 General principles

Tests for particular mutations are now mostly based on PCR. These tests are designed according to the particular type of mutation being tested (see above). One advantage of PCR is that more than one pair of primers can be included in a single PCR reaction, allowing more than one mutation to be screened in each reaction. Such a reaction is referred to as a **multiplex PCR** reaction. Specific mutations may be recognised by **allele-specific oligonucleotides** (ASOs), which anneal either to a mutant or to a wild type sequence but not both. An important principle with PCR is that tests whose interpretation depends on the presence or absence of an amplification product must have a positive control to ensure that the PCR reaction does not fail for technical reasons, such as an inactive polymerase or impurities in the DNA sample.

General approaches for recognising specific mutations are listed below:

- Large deletions may be recognised by using PCR primers in which either one or both primers anneal with the region deleted. Deleted DNA is used as a template fails to produce an amplification band. In sex-linked diseases such as DMD the absence of a band can be readily detected in affected males. In female carriers of X-linked deletions, or heterozygotes with deletions affecting autosomal genes, there will be a change in the quantity of the amplified band. This can now be detected using automated sequencing technology because there will be a reduction in the size of the peak in the trace of the output.

- Small deletions may be recognised by amplifying a region of 100 nucleotides that spans the site of the mutation. A deletion is recognised by a band shift in the product. Even differences as small as one nucleotide can be resolved by careful electrophoresis.

- Point mutations, which include both base substitutions and deletions/ insertions of only one or two nucleotides, can be recognised by PCR. The ARMS test (amplification refractory mutation system) is commonly used. In its original form two PCR reactions are carried out in parallel and the products run in adjacent lanes during electrophoresis. One primer that anneals at a site away from the site of the mutation is common to both reactions. In one reaction the other primer anneals to the sequence of the mutant allele. In the second reaction the other primer anneals to the sequence of the wild type allele. Thus a band is produced in one or other of the lanes according to whether the template carries the wild type or mutant allele. Generally the ARMS test is designed in a multiplex form that allows a number of mutations to be simultaneously screened. This requires careful design of the primers to result in well separated bands upon electrophoresis of the amplification products. Usually this test is now carried out in a modified form. Only a primer that anneals to the mutant allele is included, so that the presence of a band indicates that the mutation is present. Positive controls for the PCR reaction are included in the multiplex test. Usually two controls are used that produce bands that are either smaller or larger than any of the bands that will be produced by mutant alleles. These controls test that the PCR reaction is functioning properly, so that the absence of a mutant band can be safely interpreted as indicating that the mutation is not present.

- Sometimes a mutation will either create or destroy a restriction site. This may be simply detected by amplifying a region around the site and attempting to digest the product with the appropriate restriction enzyme.

- TREs can be screened using PCR products that anneal either side of the expansion site. Replication slippage during amplification often results in a ladder effect. This can be very useful for determining the repeat number precisely. Large amplification will sometimes result in the distance between the two primers becoming too large for the PCR reaction to work efficiently. Such cases need to be investigated by Southern blotting.

- ASOs can also be used in hybridisation assays to detect a mutation. Usually the ASOs are attached to the membrane. This is then hybridised to DNA amplified by PCR from the region being tested. A number of different point mutations can be simultaneously screened by having a number of ASOs arranged in an array. DNA chips are an extension of this idea, where the array is miniaturised to allow a very large number of ASOs to be incorporated. In the extreme, an ASO for every single nucleotide position can be included, allowing every possible point mutation to be detected.

8.2.2 Test for mutations in the *CFTR* gene

In people of European extraction, the ΔF508 mutation is the most common CF allele. The proportion of CF alleles that are ΔF508 shows a gradient across Europe, increasing from about 50% in southern European countries to 85% in Denmark and the Basque region of Spain. The average figure is about 70%. Twenty other mutations are also relatively common, having a combined frequency of about 15% (depending on country). The remaining 15% of mutations are individually uncommon, but over 800 have been characterised. The distribution of CF alleles has consequences for genetic testing for CF. Assuming a frequency for the ΔF508 allele of 70% and a combined frequency of 15% for the other common alleles, the frequency of the various classes of CF-affected individuals can be calculated as follows:

49%   ΔF508/ΔF508
21%   ΔF508/common mutation not ΔF508

21%   ΔF508/rare mutation
4.5%   common mutation not ΔF508/rare mutation
2.25%   common mutation not ΔF508/common mutation not ΔF508
2.25%   rare mutation/rare mutation

These figures will vary in different European countries as allele frequencies change. The important point is that over 90% of patients with CF will carry at least one ΔF508 allele. An initial test for the ΔF508 mutation will therefore identify most individuals who are either affected by CF or who are carriers. Testing for ΔF508 together with the 20 common mutations will identify 70% of CF cases in Europeans (more in some countries). However, detecting **compound heterozygotes** between ΔF508 and a rare mutation will be difficult. This is an important complicating factor in embryo testing, neonatal screening or carrier screening in adults. In other populations, different mutations may be more common, and testing for *CFTR* mutations must therefore be designed to take account of the genetic structure of the local population or the ethnic origin of the subject.

The ΔF508 mutation may be detected by a simple and robust PCR test that amplifies a 100-bp region spanning the site of the mutation. The 3-bp deletion results in the ΔF508 allele migrating slightly faster than the wild type allele. As a result, all three possible genotypes may be readily distinguished from each other (Figure 8.1). A multiplex ARMS test can be used to recognise the other 20 most common CF mutations found in European populations. Figure 8.2 shows the result of a multiplex test using a commercial testing kit.

Sometimes the base change creates or destroys a restriction site, which forms the basis of a very simple test for the presence of the mutation.

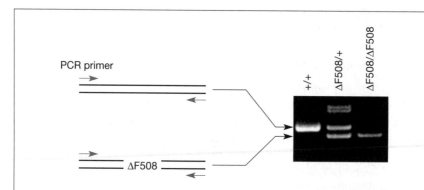

Figure 8.1 A PCR test for the ΔF508 mutation. Chromosomal DNA is amplified using primers that anneal either side of the site of the ΔF508 mutation. The amplification products are separated on a denaturing polyacrylamide gel and visualised by ethidium bromide fluorescence. The wild type allele is 3 bp larger than the mutant allele so migrates slightly more slowly. The three possible diploid genotypes are clearly distinguishable. The two additional bands in the heterozygote lane are heteroduplexes between wild type and mutant DNA strands. (Data courtesy of the North Trent Molecular Genetic Laboratory, UK.)

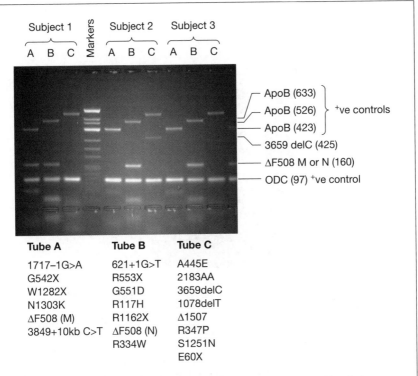

Figure 8.2 Multiplex PCR ARMs test for CF alleles. A commercial testing kit called Elucigene CF20, marketed by Cellmark Diagnostics, can detect the presence of the 20 most common CF mutations. Three multiplex PCR reactions are carried out in parallel for each subject. Each tube contains a number of separate primer pairs. Each pair is designed to anneal a particular allele to produce a band whose size is well separated from any other possible bands from the same tube. Therefore a band is produced only if the allele is present. Because the absence of the band is used as evidence that the allele is absent, it is essential to have positive controls that show the PCR reaction is functioning properly. Each tube also contains primers that will amplify two control sequences, one larger and one smaller than any of the expected products. The diagram illustrates tests on three different subjects. Subject 1 is producing bands with the primers targeted at the wild allele (*N*) and mutant (*M*) alleles at the ΔF508 site. This subject is therefore a ΔF508 heterozygote. Subject 2 is producing a band in tube B corresponding to that expected from the 369delC allele. This subject is therefore heterozygous with respect to this allele. Subject 3 only amplifies the wild type allele at the ΔF508 site, and produces no other bands. Therefore, this subject does not contain any mutations at the CF locus. (Data courtesy of Richard Kirk, North Trent Molecular Genetic Laboratory, UK.)

An example is given in Figure 8.3 for the G551D and R553X mutations, which are the third and sixth most common CF mutations worldwide. Both mutations affect the sequence GTCAAC, which is the target site for the restriction enzyme *Hinc*II. Thus the presence of one or the other is indicated by loss of this restriction site. In addition, the G551D mutation creates an *Mbo*I site, which can therefore be used as a further test to distinguish between the two mutations when loss of the *Hinc*II site is observed.

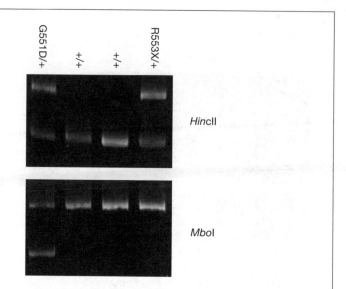

Figure 8.3 Diagnostic restriction enzyme assays for the *CFTR* G551D and R553X mutations. The wild type sequence, G<u>GTC</u>AAC is a target for *Hinc*II (underlined). This is altered to <u>GATC</u>AAC in G551D, which contains a target for *Mbo*I (underlined) but is no longer digested by *Hinc*II. In R553X the wild type sequence is altered to GGTCAAT, which is not a target for either enzyme. In the test, a fragment containing the mutation site is amplified by PCR and digested separately with each enzyme. If the individual contains either mutation, the *Hinc*II digest fails and so a heterozygote produces two bands (the wild type allele is still digested). When digested with *Mbo*I only a G551D heterozygote produces two bands because one allele is a target for this enzyme. (Data courtesy of the North Trent Molecular Genetic Laboratory, UK.)

These and other similar tests provide a simple means for identifying 80–90% of CF mutations. The remaining mutations provide a much sterner challenge because it is not practicable to develop individual tests for each of the hundreds of mutations that have been recorded. Instead it is necessary to use techniques that scan the gene to detect differences from the wild type without defining the precise nature of these differences. These techniques are reviewed below and some examples of the detection of rare CF mutations are included.

Neonatal screening for CF can be carried out by assaying for levels of immunoreactive trypsin in the Guthrie blood-spot sample. An abnormally high level may indicate CF but it could be due to other causes. Such babies would therefore be subjected to DNA tests. Because most CF cases would have at least one ΔF508 allele, screening for this mutation would detect nearly all cases. This test can also be carried out on the Guthrie blood-spot sample. A diagnosis of CF would be confirmed by continued elevation of immunoreactive trypsin levels after 28 days and by measurement of sweat electrolytes.

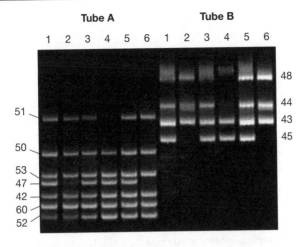

Figure 8.4 Multiplex test for deletions in the dystrophin gene. Each test consists of two tubes (A and B) containing pairs of PCR primers that amplify specific exons of the dystrophin gene, as shown at the sides. Individual number 6 has a deletion of exons 45 and 47, as judged by the missing bands in tubes B and A respectively. This probably results from a single deletion that spans exon 46, which was not tested. Individual 2 has a deletion of exon 45; the band for exon 47 is weak in this individual but was clearly visible in the original photograph. Individual number 4 apparently has a deletion of exons 44, 48 and 51. These exons are non-contiguous and so this result would be regarded with suspicion and the sample retested. The most likely explanation is that the PCR reaction failed to amplify the three largest bands in this sample. (Data courtesy of the North Trent Molecular Genetic Laboratory, UK.)

8.2.3 Duchenne muscular dystrophy

Two-thirds of DMD cases are caused by large deletions that remove one or more exons. The deletions cluster in two regions: at the 5′ end, affecting exons 3–8, and towards the 3′ end, affecting exons 44–60 (predominantly 44–50). Multiplex PCR can be used to detect loss of these exons. In this test PCR is used with a mixture of different primer pairs. Each primer pair is designed to amplify a single exon using genomic DNA as a template. The products of the PCR show a band for every exon present; a missing band is therefore diagnostic of a deletion. Figure 8.4 shows an example of this multiplex PCR test used to screen for deletions of exons in the 3′ deletion hotspot.

The remaining cases of DMD not caused by deletions are mostly nonsense or frameshift mutations that cause chain termination; these may be detected by the protein truncation test described below.

8.2.4 Trinucleotide repeat expansions

Each TRE is a separate event; however, each mutation affects the same site and so can be readily detected by a PCR reaction based on primers

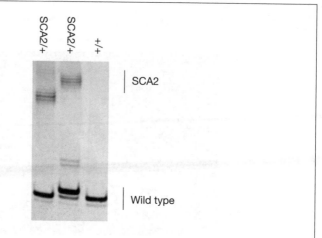

Figure 8.5 TREs in spinocerebellar ataxia type 2 (SCA2). The CAG repeat sequence was amplified using PCR primers that anneal to flanking regions. SCA2 heterozygotes have an extra band corresponding to the amplified allele. The samples have been separated on a polyacrylamide gel and visualised by silver staining. (Data courtesy of the North Trent Molecular Genetic Laboratory, UK.)

that anneal either side of the expansion site. The size of the PCR product indicates whether an expansion event has occurred. The extent of the expansion may be important in the prognosis of age of onset and severity; this information is also revealed by the expansion. Figure 8.5 shows an example, where spinocerebellar ataxia type 2 is diagnosed by such a test.

8.2.5 Multiplex ligation-dependent probe amplification

This is a technique that allows multiple targets to be tested simultaneously in a multiplex PCR reaction. It is particularly useful for detecting copy number variants such as occur in small deletions in autosomal genes. Such deletions are difficult to detect by conventional PCR, because the remaining copy of the gene will be amplified to give a signal. It can also be used to detect trisomies, point mutations and whether the target is methylated. The principle of the technique is outlined in Figure 8.1. Two oligonucleotide probes are annealed to adjacent sequences on their target DNA so they can be ligated together. The ligated probe is then amplified by PCR using primers that anneal to the end of each probe, so that amplification will occur only if the two probes have been joined together by ligation. Specificity is generated by annealing the probes at 55°C, which minimises non-specific annealing, and by using an NAD-dependent thermotolerant ligase that is active at 55°C and is intolerant of mismatches at the 3′ end of the probe. A common 5′ extension is incorporated into each probe pair so all the ligation products arising from probes with different targets can be amplified by a single primer pair after ligation. One of the probes for each target incorporates a stuffer fragment, which is a different size from the stuffer fragment of all the other probes. This means each PCR reaction generates a product which is a different size from the products of all the other PCR reactions, allowing the amplified target to be identified on the basis of its size when it is separated by capillary electrophoresis using apparatus used in DNA sequencing. One of the common PCR primers is fluorescently labelled so that the PCR product forms a peak as it passes the fluorescent detector. The area under the peak is proportional to the initial quantity of the ligated probe template, so that when one of the two target sequences is missing due to a deletion, the peak has half the area of the control from normal DNA. Similarly, a trisomic target produces a signal 1.5 times the area of the normal diploid control. Methylation can be detected when ligated probes are hybridised to the target by digestion with methylation-sensitive restriction endonucleases, such as *Hha*1, which will destroy unmethylated DNA to leave the methylated DNA intact.

Sets of probes are designed systematically to test the presence of all exons in a gene. Figure 8.6 shows the detection of exon 13 deletion in *BRCA*1, where up to 30% of mutations are exon deletions. The majority (60–70%) of DMD mutations are exon deletions and MLPA has been shown to be more efficient at detecting these deletions than other forms of PCR.

8.3 Scanning genes for unknown mutations

Section 8.2 considered some case studies where the nature of the likely mutation is known and it is possible to design specific tests to identify them. However, many genes are affected by a large number of different mutations; 10–20% of CF alleles are rare mutations. Genes such as *BRCA*1 and *BRCA*2 are affected by a large number of different mutations. It is necessary to use techniques that can scan a sample and detect differences from the wild type gene. One ever-present problem in this type of analysis is deciding whether a change detected is a 'polymorphism' or a 'mutation', i.e. whether the change has deleterious consequences for gene function.

8.3.1 Single-stranded conformational polymorphisms

Single-stranded nucleic acid molecules form extensive secondary structure in solution because of intramolecular base pairing. This secondary structure has a large effect on the rate of migration during electrophoresis under non-denaturing conditions. Even a single base change can radically alter the secondary structure and hence the rate of migration of a nucleic acid fragment. This forms the basis of the single-stranded conformational polymorphism (SSCP) technique, which allows the presence of a base change to be detected without defining the nucleotide affected. PCR is used to amplify the gene in short sections of 100–300 bp and the products are separated under non-denaturing conditions by polyacrylamide gel electrophoresis. An example is shown in Figure 8.7 where a rare mutation in the *CFTR* gene

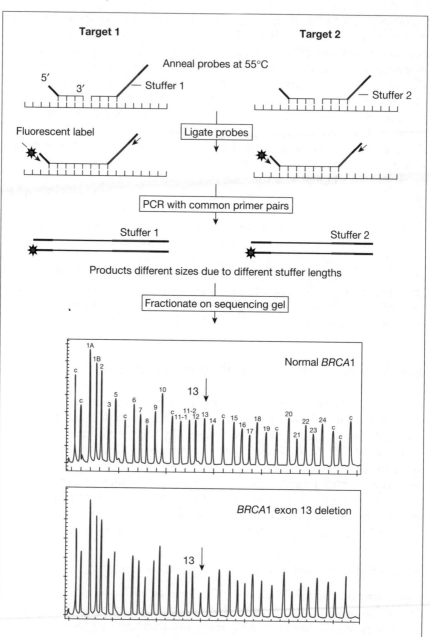

Figure 8.6 Multiplex ligation-dependent amplification. See text for explanation of the protocol. The lower panel shows the result of a diagnostic test for deletion of *BRCA1* exons. Exon 13 shows only half the signal of the control sample and therefore the patient is likely to be heterozygous for an exon 13 deletion. 'c' represents control probes annealing to non-*BRCA1* sequences. (*BRCA1* data were taken from Schouten, J.P. *et al.* (2002) *Nucleic Acids Research*, 30, e57.)

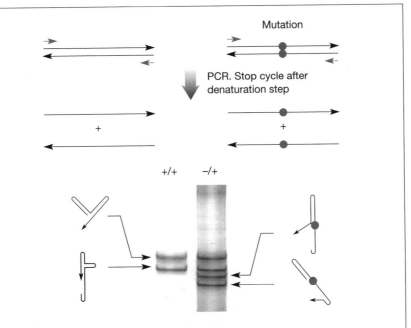

Figure 8.7 Detection of a rare *CFTR* mutation using SSCP. Reference wild type DNA and test DNA samples are amplified by PCR and the PCR reaction stopped at the denaturation stage of the cycle. The products are loaded on to a native polyacrylamide gel that allows the single-stranded molecules to form secondary structure, which affects their rate of migration. Because each strand takes up a different structure, a mutant DNA molecule produces two bands in a different position and a heterozygote will produce four bands (two wild type, two mutant). (Data courtesy of the North Trent Molecular Genetic Laboratory, UK.)

was detected. This technique is very simple and cheap, and detects 70–95% of base changes in fragments.

8.3.2 Denaturing gradient gel electrophoresis

Partially denatured DNA molecules migrate at a much slower rate during gel electrophoresis. If a molecule is subjected to electrophoresis in a gradient of denaturing agents, such as heat or chemical denaturants, it will migrate to the point where it starts to denature and then effectively stop migrating. This is called denaturing gradient gel electrophoresis (DGGE). The point at which denaturation starts is extremely sensitive to sequence and will be altered by even a single base change. Mutations may therefore be detected by a bandshift in PCR products from the test DNA compared with wild type (Figure 8.8). This method detects about 99% of mutations and is simple to operate once the conditions are fully established for each DNA fragment tested. However, designing the right conditions is not easy, and the conditions for each fragment have to be developed separately.

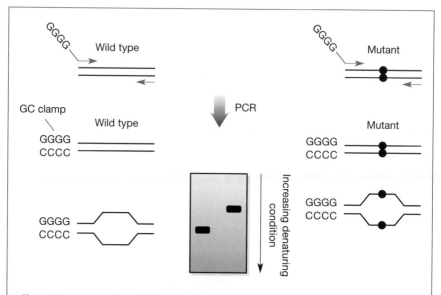

Figure 8.8 An example of DGGE. Reference wild type and test DNA are amplified by PCR. The 5' end of the upstream primer is extended with a series of C residues, resulting in a 'GC clamp' in the amplified product. The products are separated in a polyacrylamide gel with a gradient of increasingly denaturing conditions. When the duplex DNA molecules start the migration, their motion is dramatically slowed. This point is sensitive to a base pair change in sequence, so mutant and wild type DNA molecules migrate to different positions. The GC clamp is slow to denature because of the three hydrogen bonds in a GC base pair. This stabilises one end of the molecule and has been found empirically to increase the discrimination between mutant and wild type DNA molecules.

8.3.3 Mismatch cleavage detection

If a wild type DNA molecule is hybridised to a mutant sequence, at the site of the mutation there will be a mismatch in the double-stranded duplex. This mismatch may be cleaved chemically using agents such as piperidine or osmium tetroxide (chemical mismatch cleavage, CMC) or by enzymes involved in mismatch repair (enzyme mismatch cleavage, EMC). DNA is amplified from wild type and DNA to be tested. The products are separately denatured and then allowed to form heteroduplexes, before being subjected to CMC or EMC. The products are fractionated by DGGE; a mutant DNA strand is detected by the appearance of a smaller fragment (Figure 8.9). The use of chemical cleavage agents is problematic because they are highly toxic and carcinogenic. Enzymic cleavage is now well developed and is claimed to detect 99% of mutations. However, as yet it is not widely used.

8.3.4 Protein truncation test

In many genes the majority of disease-causing mutations result in truncation of the protein product. The presence of such a mutation may be

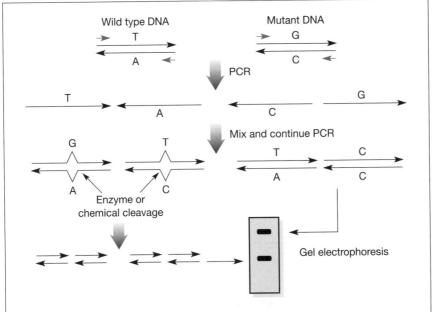

Figure 8.9 Mismatch cleavage detection. Reference wild type and test DNA samples are amplified separately by PCR for a number of cycles, then mixed and subjected to further rounds of PCR amplification, which allow heteroduplexes to form if a mutation is present in the test DNA. Chemicals (such as osmium tetroxide or piperidine) or mismatch repair enzymes are used to cleave the DNA at the site of any mismatches. Cleaved molecules are smaller and therefore migrate faster on the gel.

detected by using the product of a PCR reaction as a template for an *in vitro* transcription/translation reaction. The truncated protein from a mutant gene is detected by gel electrophoresis. The advantage of this test compared with those described above is that it will detect only changes that are biologically significant. The problem with the other methods is that many biologically neutral polymorphisms will be detected and there is no simple way of distinguishing these from disease-causing mutations. This test is likely to be the method of choice for screening for mutations in tumour suppressors such as *BRCA1*, *BRCA2* and *APC* genes, where over 90–95% of mutations are chain-terminating. An example of its use in detecting a mutation in *BRCA2* is shown in Figure 8.10.

8.3.5 DNA chips

DNA chips are a novel and still developing technology that allows the identity of every nucleotide in a test DNA sequence to be examined in a single operation. The general methods by which DNA chips or oligonucleotide microarrays are constructed and used were discussed in Section 4.9.1. Figure 8.11 provides an example of a chip that was constructed to detect mutations in exon 11 of the *BRCA1* gene. DNA chips can be

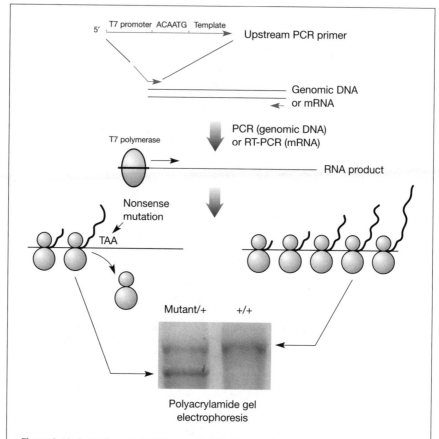

Figure 8.10 Detection of a *BRCA*2 mutation by the protein truncation test. A section of the *BRCA*2 gene was amplified by PCR. The design of the upstream primer is important. It has a 5' extension that contains the promoter sequence for the phage T7 RNA polymerase followed by a consensus eukaryote protein translation initiation sequence. The primer is designed so that the reading frame of the amplified *BRCA*2 sequence will be correctly aligned with the ATG of the translation initiation sequence. The amplified product can be transcribed *in vitro* using phage T7 polymerase and the resulting RNA translated *in vitro* using ^{35}S-labelled methionine to visualise the product by autoradiography. If the *BRCA*2 sequence contains a nonsense mutation, translation is halted prematurely and the product is truncated. This results in a faster rate of migration in the polyacrylamide gel used to fractionate the products of the *in vitro* translation. Generally, in large genes such as *BRCA*1, an exon will be amplified from genomic DNA rather than mRNA. (Data courtesy of the North Trent Molecular Genetic Laboratory, UK.)

designed to recognise specific mutations rather than examine every base pair. This may make interpretation of the results simpler because it would only identify mutations that are definitely known to be disease-causing, rather than identifying all base changes, many of which may be neutral polymorphisms. Chips have already been constructed that screen for

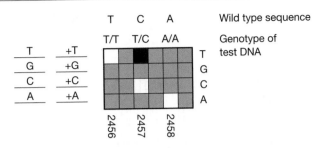

Figure 8.11 Detection of a heterozygous base substitution in the _BRCA1_ gene using a DNA chip. The left-hand part of the diagram shows the organisation of an oligonucleotide array that interrogates a single base position in the test DNA. Each oligonucleotide is 20 nucleotides long; the sequence matches the wild type sequence except that there is a substitution or insertion at residue 11 as shown. Wild type and test RNA sequences were fluorescently labelled with different coloured dyes and then hybridised separately to the chip. The composite image for three consecutive positions is shown on the right. The oligonucleotides are attached to the chip in square cells. Cells shown in white hybridised to wild type and test sequences with equal intensities, blue-coloured cells hybridised to neither target and the black cell hybridised to the test RNA but not to the wild type RNA. The test RNA sequence hybridised to the cells containing T, C and A in the oligonucleotides corresponding to the wild type sequence, but it also hybridised to the cell that contained T at position 2457, showing that one allele contained a T at this position. The individual from which the test RNA was derived is therefore heterozygous for a base substitution at position 2457. Quantitative analysis of the hybridisation signal also detected the mutation, as the wild type DNA produced a stronger signal than the test DNA at position 2457C. RNA was used because it was found to give superior results compared with DNA. The RNA was generated from PCR-amplified DNA by _in vitro_ transcription. As well as the oligonucleotides shown, oligonucleotides with various deletions were also included in each array.

known mutations in the _CFTR_ gene, the β-globin gene and the mitochondrial genome. Despite the potential of DNA chips, they are not yet used routinely in gene testing.

8.3.6 DNA sequencing

The advances in DNA sequencing technology, developed for the Human Genome Project, have made it simple and routine to resequence genes from subjects suspected of having a mutation. Figure 8.12 provides an example of a mutation in the _BRCA1_ gene being detected by automated DNA sequencing.

8.3.6.1 Chromosome tracking

When the mutation causing a genetic disease segregating in a family cannot be identified, it is necessary to track the chromosome carrying the mutation in order to diagnose the status of an embryo. This involves genotyping linked polymorphic markers and establishing the phase of markers and mutation. Highly polymorphic microsatellites are normally used. An example is shown in Figure 8.13, where a polymorphic microsatellite was used to

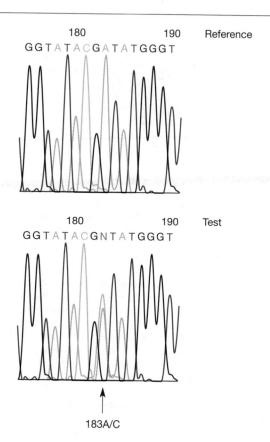

Figure 8.12 Detecting a mutation in the *BRCA*1 gene by DNA sequencing. The output from an ABI 3770 prism sequencing apparatus. Peaks corresponding to G, A, T and C are coloured black, grey, dark blue and light blue, respectively. (In reality these colours are black, green, red and blue, respectively. Printing limitations prevent the depiction of these colours in this figure.) The reference sample shows a clear green peak at position 183, showing an A is present at this position. The test sample shows both a light blue (C) and a grey (A) peak, indicating that the subject is an A/C heterozygote at this position. Note how the size of the peaks fluctuates along the length of the gel, reflecting the efficiency of chain termination at each position. However, the peak fluctuations are reproducible in the test and reference samples. For example, in both cases there is a high dark blue (T) peak at position 179 but a low red peak at position 177. At position 183 in the test sample, the expected high grey peak has reduced in size by approximately 50%. Thus, a comparison of peak height provides information as well as peak order. In this case the reduction of the grey peak at position 183 is additional evidence that a mutation has occurred. (Data courtesy of Ann Dalton, North Trent Molecular Genetic Laboratory, UK.)

track the inheritance of a chromosome carrying a Huntington's disease (HD) mutation. Although a precise test for an HD-causing mutation is available, it was not used in this case because it would have led to the diagnosis of the mother of the subject, who did not want to know her status.

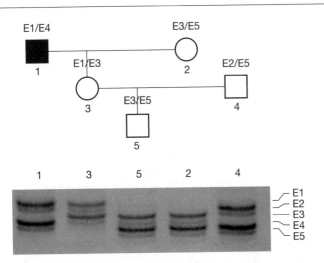

Figure 8.13 Chromosome tracking to exclude Huntington's disease. The subject (labelled 5 in the pedigree) wishes to know his HD status because his maternal grandfather (1) suffered from the disease. However, his mother (3) does not wish to know her status (so far she is disease-free). The subject must have inherited one copy of chromosome 4, which carries the HD locus, from one of his maternal grandparents. If it could be shown that the chromosome was inherited from his maternal grandmother, the risk of HD can be excluded without testing the status of the mother. Of course if the subject inherited a copy of chromosome 4 from his grandfather then the risk that he will suffer from the disease is increased from 1 in 4 to 1 in 2. The test used a polymorphic microsatellite at D4S127 that is closely linked to the HD locus, so recombination is unlikely to confound the results. The alleles revealed by this test allow the copies of chromosome 4 to be tracked through the pedigree. The position of bands produced from five alleles (E1–E5) are marked at the side of the figure. The genotypes, deduced from the data, are written above each member of the pedigree. The subject is E3/E5. The E5 chromosome was inherited from his father who has no family history of the disease. The E3 chromosome was inherited from his mother, who in turn inherited it from her mother. Therefore, the copy of chromosome 4 he inherited from his maternal grandparents comes from the grandmother, not the grandfather. We can conclude that he is not at risk of the disease. (Data courtesy of Steve Evans, North Trent Molecular Genetic Laboratory, UK.)

8.4 Summary

- Gene testing has been revolutionised by the isolation of disease genes and the amplification of small amounts of material by PCR.

- Gene testing is carried out in a number of different situations:
  - To diagnose embryos known to be at risk of genetic disease.
  - Neonatal screening.
  - As an aid to medical diagnosis.
  - To diagnose the status of an adult at risk of a late-onset disease.
  - To determine susceptibility to a complex disease.
  - To confirm the identity of a candidate gene by showing a correlation between disease and mutation.

- Some mutations, such as protein-truncating mutations, will clearly ablate protein function. Others, such as base substitutions, may be neutral polymorphisms.

- DNA or RNA may be used for the tests. DNA is easy to obtain, but the extremely large size and complex organisation of some genes may make searching for unknown mutations difficult. RNA from the right tissue may be more difficult to obtain but it will represent the gene sequence in a more manageable form.

- PCR tests have been developed to recognise the different types of mutation, e.g. large deletions, small deletions and point mutations. One commonly used test is the ARMS test, which uses ASOs as primers to test for the presence of several specific mutations in two parallel PCR assays.

- The *CFTR* ΔF508 mutation may be detected by a simple PCR test. A multiplex PCR test can be used to detect other common mutations.

- About 60% of DMD cases are caused by deletions in the dystrophin gene. These may be detected by a multiplex PCR test.

- TREs can be detected by amplifying the expansion site using PCR primers that anneal to flanking regions.

- Unknown mutations may be detected by a number of tests that attempt to find differences between the test sample and a reference wild type sequence.

- SSCP relies on the fact that the rate of migration of single-stranded DNA through a native gel is highly sensitive to secondary structure, which in turn depends on sequence.

- DGGE is based on the difference in migration rate of duplex and partially denatured DNA. The point at which a DNA molecule starts to denature is dependent on its exact sequence.

- Mismatch cleavage uses chemicals or enzymes to cleave a heteroduplex between test and reference DNA at the site of a sequence mismatch.

- The protein truncation test is based on *in vitro* transcription/translation. If the reference sequence contains a chain-terminating mutation the polypeptide produced will be smaller than the reference.

- DNA chips are, in principle, capable of detecting all possible changes in a gene sequence in a single operation. They are still at an experimental stage.

- Direct DNA sequencing for genetic testing is now routine. There remains the problem of deciding whether a change detected is a disease-causing mutation or a neutral polymorphism.

- When a family requires genetic testing and counselling for a genetic disease in which the aetiological mutation has not been identified, it is necessary to track the chromosome carrying the mutation. This is now normally done with linked microsatellite markers amplified by PCR.

Further reading

General

ANTONARAKIS, S.E. and THE NOMENCLATURE WORKING GROUP (1998) Recommendations for a nomenclature system for human gene mutations. *Human Mutation*, **11**, 1–3. Available online at http://www.interscience. wiley.com/jpages 1059-7794/nomenclature.html.

ENG, C. and VIJG, J. (1997) Genetic testing: the problems and the promise. *Nature Biotechnology*, **15**, 422–426.

Multiplex ligation-dependent amplification

SCHOUTEN, J.P., MCELGUNN, C.J., WAAIJER, R. *et al.* (2002) Relative quantification of 40 nucleic acid sequences by multiplex ligation-dependent probe amplification. *Nucleic Acids Research*, **30**, e57.

Cystic fibrosis

FERRIE, R.M., MARTIN, M.J., ROBERTSON, N.H. *et al.* (1992) Development, multiplexing, and application of ARMS tests for common mutations in the CFTR gene. *American Journal of Human Genetics*, **51**, 251–262.

Methods for unknown mutations

GROMPE, M. (1993) The rapid detection of unknown mutations in nucleic acids. *Nature Genetics*, **5**, 111–116.

JONSSON, J.J. and WEISSMAN, M.S. (1995) From mutation mapping to phenotype cloning. *Proceedings of the National Academy of Sciences USA*, **92**, 83–85.

MASHAL, R.D., KOONTZ, J. and SKLAR, J. (1995) Detection of mutations by cleavage of DNA heteroduplexes with bacteriophage resolvases. *Nature Genetics*, **9**, 177–183.

YOUIL, R., KEMPER, B.W. and COTTON, R.G.H. (1995) Screening for mutations by enzyme mismatch cleavage with T7 endonuclease VII. *Proceedings of the National Academy of Sciences USA*, **92**, 87–91.

Detection of *BRCA*1 mutations using a DNA chip

HARCIA, J.G., BRODY, L.C., CHEE, M.S., FODOR, S.P.A. and COLLINS, F.S. (1996) Detection of heterozygous mutations in *BRCA*1 using high density oligonucleotide arrays and two-colour fluorescence analysis. *Nature Genetics*, **14**, 441–447.

CHAPTER NINE

Gene therapy

Key topics

- Somatic and germline gene therapy
- Gene replacement and gene addition
- *In vivo*, *ex vivo* and *in vitro* gene therapy
- Transgenic animal models
- Vehicles for gene transfer
  - Retrovirus
  - Adenovirus
  - Adeno-associated virus
  - Liposomes and lipoplexes
  - Naked DNA
- Gene therapy for cystic fibrosis
- Progress and adverse events
  - SCID-XI
  - Chronic granulomatous disease
  - Melanoma
  - Adverse events

9.1 Introduction

In theory, genetic diseases may be treated by the introduction of the wild type gene into the cells affected by a mutation. Such an approach is called gene therapy. Following the successful isolation of the genes affected in many common monogenic disorders, there is considerable expectation that gene therapy will, for the first time, offer the prospect of a cure for genetic diseases. Moreover, the introduction of DNA molecules and oligonucleotides into cells can be used to provide novel treatments for many kinds of non-hereditary disease such as cancer. In this chapter various approaches to gene therapy are described and progress that has been made in their application reviewed. It will be seen that experiments *in vitro* and with animal models demonstrate that, in principle, gene therapy is possible. Nevertheless, the results of the first clinical trials demonstrate that successful application to human patients for the treatment of hereditary diseases is still a long way off; however, gene therapy to treat other types of disease such as cancer is giving more promising results.

9.2 Types of gene therapy

9.2.1 Somatic and germline gene therapy

There are two strategies for correcting an inherited disease by gene therapy (Figure 9.1). In **somatic gene therapy** the genetic defect is corrected only in the somatic cells of a person affected by the disease. In **germline gene therapy** a genetic modification is made to a gamete, fertilised egg or embryo before the germline has split off from the cells that will make the rest of the body. The crucial difference between these two strategies is that in the first, any genetic changes are restricted to the lifetime of the person treated, while in germline therapy any change is passed on to subsequent generations. Somatic gene therapy poses few ethical problems, but germline gene therapy represents a fundamentally new type of human activity whose consequences need to be thought through carefully before any experiments are attempted. Because of this, such experiments are currently prohibited in most countries.

9.2.2 *In vivo* and *ex vivo* gene therapy

One strategy to introduce the **transgene** (the exogenous gene) is to isolate cells so that they can be manipulated *in vitro*. This would allow the cells that have received the transgene to be selected and perhaps cultured *in vitro*. These cell clones would then be reintroduced into the patient's body. Such a strategy is called *ex vivo* **gene therapy** (Figure 9.1). Alternatively, the transgene could be directly introduced into the cells affected by the disease; this is called *in vivo* **gene therapy**.

Ex vivo gene therapy is suitable for genetic defects that affect blood cells. All blood cells are derived from pluripotent stem cells in the bone marrow. They can be isolated, cultured *in vitro* and successfully reintroduced *in vivo*. As described below, *ex vivo* therapy was used for the treatment of adenosine deaminase (ADA) deficiency, the only gene therapy experiment that has so far resulted in any sign of clinical improvement. Clinical trials using *ex vivo* approaches are currently under way for Gaucher's disease, Fanconi's anaemia, Hurler's syndrome and chronic granulomatous disease. In principle, the major group of monogenic disorders that could be treated by *ex vivo* therapy are the haemoglobinopathies. Gene

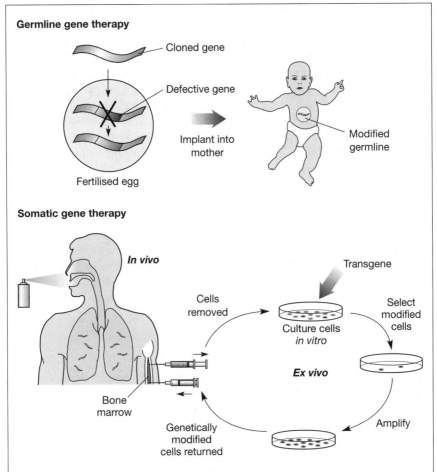

Figure 9.1 Germline and somatic gene therapy. Germline gene therapy (top) requires that the transgene is introduced into the fertilised egg or into the embryo before the germline separates from the soma. As a result, the baby is born with modified somatic and germline cells and will pass the modification on to the next generation. In somatic gene therapy the change is limited to the cells of the subject's body and is not passed on to the next generation. Somatic gene therapy can be *ex vivo* or *in vivo*. *Ex vivo* therapy involves removing cells from the patient, in this case blood cell progenitors from the bone marrow, and modifying them *in vitro*. After the modification, the cells may be cultured to increase their number before reintroduction into the patient. *In vivo* therapy involves direct introduction of the transgene into the cells of the subject's body. In this case a transgene is introduced into the cells that line the lung by means of an aerosol.

therapy for this group of diseases is complicated by the mechanisms that control the expression of both the α-globin and β-globin gene cluster. Expression of genes in these clusters requires the action of LCRs that operate at a considerable distance from the genes controlled (see Chapter 5). Increased understanding of these processes has led to recent advances in this area (see Further reading).

Clearly, *ex vivo* approaches for diseases such as CF and muscular dystrophy will not be practical. The cells of the lung epithelia divide only very slowly so it is not possible to culture them *in vitro*. Even if it were, it is not easy to imagine how the lungs of affected individuals could be readily repopulated by cells manipulated *in vitro*. The accessibility of such cells thus becomes a major factor in determining whether gene therapy is possible. In the case of diseases such as CF it will be necessary to manipulate the cells *in vivo*. This requires that the transgene is delivered to those cells affected by the disease. Thus DMD would be treated by introducing the dystrophin gene into muscle cells, CF by introducing the *CFTR* gene into the cells of the airway, and so on. In both diseases these are the cells whose malfunction causes the major life-threatening effects. Nevertheless, in both examples there are many other cell types that fail to function normally and contribute to the pathophysiology of the disease. Patients with CF may suffer, *inter alia*, from pancreatic insufficiency, gastrointestinal malformation, defects in the vas deferens and liver problems. Patients with DMD may also have heart and central nervous system defects. Even if the defects were successfully corrected in the tissues primarily affected by the disease, the disease may not be fully cured. Furthermore, many somatic cell types may not be easily accessible for treatment.

9.2.3 Gene addition or replacement

The most perfect form of gene therapy would be to completely replace the defective gene with a functioning transgene. This would allow therapy for all types of mutation, including dominant ones. Replacement would involve homologous recombination between the transgene and the chromosomal copy. This is possible as there are many situations where geneticists rely on such recombination; for example, in the generation of gene deletions in transgenic mice. However, gene replacement requires procedures to select a small minority of cells that have the alteration required. The selection procedures make use of sensitivity or resistance to drugs. *In vivo* gene therapy would need to be carried out without such selection. Although gene replacement has been shown to occur at a very low frequency in experiments with mouse embryonic stem cells, such procedures in human cells will not be technically feasible in the foreseeable future. Germline therapy will require accurate gene replacement, because this is the only way to be sure that the gene will be expressed in a completely normal fashion. This is one of the powerful practical reasons that precludes germline therapy for the present.

For somatic gene therapy, this leaves gene addition rather than replacement, where the transgene works alongside the mutated gene in the same cell. This should allow correction of the lack of gene function that occurs in recessive conditions. However, it cannot correct a dominant mutation because the mutant protein will still be produced and exert its detrimental effect. It is theoretically possible to use antisense strategies to inhibit a gene affected by a dominant mutation. This relies on producing an RNA molecule, or introducing an oligonucleotide, that is complementary to the mRNA originating from the gene. The mRNA and this **antisense mRNA** form a double-stranded hybrid that cannot be translated and is specifically degraded. Targeting to a mutant mRNA can be achieved by using an oligonucleotide whose sequence is complementary to mRNA from the mutant version of the gene.

The transgene can be integrated into a chromosome or can exist as an independent copy (**episome**). The advantage of the former is that the transgene will be stable in cell division and be inherited by progeny cells. The disadvantage is that, as noted above, the site of integration cannot be controlled. It is possible that the new gene will insert within a functioning gene, resulting in its inactivation. Alternatively, it may activate an oncogene by integrating next to it. As a consequence, the cell may become transformed into a cancer cell. This is not a theoretical speculation. Some retroviruses are known to be oncogenic by this mechanism.

9.2.4 Animal models

Development of gene therapy techniques will require a large amount of experimentation. For obvious reasons it is difficult to do most of this on human subjects. Animal models are extremely useful to demonstrate the principle and develop the technology. Some diseases have natural counterparts in animals. For example, *mdx* in mice and *xmd* in dogs (golden retrievers) are X-linked mutations that cause a disease similar to DMD. Alternatively, transgenic mice can be generated with the homolog of the human disease gene deleted. Usually these mice suffer from similar symptoms and can therefore be used to develop gene therapies. It is important to realise that there will also be important differences and the results may not always be directly transferable from the animal model.

9.3 Methods of transferring transgenes into target cells

Gene therapy relies upon methods of introducing transgenes into target cells. This means a **vehicle** is needed for its delivery. Sometimes the term 'vector' is used in this context. In this account, the word 'vector' is reserved for its normal meaning of a DNA molecule to which a gene may be spliced to mediate its expression and/or replication in the host cell. Sometimes the use of the two words will overlap; for example, a retroviral genome into which a transgene has been spliced is a vector, but the **virion** (see below) containing the recombinant genome is a vehicle. The utility of the vehicle will depend on a number of factors:

- Efficiency of delivery to the target cell.
- Specificity of delivery for the target cell (**tropism**).
- Whether the target cell needs to be dividing.

Table 9.1 Advantages and disadvantages of the various approaches used to transfer genes.

	Advantages	Disadvantages
Retroviruses	Integration results in stable modification of target cell Infect replicating cells Suitable for *ex vivo* treatment	Uncontrolled integration may have oncogenic consequences Cannot infect non-dividing cells, may be overcome by use of related lentiviruses Provoke immune response Expression of transgene becomes attenuated in long term
Adenoviruses	Infect non-dividing cells, so suitable for CF and DMD Non-integration avoids safety hazard of uncontrolled integration Efficient gene transfer, at least *in vitro*	Expression of transferred gene is transient Provoke strong immune reaction
Adeno-associated virus	Non-pathogenic Broad tropism Long-lasting expression of transgene	Difficult to raise high titres in packaging cell lines Small insert size up to 4.5 kb
Lipoplexes	Non-immunogenic, so safe Can carry large DNA molecules	Inefficient delivery Transient expression
Naked DNA	Non-immunogenic Simple to prepare and administer to muscle cells	Inefficient gene transfer Limited in cell types that can be treated Expression transient

- Whether the vehicle will provoke an immune response.
- The size of the DNA that can be carried.
- Stability and longevity of the gene in the target cell.
- Expression of the introduced gene.

Various approaches can be taken to transfer the gene to the target cell. The advantages and disadvantages of these are summarised in Table 9.1 and are discussed in more detail in the following sections.

9.3.1 Virus-based vehicles for gene therapy

9.3.1.1 Retrovirus-based gene transfer

Retroviruses (Figure 9.2) have two identical RNA genomes. Upon infection, the RNA is copied into DNA by a virally encoded reverse transcriptase. This DNA copy is then integrated into the host genome in a process that depends upon the **long terminal repeats** (LTRs) at each end of the genome

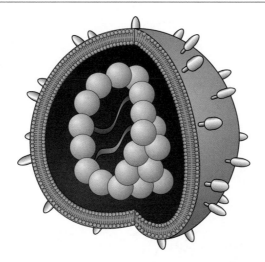

Figure 9.2 Retrovirus virion. The mature virus particle found outside the cell is called a virion. The retrovirus virion is composed of an outer lipid envelope into which are inserted envelope proteins that recognise specific proteins on the outside of target cells. Within the outer lipid layer is an inner nucleocapsid, a hollow protein structure made up of the *gag* and *pol* proteins surrounding two RNA genomes.

(Figure 9.3). The retroviral genes are transcribed from a promoter within the 5′ LTR regions. The resulting RNA molecules are packaged into the virus particles. This packaging process requires a sequence called ψ. As well as the *pol* gene encoding reverse transcriptase, the retroviral genome contains the *gag* gene that encodes a viral core protein and the *env* gene that encodes a viral envelope protein. The *env* protein recognises receptors on target cells, facilitating infection.

Figure 9.3 shows how retroviruses may be used as gene delivery systems. Because retroviruses can be oncogenic if they integrate upstream of a cellular oncogene, it is essential that any retrovirus-based gene therapy product is not contaminated by replication-competent viruses. This is achieved by using a packaging cell line in which essential retroviral genes are divided between two different DNA molecules, neither of which contains the ψ packaging sequence necessary for incorporation into the virion (the mature virus particle). A third DNA molecule carries the gene to be transferred, together with a functioning ψ sequence sandwiched between LTRs. This packaging cell line produces retrovirus particles containing the gene of interest that can infect the target cell. Once inside the target cell, the pol protein contained within the virus particle will ensure that the gene is integrated into the host genome. The LTRs are necessary for integration and also act to express the gene once integrated. Because multiple recombination events will be required to produce a functioning retrovirus in the packaging cell line, it is highly unlikely that the preparation will be

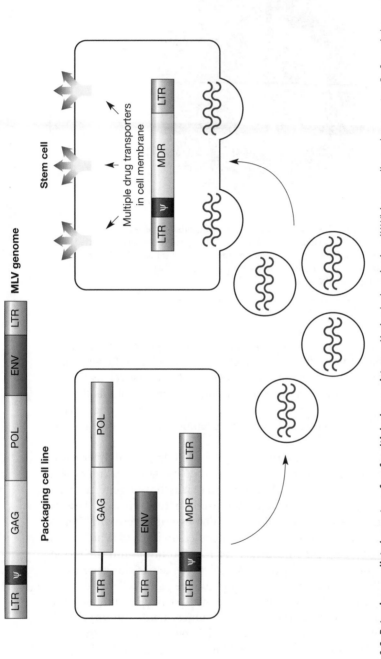

Figure 9.3 Retrovirus-mediated gene transfer of multiple drug resistance. Murine leukaemia virus (MLV) is normally used as a vector. So far as is known it is not pathogenic to humans; however, it is prudent to exclude replication-competent viruses from the preparation used for the treatment. To do this, a packaging cell line is used to produce virions containing the gene to be transferred, in this example the multiple drug resistance gene (MDR). The packaging cell line is transfected with three separate plasmids. Two produce the retroviral gene products that form the virion. Neither of these two plasmids contains a ψ packaging sequence. The third plasmid carries the MDR gene together with a ψ sequence flanked by LTR sequences, which results in it being packaged into the virion. The virion can now infect stem cells, and the MDR construct will integrate into the host genome. When expressed, the MDR-encoded protein forms a pump that removes toxic drugs from the cell.

contaminated. Nevertheless, each batch produced for clinical testing is carefully examined for any replication-competent virus particles.

The advantages of retrovirus-mediated gene transfer are its efficiency compared with other methods of gene transfer, together with the stability of the introduced gene. It is generally used for *ex vivo* gene therapy, which is based on replicating cells. Many gene therapy trials are currently in progress using retroviral vehicles for gene transfer. Nevertheless, there are significant problems with the use of retroviruses for gene therapy:

1. Retroviruses can infect only proliferating cells because the viral genome integrates into the host chromosome and can gain access to the chromosomes only when the nuclear envelope disappears at mitosis. This prevents the use of retroviral vehicles for the treatment of genetic diseases such as DMD or CF where the cells most affected are non-dividing or only dividing slowly.
2. Retrovirus-mediated gene transfer is limited in the size of DNA molecule that can be accommodated, which mostly limits its use to cDNA.
3. Retrovirus-mediated gene transfer does not specifically infect one cell type. This results in a loss of efficiency as most of the genes do not reach the target cells. Moreover, when it is used to alter the characteristics of a particular cell type, such as a cancer cell, modifying the wrong cell type may be actively counter productive. One approach for improving target specificity is to modify the viral env protein that interacts with receptors on the exterior of the host cell. This can be done by fusing a protein hormone to the env protein so that the virus may be targeted to cells displaying the receptor for that hormone. Another strategy is to replace the env protein with that of another virus. For example, in one experiment the mouse leukaemia virus (MLV) env protein was replaced with the env protein of human vesicular stomatitis virus (HSV). The engineered MLV virus infected only cells displaying the HSV receptor and no longer infected the normal target cells of MLV.
4. Retroviruses integrate at random sites in the genome. If they integrate near a cellular oncogene they may cause its activation through transcriptional activation.

Lentiviruses, such as HIV, are related to retroviruses but have evolved the capacity to replicate in non-dividing cells. Thus, in principle, viruses such as HIV could be used for gene therapy in those cases where the target cell is non-dividing. Naturally there would be considerable concern about the prospect of using HIV as a vehicle for gene therapy because it causes AIDS. Moreover, HIV infects only T cells because the env protein on the viral surface binds the CD4 protein during infection. So its natural tropism would prevent it from infecting the targets of most gene therapy regimes. The tropism has been changed by substituting the HIV *env* gene for the *env* gene of human vesicular stomatitis as described above. To lessen the risk still further, HIV vectors have been engineered to remove all viral genes except those necessary for infection of non-dividing cells. Such vectors contain only 25% of the original HIV genome. Nevertheless, there is a fear that recombination *in vivo* with related viruses could reconstitute an infectious virus.

9.3.1.2 Adenovirus-based gene transfer

Adenoviruses (Figure 9.4) are double-stranded DNA viruses that naturally infect the non-dividing cells of the respiratory and gastrointestinal tracts. They make an attractive vehicle for gene therapy because they have evolved to evade host immune mechanisms and to be highly efficient at infecting their target cells. Furthermore, once inside the target cell they replicate as episomes and so avoid the dangers of uncontrolled integration. There are at least 50 different **serotypes**. Serotypes 3 and 5 show a high degree of tropism for the respiratory tract and thus these strains have been used as vehicles for CF gene therapy. The ability of adenoviruses to infect non-proliferating cells has resulted in attempts to use them for gene therapy of DMD, where it is necessary to treat non-dividing muscle cells.

Of course it is necessary to disable the adenovirus to prevent it replicating and killing the host cell. Gene expression can be divided into early and late phases. DNA replication and expression of the genes whose products constitute the virion depend upon prior expression of early genes. The infective cycle can therefore be halted by deleting E1, an essential early gene (Figure 9.5). At the same time this makes room for the insertion of foreign DNA, up to a size of about 6 kb.

Adenoviruses are efficient at delivering DNA to their target cells. However, a major drawback is that clinical trials have shown that even

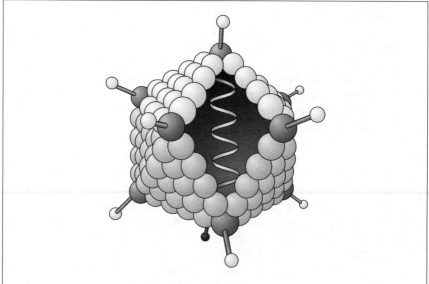

Figure 9.4 An adenovirus virion. This is an icosahedral structure consisting of a capsid made up of 12 different proteins enclosing a single double-stranded DNA molecule 35 kb in size.

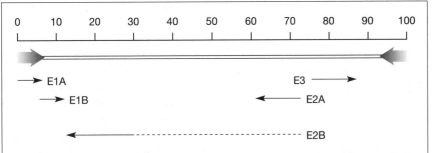

Figure 9.5 Early gene transcription in adenoviruses. The adenovirus genome is 36 kb in size, conventionally represented by a map with 100 units. Each end of the molecule consists of a 100-bp sequence that is an inverted repeat of the sequence at the other end. During the first 6–8 hours after infection, mRNA molecules representing about 25% of the genome are derived from scattered regions of the genome called E1 to E4. Each region gives rise to multiple transcripts by differential splicing, but all have the same 5′ end. The resulting overlapping mRNA molecules give rise to proteins that are subsets of one another. In some cases one mRNA molecule is read in overlapping reading frames so that the resulting proteins have common amino terminal sequences but different carboxy terminal sequences. The very first region to be transcribed is E1A; the E1A protein is required for the expression of all other adenovirus genes. Early gene transcription allows DNA replication followed by late RNA transcription that encodes the capsid proteins of the virion.

disabled forms provoke a strong immune response. This limits the dose that can be safely administered. Moreover, expression is transient so that the therapy would need to be repeated. Repeated therapy results in the immune response becoming more severe. Recently, adenovirus vehicles with deletions of other early genes such as E2, E3 and E4 (Figure 9.5) have been constructed to further reduce late gene expression which might be responsible for the immune response.

The dangers of adenovirus vectors were highlighted by the tragic death in 1999 of a patient undergoing gene therapy with an adenovirus vector. James Gelsinger was an 18-year-old who was taking part in a phase I trial to treat an inherited deficiency of ornithine transcarbamylase (OTC). Phase I trials are designed to determine whether the treatment is safe, but are not concerned with efficacy. The death occurred even though the trial was using an adenovirus vector in which both the E1 and E4 genes had been deleted, which should have reduced the harmful immune response to the virus. This event was a severe setback to gene therapy trials, especially to the use of adenovirus vectors. It also highlights the ethical issues associated with gene therapy, which will be discussed in Chapter 12.

9.3.1.3 Adeno-associated viruses

Adeno-associated viruses (AAV) are naturally replication-defective viruses that depend on other helper viruses, such as adenovirus, to provide essential functions. AAV-based systems offer a number of advantages over other virus systems: they are not naturally pathogenic, they are tropic for airway cells, they infect non-dividing cells, they show longer-lasting expression (up to 6 months in animal models) and they can show chromosome-specific integration in chromosome 19. The genome of AAV is a single-stranded DNA molecule containing two genes, *rep* and *cap*. The rep protein mediates viral integration into the specific site on chromosome 19. Cap is the structural protein of the capsid. These genes are flanked by two inverted terminal repeats that are essential to pack the DNA into the virus capsid. For gene therapy the *cap* and *rep* genes are replaced by the transgene. This construct is introduced into a packaging cell line that supplies the cap and rep proteins in *trans*. A disadvantage of this arrangement is that the specific targeting to chromosome 19 is lost as the rep protein is not present. In animal models there has been long-lasting expression of transgenes from such tissues as muscle, liver and brain. Recently, an AAV vector was used in a promising human trial to correct haemophilia A.

9.3.2 Non-viral methods of gene transfer

9.3.2.1 Liposome- and lipoplex-mediated gene transfer

In aqueous solution DNA molecules are negatively charged. The outside of most cells is also negatively charged, so DNA molecules and cell surfaces may be expected to be mutually repellent, which is thought to limit the efficiency of DNA uptake by cells. This problem may be overcome by encapsulation of the DNA in **liposomes**. Liposomes are spheres consisting of lipid molecules surrounding an aqueous interior. The lipid molecules used contain hydrophobic and hydrophilic domains. They form a bilayer in which the hydrophilic domains face outwards towards the surrounding aqueous environment and inwards towards the water-filled interior. DNA molecules can be encapsulated within liposomes. However, liposomes are typically 0.025–0.1 mm in size but plasmids are typically over 2 mm in size. Thus only a small number of plasmids can be fitted into each liposome and as a result liposomes are very inefficient vehicles for gene transfer.

This problem was overcome with specially formulated lipids where the hydrophilic domain is positively charged (cationic), which attracts both the negatively charged DNA and the negatively charged cell surface. These cationic lipids were very efficient at encapsulating DNA. The resulting lipid–DNA complexes are much more complicated than liposomes. The plasmids are encapsulated in tube-like structures, which in some ways resemble the lipid envelope that surrounds the protein capsid of some viruses. Because these structures are so different from liposomes, they were given a new name, **lipoplexes**. The main advantage of lipoplex-mediated gene transfer is that lipoplexes are non-immunogenic so they are safer than virus-based methods. In addition, lipoplexes are easy to prepare and are not limited in the size of DNA molecule that can be carried. The main disadvantage is that the efficiency of transfer is lower than virus-based methods. Nevertheless, recent trials have shown that in some cases they can be as efficient as virus-based vehicles.

During the course of experiments designed to test the efficiency of various liposome/lipoplex formulations, naked DNA was used as a negative control since, theoretically, naked DNA should not be taken up by cells. Surprisingly, it was found that naked DNA injected directly into the muscle tissue of mice was taken up and expressed by muscle cells. This is thought to occur by DNA entering the cell through small lesions in the cell membrane. This opens up a novel route for the introduction of genes into muscle cells. One possibility is to use muscle cells to produce proteins whose action is not cell-limited, such as insulin or blood clotting factors. Another idea is the use of naked DNA to express proteins that would act as vaccines against infectious diseases or even cancers.

9.4 Gene therapy for cystic fibrosis

Soon after the *CFTR* gene was cloned it was demonstrated that the wild type gene could correct the chloride conductance defect of a CF cell line *in vitro*, demonstrating that in principle it should be possible to treat the disease by gene therapy (see Chapter 5). The life-threatening consequences of the disease affect the epithelial cells that line the lung airways and thus these are the cells into which the wild type gene would have to be introduced. Clearly, it will not be possible to treat the cells by *ex vivo* therapy, so an *in vivo* method of treatment will need to be developed.

The first experiments were carried out with an adenovirus vector using transgenic mice lacking *CFTR* function as a model (CF transgenic mice). They showed that the gene could be delivered to the target cells in the airway and correct the chloride ion transport defect, for 6 months after introduction of the wild type gene in some experiments. The scene was set for gene therapy trials with human patients with CF. There are several methodological problems that need to be considered before describing the results.

First, in which part of the airway is expression most important? *CFTR* expression is highest in the submucosal gland, which may secrete water on to the airway surface. However, *CFTR* is also expressed in the epithelia that line the alveoli, so CFTR function may be important there as well. It may be that restoration of function in airway progenitor cells may be necessary for successful treatment. At present, gene delivery systems are based on aerosols, which can only access epithelial cells and not submucosal cells or progenitor cells.

Second, what proportion of cells need to be corrected for relief of the clinical symptoms? *In vitro* experiments with monolayers of CF epithelial cells suggest that restoration of chloride ion conductance in only 6% of the cells is sufficient because chloride ions can move laterally from cell to cell through **gap junctions**, which connect adjacent cells. However, sodium ion absorption is cell-limited; thus most cells will need a functioning CFTR protein if regulation of sodium ion absorption is important.

Third, how should preliminary trials be conducted? If the gene is delivered to cells deep in the lungs it will be difficult to gain access to these cells to measure whether the gene has been delivered and whether there is any evidence of *CFTR* function. Alternatively, the gene could be delivered to the nose where it is possible to sample cells to determine whether the *CFTR* gene has been delivered and, if it has, to monitor its expression and function. However, no clinical outcome can be observed because the nasal passages are not involved in the pathophysiology of CF. A compromise may be the maxillary sinus, which can be accessed to monitor the success of gene delivery and function but which is also clinically relevant.

9.4.1 Adenovirus trials

There have been a number of clinical trials based on adenovirus vehicles, but with only limited success in gene delivery and function. *CFTR* expression was variable and transitory. Increasing the dose of the virus to improve the outcome results in a strong immune response in the form of inflammation and fever. The problem may be that the cells that need to be treated are the columnar ciliary cells that line the airway. Previous *in vitro* experiments have been carried out with the underlying basal cells. The columnar cells are much more resistant to infection by adenoviruses. Perhaps this should not be surprising. Adenoviruses were chosen as vehicles because they have evolved to infect the lung. While this may be true, it is probably equally true that columnar cells have also evolved to be resistant to such infection.

9.4.2 Lipoplexes

Trials where aerosols of lipoplex-encapsulated DNA were sprayed into the nasal cavity have had some very limited success. Gene transfer was monitored by the presence of *CFTR* DNA and mRNA, and *CFTR* expression was monitored by chloride ion conductance and sodium ion absorbance. Both transfer and expression were detected, but the levels were low and did not persist for any significant length of time. The levels were comparable to those observed with adenovirus vehicles but, importantly, without the inflammatory response. These trials were conducted using the nasal epithelium, which is not medically relevant to CF and is probably an easier target than the usually infected lower airways typical of patients with CF. Moreover, although the degree to which it is necessary to remediate *CFTR* function is not known, the levels of *CFTR* expression observed are not likely to prove sufficient for clinical improvement.

Lipoplexes are taken up into the cell by **endocytosis**: the cell membrane invaginates to form an intracellular vesicle. These vesicles fuse to form structures called **endosomes**. A problem with lipoplexes is that they remain trapped in the endosome and so the DNA does not reach the nucleus. A suggestion to overcome this problem is to encapsulate an adenovirus vector in a lipoplex. This may combine the benefits of both vehicles: the negatively charged lipoplex provides an efficient means of access into the cell, while the adenovirus can escape from the endosome and deliver the DNA to the nucleus.

Although it is exciting that clinical trials for CF have been undertaken with some positive results, clinically effective treatment is still a long way off. Far more efficient ways have to be found to deliver and express the *CFTR* gene. One exciting development is the construction of human artificial chromosomes based on an α-satellite sequence as a centromere. This may provide a vector for the safe and permanent maintenance of the *CFTR* gene. Since such a vector will be capable of carrying a large DNA insert, it might be possible to transfer the entire *CFTR* chromosomal locus instead of *CFTR* cDNA. This may enable the pattern of expression to match the natural system more faithfully.

A second way forward may be treatment *in utero*, where immune tolerance to adenovirus may be greater. It may also avoid the early and possibly irreversible damage to the lung that occurs in patients with CF. One intriguing report shows that in a CF transgenic mouse, transient expression of *CFTR in utero* results in a permanent reversal of the CF phenotype. The *CFTR*-deleted mice used in the study suffer from abnormalities of the gastrointestinal tract that mimic meconium ileus, a form of intestinal blockage seen in 5–10% of new-born patients with CF. Transient expression completely prevented this abnormality so *CFTR* expression may be required only briefly during development, but not subsequently. If this is true, it may be necessary to reassess current strategies for gene therapy of CF. However, some caution would be appropriate before such a conclusion is drawn. At present it is not known how relevant this model will be to the human airway. Previous experience has shown that studies with the CF mouse model do not always transfer to the human situation.

9.5 Progress and adverse events

The initial impetus for gene therapy was the idea that genes defective in common single-gene disorders such as CF could be augmented or replaced by the introduction of the wild type gene. CF appeared to be a good candidate because the epithelial cells of the lung airways which required *CFTR* function were in principle accessible to outside intervention by use of aerosols (Figure 9.1). Clinical trials for CF have been carried out using both adenovirus and lipocomplexes to deliver the *CFTR* cDNA. The results have been disappointing. In both cases there has been some evidence of gene transfer, but expression was transitory and there was no evidence of clinical improvement.

Since the first clinical trial for gene therapy there has been a total of 1472 trials worldwide (according to the Gene Therapy Worldwide Database as of October 2008). However, only 47 of these have progressed to phase III trials, the stage which tests for efficacy on a large scale. Although the initial focus was on single-gene disorders, two-thirds of clinical trials to date have been concerned with anti-cancer treatments, 9% have been concerned with cardiovascular disease and 8% each for single-gene disorders and infectious disease. So far, there has been no product licensed for clinical use in the USA or Europe, although an anti-cancer treatment called Gendicine has been licensed in China, which we will discuss further below. So clearly gene therapy is still a developing technology. One area that is generating considerable excitement is the use of siRNA rather than the transgenic expression of whole genes. In this section we will discuss the limited number of examples where the gene therapy treatment aimed at the expression of whole genes has been shown to be clinically effective. The use of siRNA was considered as a separate topic in Chapter 7.

9.5.1 SCID-XI

Severe combined immunodeficiency (SCID) is a heterogeneous group of disorders which, as the name implies, results in a complete failure of the immune system. As a result, children suffering from this disorder are very vulnerable to infections of all kinds and have to live in a completely sterile environment. SCID-XI is an X-linked inherited disorder characterised by an early block in T and natural killer (NK) lymphocyte differentiation. This block is caused by mutations of the gene encoding the γC cytokine receptor subunit of interleukin-2, -4, -7, -9 and -15 receptors, which participate in the delivery of growth, survival and differentiation signals to early lymphoid progenitors. T cells and NK cells are derived from pluripotent haemopoietic stem cells, which are capable of developing into any of the diverse cell types found in the blood. These stem cells express a glycoprotein on their surface called CD34, hence they are called CD34+ cells. The nature of the defect makes it a good target for *ex vivo* therapy. Cells can be purified from peripheral blood or bone marrow by methods such as attaching an anti-CD34 antibody to magnetic beads. The purified CD34+ cells can then be cultured *in vivo* and transduced with a retroviral vector carrying the γC gene. Transduced cells expressing γC protein can be selected because they will have a survival and competitive advantage in the presence of a cytokine compared to the initial cells. These cells can then be reinfused back into the patient from whom the cells were isolated in the first place. Because the patient receives their own cells, there is no need for tissue matching or immunosuppressant therapy.

A gene therapy trial for SCID-XI was initiated in France, based on the use of γC cDNA inserted into a defective γC Moloney retrovirus-derived vector and *ex vivo* transduction of CD34+ cells. The treatment was successful in the 10 children that took part in the trial. T and NK cells expressing γC were detected and T, B and NK cell counts and function, including antigen-specific responses, were comparable to those of age-matched controls. Other similar trials in the UK and France also produced successful results. The first trial reported in 2000. In 2002, two of the 10 children treated in this trial developed a leukaemia-like condition. In these children the integration of the retrovirus was found to have inserted near a proto-oncogene called *LMO2*. The problem with this approach is that cells where a proto-oncogene has been activated by retroviral insertion will proliferate faster and so be at a selective advantage. It was hoped that this problem could be reduced by reimplanting fewer CD34+ cells carrying the transgene, so providing a smaller population from which the tumour cells could arise. The trial was restarted in 2005, but another child developed the leukaemia-like condition. In 2007 another of the children that had apparently been successfully treated developed leukaemia. Thus, altogether, of eight children who were successfully treated, four have developed leukaemia, one of whom has died as a result. So in principle, gene therapy has been shown to cure a monogenic disorder, but at the same time the treatment has raised serious questions about the safety of retroviral vectors. Current research is directed towards developing so-called self-inactivating (SIN) vectors in which the promoters in the LTR of the retroviruses are inactivated, which should prevent them from activating adjacent proto-oncogenes. A similar trial in London has not so far resulted in the patients developing leukaemia. So it is possible that the precise technical details of the protocol that was used in the trial may be important. For example, the way in which cytokines were used to promote the growth of transduced cells *ex vivo* before reinfusion into the patient.

9.5.2 Chronic granulomatous disease

Chronic granulomatous disease (CGD) is a severe immunodeficiency disorder that results from a failure of phagocytes, the most important of which are neutrophils, to have antimicrobial toxicity. Neutrophils kill bacteria or fungal cells by a respiratory burst of activity which produces highly reactive superoxide radicals. The superoxide is produced by NADPH oxidase which transfers electrons across membranes to react with molecular oxygen. This may operate across the plasma membrane to generate the superoxide radicals outside the cell, or across the membrane of phagosomes, an intracellular structure containing microbes which have been engulfed by the neutrophil. Patients with CGD carry mutations in one of four genes encoding the subunits of NADPH, so the neutrophils of their immune system are unable to generate this respiratory burst in response to microbial infection. About 70% of the mutations are in an X-linked gene encoding an NADPH subunit called *gp91phox*. A clinical trial was carried out reintroducing the *gp91phox* gene on a retroviral vector into CD34+ cells recovered from peripheral blood and cultured *ex vivo*. Before the transduced cells were reinfused, the patients were first subject to non-myeloablative treatment. This chemotherapy reduces, but does not completely eradicate, all the bone marrow cells of the recipient. The purpose of this treatment is to increase the chance that the reinfused cells will re-populate the patient, while still preserving some immune function. After treatment both patients showed a substantial improvement in neutrophil function and a clear clinical benefit.

Although the treatment was based on a retroviral vector, it was hoped that the treatment would not suffer from the same problem as the French

SCID trial, because expression of the *gp91^{phox}* gene should not confer any selective advantage on the transduced cells. However, this then introduces the problem that the cells transfected with *gp91^{phox}* will have to compete on equal terms with the resident stem cells – this is why it was thought necessary to reduce the number of resident CD34+ cells by the non-myeloablative treatment. The investigators monitored the locations in the genome where the retrovirus had inserted in the transduced cells. At first they detected many different sites of insertion; but with time the profile in both patients progressively changed, so that in an increasing proportion of CD34+ cells the retrovirus was inserted in the promoter of one of three genes: *MDS-EVI*1, *PRDM*16 and *SETBP*1. The significance of this observation is that *MDS-EVI*1 activation has been observed in acute and chronic myeloid leukaemias. So it is suspected that overexpression of these genes may have promoted cell growth and proliferation. This was probably one reason that the trial was successful, but it highlights the problem that the use of retroviral vectors leads to selection of cells where the retrovirus has activated genes that promote cell proliferation and perhaps eventually leukaemia. The two patients remained disease-free for 2 years with no sign of leukaemia. However, one of the patients subsequently died from a bacterial infection that was judged to be due to the return of the underlying CGD disease, not to the consequences of the gene therapy treatment.

9.5.3 Melanoma

So far, we have considered gene therapy trials which were designed to replace a defective gene. A trial for the treatment of melanoma illustrates another approach. Peripheral blood lymphocytes act as an important defence against cancer. T lymphocytes have receptors on their surface (T cell receptors or TCRs) that recognise antigens that are commonly expressed on the surface of cancer cells (tumour-associated antigens or TAAs). TCRs are composed of α and β chains, of which a relatively small number recognise a variety of TAAs that are common to a variety of different cancers. The isolation of these lymphocytes followed by *ex vivo* expansion and reinfusion back into the patient provides a promising new treatment for cancers. The problem is being able to purify enough of the T lymphocytes that are expressing a TCR that will be active for the particular cancer being treated. A way round this problem is to isolate peripheral blood lymphocytes, which can be readily isolated in large quantities, and then introduce a transgene expressing the gene encoding an appropriate TCR that recognises the cells from the cancer being treated. To test this idea, 15 patients with melanoma were treated with peripheral blood lymphocytes engineered *ex vivo* using a retroviral vector to express both chains of a TCR that recognises MART-1, a surface epitope expressed by melanoma tumour cells. Fourteen of the fifteen patients showed evidence of circulating peripheral blood lymphocytes expressing the anti-MART TCR and two of the patients showed regression of the melanoma tumour; both were judged to be essentially disease-free 20 months after treatment.

We saw above that, in the French trial for treatment of SCID-XI using a retroviral vector, three patients developed a leukaemia-like disorder. There were also some warning indications in the CGD trial about the safety of retroviral vectors. There have been other serious adverse outcomes of gene therapy treatment. The most well known is the death in 1999 of Jesse Gelsinger in a gene therapy trial to treat ornithine transcarbamylase deficiency. The death was caused by a massive inflammatory reaction to the adenovirus-based vector used in the trial. His death was a setback for gene therapy and the number of gene therapy trials markedly reduced after it occurred. In 2007 it was reported that a woman died while undergoing a gene therapy trial for rheumatoid arthritis in her knee. The trial, using an AAV vector, was designed to block TNF-α, a stimulatory cytokine which is thought to provoke the inflammation that causes the symptoms of rheumatoid arthritis. The cause of death was a fungal infection called histoplasmosis affecting her liver. As well as the gene therapy, she was undergoing drug therapy to reduce TNF levels. Both the gene therapy and the drug therapy would weaken the immune system, so it is not clear which was responsible for her death. Possibly the combination was to blame, with the gene therapy tipping her already weakened immune system over the edge. It is not thought that the AAV vector was a factor.

9.6 Summary

- Gene therapy seeks to treat genetic disease using the unmutated version of the gene affected.

- Two strategies can be employed: somatic gene therapy, which limits any genetic change to the lifetime of the person treated, and germline therapy, which affects gametes, and any changes will thus be inherited by future generations. Germline therapy is currently prohibited for ethical reasons.

- Gene therapy can act by replacing the mutated gene or adding a wild type version to work alongside it. Gene replacement requires techniques that are not yet available, so gene therapy of inherited diseases is limited to recessive mutations.

- Cells may be modified by *ex vivo* techniques, where cells are removed from the body, genetically modified and then reintroduced. *Ex vivo* therapy is suitable for modification of blood stem cells.

- Diseases such as DMD and CF must be treated by *in vivo* therapy, where the gene is introduced to the cells within the body of the patient.

- Various vehicles are used to introduce genes into cells:
  - Retroviruses introduce genes into replicating cells and are used for *ex vivo* therapy.

- ○ Adenoviruses are tropic for airway cells and are efficient at carrying genes into cells, but provoke a strong immune reaction in the patient.
  - ○ Lipoplexes encapsulate the DNA into cationic lipids. They do not induce an immune reaction, but may be less efficient than other methods.
  - ○ DNA can be directly injected into muscle cells.

- The first trials for CF gene therapy have successfully demonstrated that genes can be delivered to cells *in vivo*. However, expression was transient and not sufficiently strong to cause a clinical improvement. Adenovirus-based therapy provoked a strong immune reaction.

- There have been some successes:
  - ○ Severe combined immunodeficiency disease type XI (SCID-XI) has been treated successfully by introducing the wild type copy of the gene encoding the γC cytokine receptor using a retroviral vector.
  - ○ Chronic granulomatous disease (CGD) has been treated by reintroducing the *gp91^{phox}* gene on a retroviral vector into CD34+ cells recovered from peripheral blood and cultured *ex vivo*.
  - ○ Melanoma has been successfully treated by engineering peripheral blood lymphocytes to express T cell receptors that recognise antigens on the melanoma cells.

- Unfortunately, there have been several incidences of patients dying during gene therapy trials.
  - ○ A teenage boy called Jesse Gelsinger died in a gene therapy trial to treat ornithine transcarbamylase deficiency.
  - ○ Four of the eight children treated for SCID-XI developed a leukaemia-like illness as the retrovirus integrated near oncogenes, resulting in their activation.
  - ○ A woman died during treatment for rheumatoid arthritis in her knee. The trial was based on blocking the stimulatory cytokine TNF-α that induces the inflammation characteristic of rheumatoid arthritis. It was possible that the gene therapy, in combination with drug therapy that also targeted TNF-α, weakened her immune system so that she succumbed to a lethal fungal infection.

Further reading

General

BANK, A. (1996) Human somatic gene therapy. *Bioessays*, **18**, 999–1007.

SOMIA, N. and VERMA, I.M. (2000) Gene therapy: trials and tribulations. *Nature Reviews Genetics*, **1**, 91–99.

THOMAS, C.E., EHRHARDT, A. and KAY, M.A. (2003) Progress and problems with the use of viral vectors for gene therapy. *Nature Reviews Genetics*, **4**, 346–358.

Various authors (1997) Special report: making gene therapy work. *Scientific American*, **276**, 79–103.

These four papers give a general introduction to the principles of gene therapy and the different viral and liposome vectors that can be used.

Cystic fibrosis

BOUCHER, R.C. (1996) Current status of CF gene therapy. *Trends in Genetics*, **12**, 81–84.

ENNIST, D.L. (1999) Gene therapy for lung disease. *Trends In Pharmacological Sciences*, **20**, 260–266.

WAGNER, J.A. and GARDNER, P. (1997) Towards cystic fibrosis gene therapy. *Annual Review of Medicine*, **48**, 203–216.

Applications of gene therapy

CAVAZZANA-CALVO, M. HACEIA-BEY, S., BASILE, C.D. *et al.* (2000) Gene therapy of human severe combined immunodeficiency (SCID)-XI disease. *Science*, **288**, 669–672.

EDELSTEIN, M.L., ABEDI, M.R. and WIXON, J. (2007) Gene therapy clinical trials worldwide to 2007 – an update. *Journal of Clinical Medicine*, **9**, 833–842.

A comprehensive review of all the gene therapy trials to date, with sections describing the way that the different vectors have been used and the different disorders that have been addressed.

MORGAN, R.A., DUDLEY, M.E., WUNDERLICH, J.R. *et al.* (2006) Cancer regression in patients after transfer of genetically engineered lymphocytes. *Science*, **314**, 126–129.

The melanoma trial.

OTT, M.G., SCHMIDT, M., SCHWARZWAELDER, K. *et al.* (2006) Correction of X-linked chronic granulomatous disease by gene therapy, augmented by insertional activation of *MDS1-EVI1*, *PRDM16* or *SETBP1*. *Nature Medicine*, **12**, 401–409.

Adverse events

HUGHES, V. (2007) Therapy on trial. *Nature Medicine*, **13**, 1008–1009.

Describes the issues involved in the death of a woman participating in the rheumatism gene therapy trial.

Useful web sites

www.wiley.co.uk/genmed/clinical/

The home page of the Gene Therapy Clinical Trials Worldwide, an online database which gives a continually updated record of all the gene therapy trials.

http://www.esgct.org/announcements.cfm

Provides updated information on deaths in the French SCID trial and the death in the rheumatism trial.

CHAPTER TEN

Human population genetics and evolution

Key topics

- The Hardy–Weinberg distribution
- Selection
- Drift
- Migration
- Effective population size
- Population bottlenecks and founder effects
- Phylogenetic trees
- Polymorphisms in protein-encoding genes
- Mitochondrial polymorphisms
- Amplification of DNA from Neanderthal fossils
- Y chromosome polymorphisms
- Out-of-Africa and multiregional hypotheses of human origins
- Population variation in disease frequencies

10.1 Introduction

Apart from monozygotic twins, the genetic constitution of each individual is different. What sets us apart from each other is not that we have different genes – we all have copies of the same genes – but rather that these genes may take different **allelic** forms. A gene or locus is said to be **polymorphic** if there is more than one allelic form in a population and the rarer allele still occurs at a significant frequency – usually taken to be more than 1%. In some cases, different alleles may produce different phenotypes, as in mutations responsible for monogenic disorders; in others the variation may not have any discernible effect on the phenotype and these are said to be **neutral**. The study of these polymorphisms within and between populations is important for two reasons. First, they can be used to reconstruct recent human evolution and the histories of population movement and growth since the emergence of anatomically modern humans. Second, as we saw in Chapter 6, the genetic structure of human populations has a direct bearing on the design and interpretation of experiments that use population-based methods to search for risk alleles of common complex diseases.

10.1.1 Types of variation studied

Polymorphisms of nuclear genes encoding proteins were the first type of variation to be studied in human populations. Subsequently, other types of variation have become increasingly important. As in the case of genetic maps, neutral molecular polymorphisms such as RFLPs, *Alu* elements, microsatellites and minisatellites have been extensively used. Polymorphisms affecting the mitochondrial DNA (mtDNA) and the Y chromosome have proved particularly important because they are transmitted without recombination. Thus when two molecules are compared, each retains a record of all the changes that have occurred since they diverged from a common ancestor. The Human Genome Project has provided a novel source of variation in the form of single nucleotide polymorphisms (SNPs). These provide a detailed picture of the variation in individuals and populations. We shall see that rather than each SNP being informative on its own, the real benefit lies in studying linked haplotypes. We shall consider the studies using each of these forms of variation and consider how they have increased our understanding of human evolution. In the case of SNPs we shall also consider issues that affect their usefulness as tools to search for risk alleles involved in complex disease.

10.2 Basic principles in human population genetics

10.2.1 The Hardy–Weinberg distribution

Consider an autosomal gene that has two alleles, say *A* and *a*. These alleles can combine in different ways to produce three different genotypes: *AA*, *Aa* and *aa* respectively. In a population at equilibrium (that is, where the gene frequencies are not changing), the distribution of these genotypes is described by the Hardy–Weinberg equilibrium which states that the distribution of genotypes is as follows:

$AA = p^2$ where p = the frequency of the *A* allele
$Aa = 2pq$ where q = the frequency of the *a* allele
$aa = q^2$.

This distribution is valid only if a number of conditions are met:

- Mating is random.
- There is no selection for or against any of the genotypes.
- There is no migration.

In human populations these conditions are rarely completely satisfied; in particular, mating is rarely random. Humans tend to mate with partners from similar religious, cultural and socioeconomic backgrounds. We shall consider the effects of migration and selection below. Nevertheless, in the absence of obvious factors, such as a clear selection for one of the genotypes or large-scale migration, allele frequencies do generally conform to that described by the Hardy–Weinberg distribution.

The usefulness of the Hardy–Weinberg distribution is that if we measure the frequency of one of the genotypes, it allows the other two to be calculated. For example, the average frequency of cystic fibrosis (CF) in European populations is about 1 in 2000. Because CF is an autosomal recessive condition, the value of q^2 in the Hardy–Weinberg distribution is 1/2000. This allows the following simple calculation:

$q^2 = 1/2000$
$q = 1/44$
$2pq = 1/22$ (approximating p to equal 1)

$2pq$ is the frequency of heterozygotes, so, starting with the observed frequency of the disease, we can calculate the frequency of carriers. We can perform a similar calculation with autosomal dominant traits, but this time the observed incidence of the disease is $2pq + q^2$.

10.2.2 Changes in allele frequency

Allele frequencies do not always remain constant. If they did, then evolution could not have occurred, because evolution requires such changes. New alleles can be created only by mutation. The classic neo-Darwinian theory of evolution states that alleles change in frequency in response to **selection**, either for advantageous alleles or against deleterious alleles. However, selection is not the only mechanism that operates – migration and random drift of selectively neutral alleles also act to change allele frequencies.

10.2.2.1 Selection
Selection is a consequence of alleles that alter the **Darwinian fitness** of the organism, defined as the number of its offspring reaching reproductive maturity. The rate of change in the frequency of an allele in response to selection can be calculated from its selective coefficient, which is the fractional difference between the Darwinian fitness of an organism carrying the allele and the Darwinian fitness of a reference genotype. Selection is a powerful force when it operates against strongly deleterious alleles from the population, such as those responsible for monogenic disorders. If such a disorder has a selection coefficient of 1, i.e. no affected individuals have children, then selection will ensure that the frequency of homozygous individuals affected by an autosomal recessive disorder will be equal to the frequency with which new alleles are generated by mutation. There is an important corollary to this conclusion: if the frequency of a genetic disease

in a population is higher than would be consistent with a reasonable rate of mutation, then some factor must be operating to maintain the allele frequency. We shall consider several such situations in this chapter.

The essence of the theory of Darwinian evolution is that favourable mutations will increase in frequency due to selection. The rate at which a beneficial allele would spread through a population can be calculated from the selection coefficient. One of the best known examples of positive selection in human populations is the heterozygous advantage in areas where malaria is endemic conferred by heterozygosity for sickle cell anaemia and other mutations that affect haemoglobin (discussed in more detail below). The experimentally measured selection coefficient for carriers of such alleles is about 0.1.

Allele frequencies of genes involved in resistance to infectious diseases such as immunoglobulins are highly variable in different parts of the world, which is assumed to reflect different histories of exposure to infectious diseases. As we shall see below, modern humans probably emerged from Africa about 100 kya. If selection had been operating throughout this time, the value of selection coefficients operating on genes of the immune system would be in the order of 0.01.

10.2.2.2 Random drift

Mathematical analysis by geneticists such as Fisher and Haldane showed that the effects of selection on Mendelian genes provide a satisfactory mechanism for evolution. This became known as the neo-Darwinian synthesis, and for a while it was thought to be the only mechanism responsible for changes in allele frequency. An important implicit assumption of the neo-Darwinian synthesis is that all, or nearly all, genetic variation is subject to selection – an allele is either advantageous or disadvantageous; it is rarely neutral. During the 1960s, the molecular analysis of gene structure and evolution started to reveal a number of facts that were not readily explicable in terms of selection. Two of the most important observations were:

1. At the level of the amino acid or nucleotide sequence, most genes were much more variable than predicted by the neo-Darwinian theory. This fact emerged during the study of protein polymorphisms revealed by electrophoresis. At first it was hotly contested, but the advent of DNA sequencing has shown that this statement is undeniable.
2. The rate of nucleotide or amino acid substitution in a gene was apparently constant in a wide variety of genes examined. This gave rise to the concept of a molecular clock (Figure 10.1).

The first observation was inconsistent with the neo-Darwinian view, because if few alleles were selectively neutral, selection should ensure that the most advantageous allele became fixed in the population. Most genes should therefore be identical. Polymorphisms should be rare and observed only when the frequency of a new and advantageous allele was being increased by selection. The second observation is surprising because it

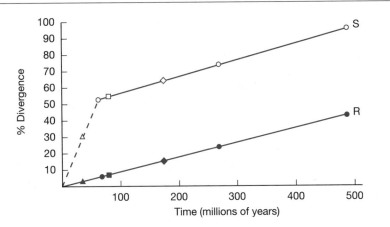

Figure 10.1 The molecular clock of globin evolution. The sequences of α- and β-globin genes were used to measure the divergence between pairs of globin genes. Both silent substitutions (filled symbols) and replacement substitutions (open symbols) accumulate at a constant rate. At first the rate of silent substitutions was 10 times that of replacement substitutions, in agreement with the neutral theory of evolution. Eventually all possible selectively neutral substitutions that could occur had occurred. Further substitutions were subject to the same selection pressures as replacement substitutions, and the rate at which substitutions accumulated slowed down. The data were obtained from following comparisons which can be dated from the archaeological record (circles): intraspecies comparisons of α- and β-globin genes that were presumed to have diverged 500 mya, comparisons of α- and β-globin genes of mammals with chicken (270 mya), and comparison of α- and β-globin genes of different mammals (85 mya). Further data for which archaeological dates are not available were obtained from comparison of δ versus β (triangles), γ versus ε (squares) and δ or β versus γ or ε (diamonds). For these data the replacement substitutions were plotted on the line R and the silent site divergences plotted directly above. The amount of divergence (corrected for multiple substitutions) was plotted against the time since divergence. Line R passes through the origin. Because the line S does not pass through the origin, and there is no independent verification of the time of the β/δ divergence, it is plotted as a dashed line. (From *Cell*, **21**, 653–68 (Efstratiadis, A. et al. 1980), with permission from Cell Press.)

would be expected that proteins would evolve to a configuration that optimised their function, and that subsequent changes would be selected against. It is difficult to imagine how, over long periods of evolution, new alleles could arise and be selected for with the same selection coefficient.

These observations led Motoo Kimura to propose what is now known as the **neutral theory** of evolution. The foundation of this theory is that many, or even most, alleles are selectively neutral. Allele frequencies change because of random sampling effects between generations. This can be explained by a simple analogy. Consider a bag containing equal numbers of white and red balls. If ten balls were randomly withdrawn from the bag, it would not be surprising if the sample consisted of four red and six white balls. The frequency of red balls has thus changed from 0.5 to 0.4. However, if we withdrew 1000 balls we would be very surprised to find that the sample contained 400 red and 600 white balls. If the different coloured balls represent two alleles of a gene, two conclusions are evident from this analogy. First, random sampling will result in a change in allele frequency from generation to generation. Second, the smaller the population size, the larger the changes are likely to be.

At every generation the frequency of an allele will fluctuate because of this sampling effect, until either the frequency of an allele becomes 100% (fixation), or it disappears completely from the population (extinction). One of these two outcomes is logically inevitable, but the time it takes depends on population size, occurring more rapidly in smaller populations. Clearly, fixation is a more likely outcome for common alleles, and extinction more likely for rarer alleles. Therefore, the most likely fate for a new allele introduced by mutation is that it will become extinct. However, because a large number of alleles are continually being created by mutation, fixation of some of them will occur by random drift.

It is important to understand that the neutral theory does not state that all alleles are neutral. First, rare advantageous alleles do exist and will increase in frequency as a result of selection. However, they are likely to be rare and make little contribution to overall variability. More importantly, many substitutions will be deleterious. Kimura showed that the rate of evolution can be described by the following equation:

$$K = v_T f_o$$

where K = rate of nucleotide substitutions per year, f_o = fraction of new alleles that are selectively neutral and v_T = mutation rate.

This equation infers that the rate of evolution of a gene, or part of a gene, will be constant and will depend on two variables, the mutation rate and the fraction of selectively neutral mutations. The validity of this equation is readily tested by comparing the rate of nucleotide substitutions in situations where we may be confident about whether or not the result is likely to be selectively neutral. For example, Figure 10.1 compares the rate of two different types of nucleotide substitution that have occurred in the evolution of the globin genes. **Silent substitutions** do not result in a change in amino acid sequence, because they result in a **synonymous** codon; that is, a codon that encodes the same amino acid. **Replacement substitutions** result in a codon that encodes a different amino acid. Clearly, silent substitutions are more likely to be selectively neutral than replacement substitutions. Figure 10.1 shows that this is indeed the case. The initial rate of silent substitutions is 1% divergence per million years, while replacement substitutions occur at about one-tenth of this rate. Silent substitutions in coding regions may still be subject to some selection pressure; for example, the secondary structure of the mRNA may be altered. Sequences that are unlikely to be subject to any selection, such as pseudogenes, evolve at a rate of 2% divergence per million years.

Genetic drift and selection are therefore factors that act together. While random drift ensures that substitutions occur at a constant rate, selection controls that rate. Neutral alleles evolve at a rate determined by the mutation frequency, but the protein sequences evolve at a lower rate, determined by the capacity of the protein to accommodate amino acid substitutions. Different proteins evolve at different rates as a result of this constraint. Even different parts of proteins may evolve differentially; for example, the active site of an enzyme is likely to be more highly conserved than other parts of the protein. A last point to note is that selection and drift may operate simultaneously, so that rare advantageous alleles may still become extinct by genetic drift.

10.2.2.3 Migration

The effects of drift and selection described above apply formally to isolated populations. Few populations are entirely isolated – immigration/migration between adjacent populations will result in gene flow that will affect their genetic structure, in particular mitigating the effects of drift. Gene flow between populations as a result of migration is called **admixture**.

10.2.3 Effective population size

The neutral theory of evolution predicts that if alleles in a population are completely neutral and frequencies change only through drift, then the number of alleles in the population, known as the population mutation parameter, θ, is related to the effective population size, N_e, according to the equation:

$$\theta = 4N_e\mu$$

where μ is the mutation rate per generation at each one of an infinite number of sites. This equation is valid for autosomal loci. Because there are only three X chromosomes for every four autosomes in the population, the equation for X-linked loci is $\theta = 3N_e\mu$. Similarly, for mtDNA the equation is $\theta = 2N_e\mu$.

The effective population size, N_e, is an abstract concept that describes the size of an ideal randomly breeding population in which allele frequencies would drift at the same rate as that observed. In real populations there will be many individuals who are not breeding; for example, children and adults past their reproductive age. For this reason, and the fact that real populations do not breed randomly, N_e is about one-half the actual or census population size. N_e also describes the long-term population size. If the population size fluctuates over time, N_e is approximately the harmonic mean of the population size, which means that N_e tends to be closer to the historical minimum rather than maximum population size.

10.2.3.1 Long-term effective population size of human populations

The importance of the equation above is that both θ and μ can be experimentally measured, thus it is possible to estimate the long-term value of N_e. Such estimates have been made from measurements of nuclear protein polymorphisms, minisatellites, microsatellites, mtDNA and Y chromosomes. All the estimates are in remarkable agreement that the long-term effective human population size is about 10 000 individuals. This figure is one of the arguments against the multiregional hypothesis of human evolution, because if this number of humans was distributed worldwide, local populations would be too small to be viable.

10.2.3.2 The mode of population expansion can be discerned in the structure of present populations

The modern human population size is much larger than 10 000, so there must have been an expansion from a very small population. If a population expands, most lineages, when traced backwards in time, would remain separate after the population expansion and then converge rapidly to a common ancestor who lived just before the expansion. The genealogy of the population is said to have a star-like morphology. The differences between two randomly chosen individuals will be due to the mutations that have occurred since the expansion, because the genealogies were separate for only a short time before the expansion. Mathematical analysis shows that the number of differences between pairs of randomly chosen individuals will follow a smooth Gaussian distribution. If the mutation rate is known, the modal number of mutations can be used to calculate the number of generations since the expansion. Estimates from mtDNA place the expansion to about 100 kya. Estimates from the Y chromosome place the expansion more recently, at 50–60 kya.

10.2.3.3 The time to coalescence is proportional to the effective population size

All individuals in a population must ultimately trace their lineage back to a common ancestor, often referred to as the most recent common ancestor (MRCA). This process is known as **coalescence**. Coalescence takes longer if there are more lineages, so the coalescence time is proportional to the effective population size. The effective population sizes of autosomes, X chromosomes, mtDNA genomes and Y chromosomes are in the ratio of 4:3:2:1. As we shall see below, this is reflected in the coalescence times of an autosomal locus such as β-globin (800 kya), mtDNA (150–300 kya) and the Y chromosome (60 kya).

10.2.4 Population bottlenecks

Human history is marked by large reductions in population size caused by famine, epidemics, climate change and wars. Once the cause of the population reduction is removed, population numbers will recover. Such an event is called a **population bottleneck**. Population bottlenecks reduce genetic variation because while the population is small the effect of drift on allele frequencies will be large, causing some alleles to become extinct, thus reducing variation. Population bottlenecks therefore leave a lasting imprint on a population and will affect the conclusions of a number of different types of investigation. For example, the reduction in genetic variation due to a population bottleneck may lead to an underestimate of the effective population size. As illustrated in Figure 10.5 and discussed further below, population bottlenecks also affect the degree of linkage disequilibrium between closely linked loci.

10.2.4.1 Founder effects

A small group that splits off from a central population and moves into a geographically isolated area and then expands is called a **founder** population. For reasons that are similar to population bottlenecks this reduces genetic variability. Although normally thought of in terms of physical migration, a founder group can also arise if it becomes reproductively isolated from the main population because of culture, religion or language. For example, Ashkenazi Jews descend from a founder group that was isolated by religion and culture rather than geography. The effect of drift when the population is small means that the population descended from a founder group may have a quite different genetic structure than the ancestral population. In practical terms this means that they often have a much higher frequency of genetic diseases than the main population because particular disease alleles attain high frequency in the founding group. Examples of this are discussed below.

10.2.4.2 Phylogenetic trees

Phylogenetic trees describe the evolution of a set of genes, populations or species (elements) from a common ancestor. The order of branching of the tree reflects the order in which the different elements split off during evolution. Trees may be rooted or unrooted (Figure 10.2). An unrooted tree

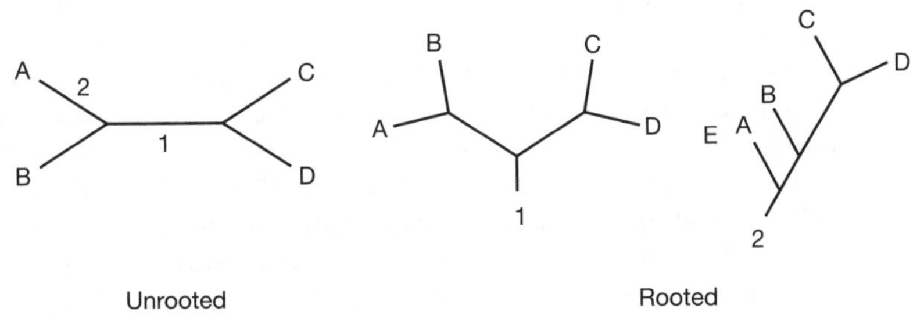

Unrooted   Rooted

Figure 10.2 Rooted and unrooted trees. The unrooted tree on the left describes the similarity relationships between four elements A, B, C and D. However, it does not specify the order in which they branched during their evolution from a common ancestor. To do this the root must be determined. Both of the rooted trees on the right are consistent with the unrooted tree on the left. The number at the base of each rooted tree specifies the position of the root indicated by the numbers in the diagram of the unrooted tree. There are three other possible rooted trees that are not shown in the diagram. The root is determined by reference to an outrider that is more dissimilar to any of the four elements than they are to each other. This allows the deduction that the outrider was the first to branch the lineage from the common ancestor. The element to which the outrider is most similar places the root.

describes the relationships between the elements in the tree without specifying the ancestral state. A rooted tree specifies which element is ancestral,
usually by reference to an outside element that is more different to the
elements in the tree than they are to each other.

There are many ways to construct phylogenetic trees and there has been
an often acrimonious debate as to which is the most accurate. One point of
general agreement is that it is never possible to be sure that a deduced phylogenetic tree is actually the true tree. Basically, there are two approaches
to constructing trees. Numerical methods quantify differences between the
elements of the tree and place the elements that are most similar on adjacent branches. **Cladistic** methods seek to identify the tree that depicts the
evolution of present-day elements from a common ancestor with the least
number of events. A commonly used method of this type is called maximum
parsimony. This method requires a large number of possible trees to be
examined and is therefore very computer-intensive. The number of possible
trees that could exist is usually much larger than can be feasibly examined,
so only a subset of the possible trees is tested and no unique tree is
identified. It is therefore necessary to carry out a statistical test, called boot-
strap analysis, to measure the confidence that can be placed in the result.
This involves carrying out the analysis on repeated occasions; each time a
small number of data are randomly replaced with other data chosen from
the data set. The percentage number of times the computer generates the
same tree is a measure of confidence in the result.

10.3 Recent human evolution

The evolution of anatomically modern humans has generated a fierce controversy. The fossil record shows that the lineage that gave rise to modern
Homo sapiens started with *Homo habilis* in Africa approximately 2.5 million years ago (mya). This was followed by *Homo erectus* who lived 1.3 mya
to 300 thousand years ago (kya), and spread from Africa to Europe and
Asia. The first members of the species characterised as *Homo sapiens* are
found in Africa dated approximately 400 kya and had spread to Europe by
300 kya. These first human populations are recognisably different from
modern humans and are thus called archaic. In Europe, archaic humans
formed a subspecies called *H. sapiens neanderthalensis* or Neanderthal
man. The last Neanderthals disappeared from Europe 30 kya. Anatomically
modern humans are first found in Africa just over 100 kya. Their remains
have been found in Israel dated 100 kya, but there is no evidence of
modern humans in the rest of the world for another 30 000 years. Remains
of modern humans are found in China dated 67 kya, Australia 55 kya,
Europe 40 kya and America 35–15 kya.

The controversy centres on the evolution of anatomically modern
humans from *Homo erectus* and archaic human populations. There are several different theories describing the emergence of modern humans. Some
palaeontologists recognise similarities in the bone structure of modern

populations that lived in the same region previously. This gives rise to the multiregional hypothesis that proposes that *H. erectus* populations evolved into modern humans independently in separate geographical regions of the world. To avoid the problem of multiple independent origins of favourable characteristics, which would be inherently unlikely, it is proposed that there was gene flow between the populations so that a favourable trait that arose in one part of the world could spread to all human populations. However, this gene flow would have to be very limited, otherwise the special characteristics of each regional population would be lost.

Not all palaeontologists agree with the multiregional hypothesis. There is also evidence from the fossil record for a contrasting view that modern humans originated in Africa from which they spread to displace archaic human populations. The initial population of modern humans consisted of about only 10 000 individuals and probably displaced the resident archaic human populations without interbreeding. The genetic data which we review below generally favour this hypothesis. However, a key point of interest is whether there really was no interbreeding with archaic populations as they were displaced, so that the archaic populations have not made any contribution to the modern human gene pool. Sequencing DNA from Neanderthal bones has allowed this question to be directly addressed.

10.3.1 Nuclear polymorphisms

Allele frequencies in polymorphisms of protein-encoding genes often vary in different population groups throughout the world. The first observation of this kind was made in 1919 with ABO blood groups (Figure 10.3). Subsequently, systematic measurements have been carried out using a wide variety of polymorphisms that can be detected at the protein level usually by electrophoresis, such as blood groups, serum proteins, human leucocyte antigens, enzymes and other readily detectable proteins. A large and comprehensive study utilising such markers has been carried out by Luca Cavalli-Sforza and colleagues utilising 120 allele frequencies in 42 populations worldwide. In this context 'populations' refer to populations that are probably still similar to those that existed in 1492 before the mass migrations of the last 500 years.

The different allele frequencies in the populations were used to reconstruct population histories by constructing phylogenetic trees. Figure 10.4. shows a phylogenetic tree of human populations worldwide. The first branch of the tree is between Africans and non-Africans. This is a major conclusion that suggests modern humans originated in Africa and thereafter spread to the rest of the world. The second branch of the tree shows that non-Africans split into Northern Eurasian and South East Asian branches, probably about 60 000 years ago. The Northern Eurasian branch gave rise to Caucasian, Northern Asian and American peoples. The South East Asian branch gave rise to the populations of New Guinea, Australian and Polynesian as well as South East Asians.

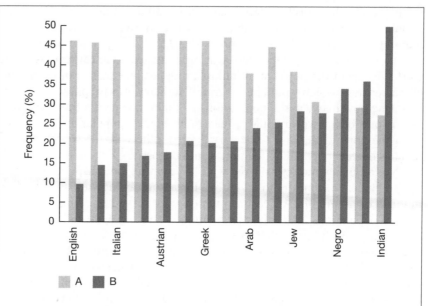

Figure 10.3 Worldwide variation in A and B blood groups. The ratio of people with blood group A to those with blood group B varies from 4.5:1 in the English population to 0.5:1.0 in the Indian population.

10.3.2 Mitochondrial DNA

Measurements of mtDNA evolution are usually based on the sequence of the two regions called *HVR*1 and *HVR*2 in the control region (see Figure 1.7) which are hypervariable and show greater variation than the rest of the molecule. Variation in mtDNA has proved extremely useful in studying recent human evolution. There are a number of reasons for this:

1. The mitochondrial genome is maternally inherited and there is no recombination between paternal and maternal genomes. This makes the construction of phylogenetic trees straightforward because, in the absence of recombination, the number of nucleotide differences between two mitochondrial genomes is a direct reflection of the history of these molecules since they originated from a common ancestor.
2. *HVR*1 and *HVR*2 DNA mtDNA evolves 5–10 times more rapidly than nuclear DNA. This makes it suitable to study recent evolution, because a sufficient number of mutations will have arisen to permit the analysis of population relationships.
3. The phylogenetic analysis was carried out by the maximum parsimony method, which many consider to be a superior method because it is cladistic; that is, the course of evolution is deduced directly from the difference in qualitative characters (see above).

Mitochondrial DNA studies do, however, suffer from two major weaknesses:

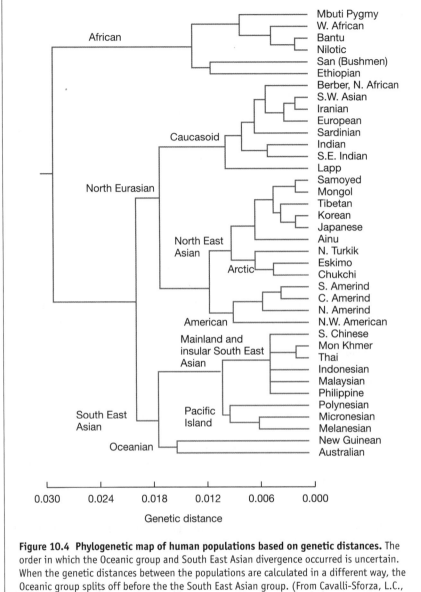

Figure 10.4 Phylogenetic map of human populations based on genetic distances. The order in which the Oceanic group and South East Asian divergence occurred is uncertain. When the genetic distances between the populations are calculated in a different way, the Oceanic group splits off before the the South East Asian group. (From Cavalli-Sforza, L.C., Meonozzi, P. and Piazza, A. *The History and Geography of Human Genes*. © 1994 Princeton University Press. Reprinted by permission of Princeton University Press.)

1. Because the mitochondrial genome is inherited without recombination, selection that applies to any part of the molecule, whether it is positive or negative, will affect the frequencies of all alleles carried by the genome. This gives rise to the concept of a selective sweep, in which a whole mitochondrial genotype is lost from a population because of selection against an allele at one locus. This will of course lead to the

loss of other genotypes which in themselves are neutral. Of course, the reverse can apply, leading to the selection for a whole mitochondrial genome because of selection for a single allele.

2. The mitochondrial genome is only a small part of the whole genome and is inherited independently of the nuclear genome through the female line. Stochastic processes play a very important part in evolution, especially when populations are small. Thus any reconstruction of human evolution based on the mitochondrial genome may not be representative of the evolution of the whole genome.

The study using mitochondrial DNA polymorphisms was published by Cann, Stoneking and Wilson in 1987. It showed that the lineage of all extant mtDNA genomes can be traced back to a single common ancestor who lived in Africa about 150 000–300 000 years ago. The origin of the most recent common mitochondrial ancestor in Africa is based on three interrelated lines of evidence:

1. Phylogenetic analysis produces trees with two branches. One branch contains only African populations, the other contains African and non-African populations. The simplest interpretation of this observation is that the ancestor was African. This phylogenetic tree drawn up was controversial for a while, because it was shown that other types of tree would also be consistent with the data. Further work showed that the initial conclusion was correct.

2. All mitochondrial lineages found outside Africa are a subset of lineages found within Africa. By far the simplest explanation for this observation is that non-African populations are derived from a group that migrated from Africa.

3. Diversity, measured as the number of different mutations in the African populations, is much higher than non-African populations. Normally the population that has the greater genetic diversity is ancestral to a population with less diversity.

The single common ancestor has been popularised as the 'African Eve'. This appellation has led to the popular misconception that all humans descend from a single woman. Although all human mitochondrial lineages trace back to a single individual woman, this does not mean that she was the only woman alive at the time when she lived. Women with different mtDNAs were alive at the same time. In each subsequent generation some lineages will be lost, and eventually all lineages but one will be lost. The woman in the ancestral population with this genotype is the so-called mitochondrial Eve; all other mitochondrial genotypes that existed in the other women alive at the same time have become extinct. Note that there is no logical reason why the mitochondrial ancestor has contributed any of her nuclear genes to present-day populations. So the rest of the genome could have had a quite different history. These arguments illustrate why it is unsafe to rely solely on mitochondrial data to reconstruct human population history.

Just as the mitochondrial genome is transmitted through the female line, the Y chromosome is transmitted only from fathers to sons. The human Y chromosome has a complex structure. About 5% of the chromosome recombines with homologous sequences on the X chromosome. This is called the pseudoautosomal region (PAR). The remainder of the chromosome does not recombine with any other chromosome and is called the non-recombining region of the Y chromosome (NRY). The NRY is 60 Mb in size, of which only 35 Mb is euchromatic. For the most part the euchromatic region of the NRY consists of highly repetitive DNA rich in transposons. However, there are a small number of expressed genes that fall into three classes. There are eight class I genes that have homologs on the X chromosome and are widely expressed throughout the body. There are also eight class II genes; they are expressed only in the testes and do not have homologs on the X chromosome. The five class III genes have homologs on the X chromosome but show tissue-limited expression, which in three cases is confined to the testes. The *SRY* gene that determines male sexual development is one of the class III genes. It is expressed only in the embryonic bipotential gonad and in the adult testes. Its homolog on the X chromosome is called *SOX3* and is transcriptionally active, and so presumably it fulfils some other essential function in both males and females.

The Y chromosome shows very little sequence variation. In part, this is due to its low effective population size, which can be considered to be only one-quarter that of an autosomal chromosome for the following reason. Suppose the human population is represented by a male and a female. Between them there are four copies of every autosome, three copies of the X chromosome, but only one Y chromosome. Population genetics theory shows that genetic variation is dependent on the effective population size, thus there will be less variation on the Y chromosome than the autosomes. Despite this relative lack of variation, many SNPs and polymorphic microsatellites have now been identified in the NRY. Comparison with the sequence of chimpanzee, gorilla and orangutan genomes allows the presumed ancestral allele to be identified as opposed to the **derived** allele, which arose by mutation during modern human evolution. By sampling the Y chromosomes from different populations, it is possible to construct a phylogenetic tree of Y chromosome evolution in modern humans similar to that constructed from mtDNA genotypes. To do this, the different individual Y chromosome haplotypes are grouped together into haplogroups. Within each haplogroup there are many different haplotypes, but they are grouped together because they all carry the derived allele at a polymorphic site, which arose by mutation in a common Y chromosome ancestor of all members of the haplogroup. Altogether, 17 haplogroups (A–R) have been defined and the phylogenetic tree, which represents the minimum number of steps to generate these haplogroups, is shown in a simplified form in Figure 10.5. A particularly useful aspect of the Y chromosome tree is that allele frequencies show greater geographical differences than other genetic

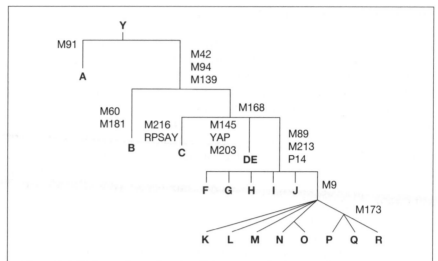

Figure 10.5 Phylogenetic tree based on Y chromosome haplotypes. The most recent common ancestor of all human males is labelled Y. Successive mutations occurred in his descendants, labelled M91, RPSAY, etc. Certain key mutations indicated allow the order in which the haplogroups arose during the descent of present-day populations from the ancestral Y male to be deduced.

trees. This is thought to be because of a phenomenon known as patrilocality, which means after marriage that the wife tends to move to live in the husband's house so that, over long periods, women's genes tend to become more widely dispersed than men's. The greater geographical differentiation means that the geographical location of the different haplogroups in conjunction with the position of the haplogroup on the Y chromosome tree can be used to trace the route of the ancient migrations of man from his origins in Africa.

The deepest branch of the tree is between haplogroup A and all the other groups. Haplogroups B–R all contain the derived alleles M42, M94 and M139, which are not present in haplogroup A. Conversely, haplogroup A carries the derived allele M91 which is not present in any of other groups. The common ancestor of haplogroup A and the other haplogroups is denoted as 'Y', and can be thought of as the Y chromosome Adam. With the exception of a single individual, haplogroup A is only found in a minority of Africans, mainly in Sudanese, Ethiopians and Khosians, and not in any non-African populations. The exception lives in Sardinia and was probably the result of a recent event. Men carrying haplogroup B, the next to branch off the tree, are also found living exclusively in Africa in similar locations to haplogroup A. This places the root of the tree in Africa, consistent with the African origins of modern humans deduced from the trees derived from protein and mtDNA polymorphisms.

A critical node in the tree is defined by the M168 allele. All non-African men and the majority of African men are descended from the individual who carried this mutation. Haplogroups C, D and E branch off the tree directly after M168. Haplogroup C men are mainly found in Africa and have largely displaced the older A and B haplogroups, possibly as a result of the migration of iron-working Bantu men from West Africa about 2500 years ago. Haplogroup C is common among Australian Aboriginals. Modern human remains have been found in Lake Mungo in Australia which are 60 000 years old and are the oldest non-African remains of modern humans. Haplogroup C may be the vestigial traces of the first migration out of Africa which followed a coastal route to Australia without venturing inland. However, this interpretation should be treated with caution because haplogroup C is also common in India, so the Australians carrying haplogroup C may have descended from a later migration from India.

All the non-African men in the world apart from haplogroup C descend from a man who carried the M89, M213 and P14 mutations. This individual probably lived in Iran about 40 000 years ago. The remaining haplogroups radiate from this individual in a starburst. The number of mutations that have accumulated in descendants of Y provide the means to estimate the time to the most recent common ancestor of Y chromosomes. This is estimated to be about 60 000 years ago, subject to the uncertainties concerning the rate of the molecular clock. This date is more recent than the time to the most recent mitochondrial ancestor, estimated to be about 200 000 years ago. This discrepancy may be explained by the smaller effective population size of Y chromosomes, which population genetic theory would predict would result in a shorter coalescence time.

Another interesting use of Y chromosome variation is to show that there is a correlation between surnames and inheritance of the Y chromosome. According to Jewish tradition, descendants of Aaron, the brother of Moses, were selected to serve as priests and all have the surname Cohen. Study of Y chromosome microsatellites show that males of both Ashkenazi and Sephardic Jews with the surname Cohen have strikingly similar Y chromosomes. Moses was a member of the tribe of Levi; but other Levite males have diverse Y chromosomes. The Cohen Y chromosomes, although similar, are not identical; this is due to mutation at the microsatellite loci. The rate of mutation can be used to date the most recent common male ancestor of Cohens, who by tradition would be Aaron. The result is that there have been about 100 generations since Aaron. Assuming 25–30 years per generation, this corresponds to 2500–3000 years before present; that is, during or shortly before the first Temple period, which ended when the temple was destroyed in 586 BC, consistent with Jewish tradition. Another study showed that all men in the UK with the surname Sykes have very similar Y chromosomes, suggesting that there was once a man called Sykes from which all men with the surname Sykes have descended.

The correlation of Y chromosome genotype and surnames is stronger for uncommon surnames; it is much less strong for common surnames such as Smith that probably had multiple independent origins. One study looked at 150 men and matched each with another man with the same surname randomly chosen from the electoral register. The surnames of the pairs were ranked according to the frequency of the surname in the general population. Overall it was estimated that only 24% of the pairs shared the same Y chromosome ancestry. This figure included pairs where the Y chromosome haplotype was not identical, but was sufficiently similar to be consistent with a shared ancestry when mutation was taken into account. Of 16 pairs with absolutely identical haplotypes, all but one pair had a surname whose population frequency was in the lower half of surname frequencies.

10.3.4 Amplification of DNA from Neanderthal fossils

Studies of protein polymorphisms, mtDNA and Y chromosomes all generally support the idea that modern human populations worldwide are all derived from a group that migrated out of Africa between 60 and 100 kya. However, it is difficult to rule out that there was a low level of interbreeding between archaic and modern human populations and that a small part of the modern human genome is derived from the genome of archaic humans. Recently it has become possible to test this idea directly because DNA has been amplified from the bones of Neanderthals, the group most likely to have exchanged genes since the archaeological record shows that they coexisted with modern humans in Europe.

DNA from Neanderthal bones can be amplified by PCR using primers designed from the human sequence. Because the DNA is very old and present in small quantities, this process is subject to a number of difficulties:

- Nearly all fossil bones are contaminated with DNA from modern humans who handled the DNA, along with diverse types of DNA that are found in the environment such as that from bacteria and fungi. It is particularly difficult to distinguish sequence from modern human DNA from Neanderthal DNA because the sequences are expected to be very similar. Bone material from beneath the surface can be used to minimise human DNA contamination. DNA extraction and PCR runs have to be carried out in laboratories where human DNA has not been handled and under conditions that prevent contamination from the operatives carrying out the experiment.

- DNA gradually spontaneously degrades in ancient bones. Remaining fragments are usually less than 50 bp long. Before attempting to extract and amplify DNA from a particular bone, its amino acid content can be used to decide whether it is likely to contain a useful amount of Neanderthal DNA: more than 20 ppm amino acids suggests that sufficient biological material has survived for DNA to be extracted.

- As well as degradation, chemical modification can alter the sequence changes. Cytosine is deaminated to uracil, which base pairs with thymine rather than guanine. This results in a C-to-T transition in the sequence. The amount of chemical modification can also be judged beforehand from the degree of amino acid racemerisation.

The first time that DNA was successfully amplified was from the original Neanderthal specimen found in Feldhofer in western Germany. To maximise the chances of seeing differences between the Neanderthal and modern human sequences, two hypervariable regions were amplified, called *HVR*I and *HVR*II. In the combined 600 bp from the two regions, modern humans differ from each other at an average of 11 nucleotides, whereas modern humans differ from the Neanderthal sequence by an average of 35 nucleotides. Thus the difference between the human and Neanderthal sequences is three times the range of variability found within modern human populations. Subsequently, mitochondrial DNA has been analysed from 15 separate Neanderthal specimens isolated from geographically dispersed sites and very similar results were obtained. The fact that mtDNA sequences from geographically dispersed sites gave similar results argues against the initial result coming from an individual who happened by chance to come from outside the normal range. If Neanderthals contributed to the human gene pool, it would be expected that the Neanderthal mtDNA would be most similar to that of modern European populations. However, there was no evidence that this is the case; the Neanderthal sequences were equally different from mtDNA sequences from European populations as they were from mtDNA sequences from Asian and African populations. These results suggest that there has been no contribution of the Neanderthal mtDNA sequence to the modern human mitochondrial sequence. This may indicate that there was no gene flow from Neanderthal to modern humans, but it is important to bear in mind that the mtDNA is not representative of the whole genome. To confirm the absence of gene flow, it is necessary to be able to compare nuclear sequences in Neanderthal DNA and modern humans.

The powerful massively parallel sequencing technologies described in Chapter 4 can be applied to sequencing DNA fragments in Neanderthal bones. DNA fragments from the bones are cloned into vectors using primers that have an additional sequence that, at a later stage, can be used to confirm that a DNA sequence was derived from the DNA extracted from the bone and not a contaminant introduced after the initial cloning. The whole library is then sequenced using 454 pyrosequencing (see Chapter 4) to produce a huge database in which each DNA fragment extracted has been sequenced many times. This allows an accurate consensus sequence to be generated based on many independent reads. The task then is to identify which sequences are genuinely Neanderthal and which are modern human contaminants. One helpful indicator of the amount of human contamination is based on known differences in the mitochondrial DNA HVR regions. DNA is amplified and sequenced using primers that flank HVR regions with known differences. The proportion of reads that produce the modern human mitochondrial sequence, as opposed to the known Neanderthal sequence, gives an estimated amount of contamination from modern human DNA. This information can be used to pick a sample showing a low level of human DNA contamination.

This methodology has allowed the whole of the Neanderthal mitochondrial genome to be sequenced. DNA was amplified from samples of Neanderthal bone from Vindija, in Croatia. Mitochondrial sequences were identified by comparing the sequences with the sequence of the modern human mitochondrial genome. These were then assembled into a complete sequence with 35-fold coverage. When sequences which are known to differ between the human and Neanderthal HVR regions were examined, 229 out of 230 reads were the Neanderthal sequence, giving a high degree of confidence that the rest of the assembled sequence was genuinely from Neanderthal mtDNA.

Altogether, the Neanderthal genome was 16 565 bp in size. This is 4 bp shorter than the modern human genome owing to an identified CACA deletion. Apart from this deletion there are 206 differences between the Neanderthal sequence and the reference human sequence. The maximum number of differences that has been observed between pairs of human mitochondrial genomes is 118. This shows that the Neanderthal falls well outside the range of variation of mtDNA sequence seen in modern human populations. The rate at which mutations have occurred in mtDNA during primate evolution can be estimated by comparing human and chimpanzee mtDNA sequence, and assuming that the MRCA of chimpanzees and humans lived 6 million years ago. Using this molecular clock it is estimated that Neanderthal and human mtDNA diverged 660 kya. This is two to three times longer than the time when the MRCA of modern humans lived. An intriguing observation was that when the modern human and Neanderthal mtDNA protein-encoding sequences were compared, using the chimpanzee mtDNA sequence as a reference, the rate of non-synonymous amino acid substitutions in the Neanderthal lineage is higher. This suggests that purifying selection to remove deleterious mutations was apparently less strong during Neanderthal evolution. An interesting explanation of this observation is that the effective population size of Neanderthals was very small.

Analysis of mtDNA suggests Neanderthals made no contribution to the modern human gene pool. However, as we noted above, it is not safe to base this conclusion solely on mtDNA because it might not be representative of nuclear DNA. However, the technology used to sequence the Neanderthal mitochondrial genome can also be used to generate nuclear DNA sequences. In order to do this, it is necessary to use a Neanderthal bone material that is free of contamination with modern human DNA. This is very difficult as most Neanderthal bones were discovered before DNA sequencing was possible or even thought of as a possibility and so appropriate precautions were not taken. Thus they have not been handled in a way that would prevent this contamination. As we discussed above, the extent of modern human contamination can be measured by the proportion of the recovered DNA that shows the modern human sequence instead of the Neanderthal sequence in mitochondrial HVR. Neanderthal nuclear DNA is being recovered that is apparently free of extensive human contamination and the sequence of nuclear DNA is being determined in this way. There is a really exciting prospect that it may be possible to generate the sequence of the whole Neanderthal genome. Initial results from nuclear sequencing of

Neanderthal DNA from different laboratories were contradictory. However, the cause of the problem has been traced to contamination with modern human DNA. There is now agreement that the nuclear Neanderthal DNA sequence broadly supports the conclusions from mtDNA sequencing; that is, that the lineages leading to modern humans and Neanderthals diverged between 600 and 700 kya, which is about three times the earliest date for the common ancestor of all modern humans. While it seems that Neanderthals did not make a large contribution to the modern human gene pool, it is difficult to exclude the possibility that that they made a small contribution, and whether they did or not is still a matter of active research and has generated considerable controversy.

10.4 Disease frequencies in different populations

We saw above that selection would normally act to restrict the frequency of alleles that cause severe monogenic diseases to be close to the mutation rate. There are a number of diseases where in certain populations this is manifestly not the case. Table 10.1 lists some examples. There are two main reasons why disease frequencies in one population may be higher than expected – heterozygous advantage and founder effects/inbreeding.

Table 10.1 Examples of elevated frequencies for monogenic diseases in particular populations. The low frequency of type 1 (insulin dependent) diabetes in Japan is given by way of comparison to the high level in Scandinavia.

Disease/allele	Population	Frequency/10^3 population	Possible cause
Sickle cell disease	Africans	10–20	Heterozygous advantage
Thalassaemia	Africans/Mediterraneans/ South East Asia	10–20	Heterozygous advantage
Cystic fibrosis	N. Europeans	0.5	Heterozygous advantage?
Tay–Sachs disease	Ashkenazi Jews	0.17–0.4	Founder effect or heterozygous advantage?
Gaucher's disease	Ashkenazi Jews		Founder effect
BRCA1 185delAG	Ashkenazi Jews	10	Founder effect
BRCA2 617delT	Ashkenazi Jews	10	Founder effect
Porphyria	South African (white)	3	Founder effect
MLH1 (HNPCC susceptibility)	Finnish	14 families	Founder effect
Polydactyly	Amish community (US)		Founder effect
Type 1 diabetes	Scandinavia	2	Environment and
	Japan	0.03	genetic effects

10.4.1.1 The haemoglobinopathies

Heterozygous advantage occurs when an allele that is deleterious as a homozygote is advantageous as a heterozygote. This results in a **balanced polymorphism** where selection against the allele in the homozygous state is balanced by selection for the allele in the heterozygous state. The high frequency of haemoglobinopathies in countries where malaria is endemic is a balanced polymorphism. The geographical distribution of malaria correlates with the frequency of sickle cell anaemia and the thalassaemias. Haemoglobin S (HbS) is responsible for sickle cell anaemia (see Chapter 4). HbS differs from the wild type protein by a substitution of glutamate by valine at residue 6 in the β-globin polypeptide. HbS crystallises at low oxygen tensions and as a result the red blood cells collapse into a sickle shape. HbS heterozygotes are clinically normal but are resistant to malaria for reasons which are not currently known. The allele frequency of HbS has equalised to 0.1 in a number of different geographic locations. This exactly corresponds to the expected value calculated from the observed Darwinian fitness of the three different genotypes.

There are four African regions where the allele frequency is at a maximum: Bantu, Benin, Senegal and Saudi. All HbS alleles are the result of the same nucleotide substitution. However, the linked haplotype is different in each region and corresponds to a haplotype that is locally common in each case. This suggests that the mutation occurred on four separate occasions, although it is possible that it occurred once and has been introduced to the four different genetic backgrounds by migration.

Heterozygous advantage is also likely to be responsible for the elevated frequencies of α and β thalassaemias in areas where malaria is endemic. β thalassaemia is more common in Europe and Africa and less common in Asia. Malaria used to be endemic in Mediterranean countries, although it has now been eradicated. This explains the high frequency of β thalassaemia found around the Mediterranean. Indeed, thalassaemia is derived from the Greek word *thalassa*, meaning sea.

10.4.1.2 Cystic fibrosis

Among Northern Europeans the incidence of CF is about 1 in 2000. On average, 70% of CF mutations are the ΔF508 allele. This allele shows linkage disequilibrium with three nearby microsatellites, which allows the haplotype of the founder chromosome to be deduced. Based on the number of mutations that have occurred in the linked microsatellites, there is a controversial estimate that the mutation occurred 52 000 years ago. The ΔF508 founder haplotype is rare in modern European populations; the next two most frequent CF mutations also occurred on a chromosome with the same haplotype, which is rare in Europe. This suggests that the common CF mutations occurred in a population that had a different genetic background from modern Europeans. The most likely scenario is that they occurred in the first Paleolithic human population in Europe, after the founder group

had split off from the main stock of modern humans expanding out of
Africa. Eventually, as we saw above, this population was absorbed by the
expansion of Neolithic farmers from the Middle East.

Why has the ΔF508 mutation maintained such a high frequency? Founder
effects could have been responsible for a high frequency initially, but the
minimum age of the mutations means that selection against the homozygote
would long ago have reduced the allele frequency below the present figure of
0.02. The likely explanation is heterozygous advantage. It is possible that
reduced CFTR function in the heterozygote protects against infantile diarrhoea, because it could reduce water loss across the intestinal mucosa.
Recently, there has been support for this hypothesis from experiments in mice.
If this is the explanation, it is interesting that more than one CF mutation
occurred and was selected for in European Paleolithic populations, but similar
mutations were not established elsewhere in the world. Perhaps this reflects
differences in the environment or lifestyle of the early Europeans.

10.4.2 Founder effects

Founder effects occur when a small group, separated from the main stock,
expands in number. As a result, disease alleles present by chance in the
founder population may have increased in frequency under the influence of
drift and the increased inbreeding that would occur in small isolated
groups. This can result in local populations that have a particularly high
frequency of a monogenic disorder that is rare elsewhere. Sometimes these
populations originated by migration of founder groups in historical times
and the identity of the individuals who carried the disease allele is known.

10.4.2.1 The Ashkenazi Jews

Table 10.1 shows that a number of different diseases are more common in
Ashkenazi Jews than elsewhere and are maintained at a higher rate than
can be explained by mutation. It has been suggested that the high rate of
Tay–Sachs disease is a result of carriers being more resistant to tuberculosis,
which was endemic in the tenements of Eastern Europe where Ashkenazi
Jews lived. An alternative explanation is that the high frequency of all of
these diseases is explained by a **founder effect**. The *BRCA*1 185delAG and
*BRCA*2 617delT mutations are extremely rare outside this population group
and so are almost certainly founder mutations.

The Ashkenazi Jews originated in the vicinity of Strasbourg in eastern
France in the early Middle Ages. They expanded eastwards at the time of
the Crusades and subsequently to America in the late nineteenth and early
twentieth centuries. During this time their culture and religion operated to
maintain an effective reproductive isolation from neighbouring populations.
From the time of their migration away from Strasbourg to the present day,
their population number would have increased several thousand-fold. They
thus conform to a classic founder population. Selection against homozygotes would not have had sufficient time to operate to reduce the frequency
of deleterious alleles. Furthermore, the *BRCA*1 and *BRCA*2 mutations

Human population genetics and evolution

would have limited effect on fitness since they have their effect for the most part after childbearing age.

10.4.2.2 Complex disease

Table 10.1 also shows that a complex disease such as type 1 diabetes mellitus is unevenly distributed in different ethnic populations. Part of this variation could be environmental in origin. However, we saw in Chapter 6 that the HLA alleles *DR3* and *DR4* contribute about 40% of the genetic risk. These alleles themselves show variation in allele frequency between populations, so susceptibility to common diseases may vary between populations because of their different genetic structures. It may be that different alleles at the HLA locus are selectively neutral and that ethnic variation reflects the effects of genetic drift. Alternatively, they have been selected for because they conferred resistance to a disease that was historically more common in certain areas.

In the case of diabetes, the situation is even more complex. The *DR3/DR4* alleles do not themselves cause the susceptibility to diabetes. They are in linkage disequilibrium with the real culprits. So in people of Northern European extraction, the *DR3/DR4* alleles act as markers of the chromosomes that carry the susceptibility alleles. In other populations, the susceptibility alleles show different linkage configurations and the association between *DR3/DR4* and diabetes is not so marked.

10.5 Summary

- Present variation within and between human populations allows the reconstruction of the past history and evolution of human populations.

- A number of genetic observations suggest that present-day human populations are derived from a small population of about 10 000 individuals that lived in Africa about 100 kya.

- Genetic variation in *Homo sapiens* is very low by comparison to other species such as chimpanzees. This lack of variation is seen whether nuclear polymorphisms, mtDNA or Y chromosome sequence is studied. This low level of variation suggests that the long-term effective population size of humans is about 10 000 individuals. A rapid expansion in population numbers occurred 50–100 kya. There was probably a severe bottleneck in the European lineage.

- About 90% of the genetic variation occurs within rather than between populations, consistent with a recent expansion from a single ancestral population. The greatest amount of variation occurs within African populations, suggesting an African origin.

- Phylogenetic trees constructed from nuclear polymorphisms, mtDNA and Y chromosomes all have their deepest root in Africa, suggesting an African origin.

- PCR amplification of mtDNA from fossilised Neanderthal specimens indicates that they were not ancestral to present-day human populations. Using massively parallel sequencing technologies it might be possible to derive the sequence of the whole Neanderthal genome.

- Many genetic diseases are present in certain populations at rates which greatly exceed the level at which new alleles are introduced by mutation. This can occur for two reasons – heterozygous advantage and founder effects.

- Haemoglobinopathies are present at a high level where malaria is endemic, because carriers are more resistant to the disease. As a result, selection against alleles in the homozygote is balanced by selection for the allele in heterozygotes.

- Cystic fibrosis is much more frequent in Europe than other areas. The most frequent allele, $\Delta F508$, was present in mesolithic populations absorbed by the advance of Neolithic farmers. It is possible that the high frequency of CF alleles is also due to heterozygous advantage.

- Many populations founded by small groups of migrants have elevated frequencies of otherwise rare diseases. The Ashkenazi Jews have elevated frequencies of several different diseases that probably arose through founder effects. Some argue that Tay–Sachs disease is an example of heterozygous advantage.

- Susceptibility to complex diseases such as diabetes is also affected by population variation in allele frequencies. These may reflect variation of previously neutral alleles owing to random genetic drift, or the susceptibility alleles may influence resistance to infectious diseases.

Further reading

General aspects of population genetics
CAVALLI-SFORZA, L.L., MENOZZI, P. and PIAZZA, A. (1994) *The History and Geography of Human Genes*. Princeton University Press, Princeton, NJ.

Detailed description of the theory of population genetics and review of nuclear polymorphisms.

HARPENDING, H.C., BATZER, M.A., GURVEN, M. *et al.* (1998) Genetic traces of ancient demography. *Proceedings of the National Academy of Sciences USA*, **95**, 1961–1967.

Provides a detailed mathematical description of how past demographic changes can be deduced from the genetic structure of present-day populations.

JORDE, L.B., BAMSHAD, M. and ROGERS, A.R. (1998) Using mitochondrial and nuclear DNA markers to reconstruct human evolution. *Bioessays*, **20**, 126–136.

PAABO, S. (1999) Human evolution. *Trends in Cell Biology*, **9**, M13–M16.

Two general reviews of what is known about recent human evolution.

STONEKING, M. (2008) Human origins. The molecular perspective. *EMBO Reports*, **9**, S46–S50.

A review of the different theories of human evolution and the contribution of molecular genetics to resolving the controversies.

WOLPOFF, M.H., HAWKS, J. and CASPARI, R. (2000) Multiregional, not multiple origins. *American Journal of Physical Anthropology*, **112**, 129–136.

Reviews the arguments for the multiregional hypothesis.

CANN, R.L., STONEKING, M. and WILSON, A.C. (1987) Mitochondrial DNA and human evolution. *Nature*, **325**, 31–36.

The original paper that described the out-of-Africa evolution of early humans based on mtDNA genealogies.

Y chromosome evolution
JOBLING, M.A. (2001) In the name of the father: surnames and genetics. *Trends in Genetics*, **17**, 353–357.

A review of the correlation between Y chromosome genotypes and surnames.

JOBLING, M.A. and TYLER-SMITH, C. (2003) The human Y chromosome: an evolutionary marker comes of age. *Nature Reviews Genetics*, **4**, 598–612.

Gives a detailed description of the molecular architecture of the human genome and a consensus Y chromosome genealogy drawn from a number of different studies.

KING, T.E., BALLEREAU, S.J., SCHURER, K.E. and JOBLING, M.A. (2006) Genetic signatures of coancestry within surnames. *Current Biology*, **16**, 384–388.

Research paper that shows that the correlation between surnames and Y chomosomes holds only for rare surnames.

SYKES, B. and IRVEN, C. (2000) Surnames and the Y chromosome. *American Journal of Human Genetics*, **66**, 1417–1419.

Men with the surname Sykes share the same Y chromosome.

THOMAS, M.G., SKORECKI, K., BEN AMI, H. *et al.* (1998) Origins of Old Testament priests. *Nature*, **394**, 138–140.

Shows that Jewish males with the surname Cohen descended from a single male ancestor who lived 2500–3000 years ago.

Neanderthal DNA sequencing
GREEN, R.E., KRAUS, J., PTAK, S.E. *et al.* (2006) Analysis of one million base pairs of Neanderthal DNA. *Nature*, **444**, 330–336.

GREEN, R.E., MALASPINES, A.S., KRAUSE, J. *et al.* (2008) A complete Neanderthal mitochondrial genome sequence determined by high-throughput sequencing. *Cell*, **134**, 416–426.

NOONAN, J.P., COOP, G., KUDARAVALLI, S. *et al.* (2006) Sequencing and analysis of Neanderthal genomic DNA. *Science*, **314**, 1113–1118.

CHAPTER ELEVEN

DNA profiling in forensic criminology

Key topics

- Current markers used for DNA profiling
- DNA databases
- Difficulties in the legal interpretation of DNA profiles
  - Statistical issues
  - Sample contamination
- Oversights of forensic DNA profiling

11.1 Introduction

The multitude of SNPs, polymorphic minisatellite and microsatellite loci means that the genome of every individual is unique and so can be used as a means of personal identification. This was first demonstrated by Alec Jeffreys, a geneticist working in Leicester, UK. The original demonstration was based on using Southern hybridisation to visualise the variation in the length of minisatellites. The first experiments used a core sequence common to many minisatellites and produced complex patterns that were difficult to interpret apart from the obvious except for one important respect – the result from any two individuals was clearly different, i.e. the banding consti-tuted a genetic fingerprint. Using single locus probes produced simpler, more interpretable patterns. The technique was quickly applied in an immi-gration case to prove to the Home Office that a Ghanaian boy really was the son of a couple living in this country and to resolve a paternity dispute.

The first application to forensics was in 1986. Two murders of young women had been committed in a village called Enderby, near Leicester in the UK. In both cases the victim had also been raped. The police had arrested a suspect who had confessed to the crimes. In fact DNA

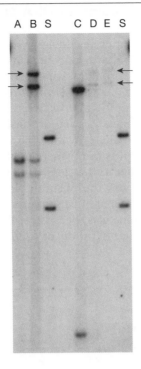

Figure 11.1 The first application of DNA fingerprints in forensics. Autoradiogram resulting from a single locus minisatellite probe to analyse DNAs as follows. A, From root hairs of the first victim; B, mixed semen and vaginal fluid from the first victim; C, blood from the second victim; D, vaginal swab from the second victim; E, semen stain from the second victim's clothes; S, blood from the first suspect. Arrows show semen alleles common to both murders that do not originate from either victim, but also do not match the alleles from the suspect. (Reprinted by permission from Macmillan Publishers Ltd: from Jeffreys, A. (2005) *Nature Medicine*, **11**, 1035–1039.)

fingerprinting showed he was innocent (Figure 11.1). The real perpetrator was caught because he tried to evade giving a sample when police asked all local men to volunteer to have their DNA tested. The case established that the technique could be used to identify criminals by tiny amounts of biological material they leave behind at the crime scene. This concept has since revolutionised forensic science and has become an essential tool in the fight against crime. It is also used in other areas such as establishing paternity and identifying victims of war crimes and mass disasters. Jeffreys' original technique has now been replaced by a technique which uses PCR to amplify microsatellites producing simple quantifiable patterns. It is now called DNA profiling rather than fingerprinting. This chapter describes DNA profiling and how it is used in forensics and other applications, including databases against which evidence from crime scenes can be checked. The evidence used in court to convict an accused person must be of the highest standard. We will discuss the difficulties which are encountered and the precautions which must be taken to meet this standard.

11.2 Current markers used for DNA profiling

The original DNA fingerprinting technique involved the isolation of a substantial quantity of DNA and a labour-intensive Southern blot procedure using P-32-labelled probes. It could only be carried out by specialised laboratories in a limited number of cases. An important technical advance was the adoption of PCR to amplify microsatellites or short tandem repeats (STRs). The number of repeats in each of these loci is highly variable so that the likelihood that two alleles will be the same is low. To genotype a DNA sample, multiplex PCR is used where a mixture of primers amplify a number of different STR loci in the same tube (Figure 11.2). The primers are fluorescently labelled and the position where they anneal carefully chosen so that the size of the amplified products can be conveniently separated on automated sequencing gels. Because of the high resolution afforded by the sequencing gels, the migration of each product can be precisely

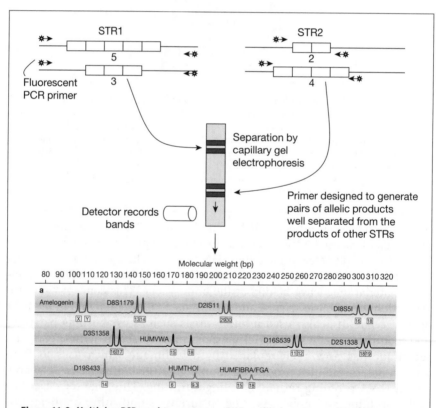

Figure 11.2 Multiplex PCR used to generate a DNA profile based on multiple short tandem repeats (STRs). Primers are chosen so that amplified DNA from the two alleles of each STR form a pair of peaks that are well separated from the peaks of the other STRs when the mixture of DNAs is subject to high-resolution electrophoresis.

quantitated. Each locus produces a pair of peaks, one from each of the polymorphic alleles. The migration of each peak determines the number of repeats in that allele, so that the whole electropherogram yields a digital profile that can be stored in a computerised database. PCR amplification of STRs is also much more sensitive, so that only about 10 ng of DNA is required compared with the 250 ng required for the original Southern blot experiment.

Using a limited number of loci means that the pattern produced from each individual is not necessarily unique, so the result is now referred to as a DNA profile rather than a DNA fingerprint. The odds that any two different DNA samples would produce the same profile are the product of the odds that each individual STR locus is identical, which in turn depends on the different allele frequencies in the population. The odds of a spurious association reduce as the number of STRs profiled increases. The actual number used varies between the different national jurisdictions. In the UK a panel called SGM+ is used which consists of 10 STRs, which is estimated to produce a spurious match probability of less than 10^{-13}. In the USA a panel called FBI CODIS consists of 13 STR loci, which has an even lower match probability. To aid international exchange of data, a reduced set of seven core loci is used in Europe by Interpol.

As well as the autosomal STR loci, all panels also include a pair of primers that amplify a sequence from the loci encoding amelogenin, which distinguishes the sex of the DNA sample. Amelogenin is an extracellular matrix protein found in tooth enamel. One copy of the encoding gene, *AMELX*, is found on the X chromosome and another copy, *AMELY*, is found on the non-recombining region of the Y chromosome. *AMELX* contains a 6-bp deletion in intron 1. The primers used in the panels amplify across this region so that the amplified product from *AMELY* is 6 bp larger than that amplified from the *AMELX*. This produces two peaks on the electropherogram from males, whereas females only produce a single peak.

Most serious crimes are committed by men, so often forensic investigation will require the male DNA to be distinguished from the female DNA of the victim. In rape cases, this is achieved by a technique called differential lysis, which separates out the male sperm DNA. However, this may not be possible if the rapist has been vasectomised, wears a condom or does not naturally produce sperm (azoospermic). In these cases, traces of the male DNA must be profiled in the presence of a great excess of the victim's DNA. STRs on the Y chromosomes can be used in this situation. Because the Y chromosome is non-recombining, variation is only produced by mutation (see Chapter 9). As a result, the number of different haplotypes is limited and the chances of a match are much higher than would be the case with independently segregating autosomal STRs. Mitochondrial DNA is present at 2000 copies per cell, so it is much easier to amplify from small amounts than autosomal loci. This makes it useful in circumstances where the amount of DNA is limited because of degradation. However, like the Y chromosome, mitochondrial DNA is non-recombining so the number of different haplotypes is reduced compared with autosomal loci.

11.3 DNA databases

The electropherogram produced by panels of STR loci such as the SGM+ can be summarised by an indicator for gender and a set of 20 numbers describing the number of repeats at each of the two alleles at the 10 auto-somal loci. This makes the information easy to store and search on a central DNA database. When a crime has been committed that yields a DNA profile, the profile can be searched against this database. Such databases have now been established in many countries. The size of the database and criteria for inclusion of samples varies. The UK has the largest database because it has the most liberal criteria for inclusion: anyone arrested for a recordable offence is required to provide a sample, usually a mouth swab, regardless of whether they are found innocent or even charged and regard-less of whether the offence is sufficiently serious to warrant a prison sentence. In addition, samples are kept from volunteers who give samples as part of a particular investigation, provided they give their written permis-sion, which, once given, is irrevocable. Such samples could be taken from crime victims or for elimination purposes from members of a population group in respect of a specific crime. At the end of 2006 there were 4 million subject samples and 385 000 crime scene samples in the UK database and it was growing at a rate of 700 000 per year. When a sample from the scene of a violent crime is compared with the database, there is a 52% chance that it will produce a match. There are many examples of murder, rape and other violent crimes being solved in this way. Some examples are given in Box 11.1. As of June 2008 The US National DNA database contained six million offender profiles and 225 000 crime scene profiles. In total, it had made 71 000 matches.

DNA databases provide the opportunity to reopen old cases where DNA can be recovered from archived evidence. Normally this requires a match to be found in the DNA database; however, in exceptional circum-stances a match can be searched for among close relatives (called family searching). For example, a sibling of the perpetrator would be expected to match 50% of the profile at a crime scene. Thus, if such a close match is found in the database, the relatives can become suspects even though their DNA is not in the database. The conviction of the Yorkshire 'shoe rapist' illustrates both these points (see Box 11.1).

The liberal criteria for inclusion in the UK DNA database is controversial. It means that 5% of the total population, and perhaps 30% of young men, are classified as potential criminals even if their inclusion arises from minor offences. However, the examples given in Box 11.1 illustrate how DNA samples provided after driving offences led to conviction for much more serious offences. Alec Jeffreys argues that the whole population should be profiled and stored on the database to prevent discrimination against cer-tain population groups who have more contact with the police and criminal system. Another controversial issue is whether the DNA itself can be stored to retrieve further information. For example, some SNP allele frequencies and haplotypes are specific to particular ethnic groups, so it might be possible

BOX 11.1: VIOLENT CRIMES SOLVED BY DATABASE MATCHING

Case 1 Raped in her own home

Early on a Sunday morning in the north west of England a young woman was raped at knifepoint in her own house by a man who entered by an open downstairs window. Around the same time, on the same housing estate, two other women fought off a man who entered their homes. Clearly, a dangerous rapist was active who probably lived locally. PCR profiling of semen samples produced mixed profiles, one of which came from the woman's own DNA. After subtracting this profile, the residual partial profile was used to search the National DNA database, producing two matches. One of these was to a young man who lived locally and his fingerprints matched those found at the scene of the crime. Further work on the partial profile obtained from the semen specimen improved the profile to the point where the other match to the database could be eliminated. The man was charged, convicted and sentenced to 10 years in prison – a sentence that would have been longer if he had been over 18.

Case 2 Operation Nightingale catches the Yorkshire shoe rapist

Between 1983 and 1986 a series of rapes was committed in the Dearne Valley, near Rotherham, Yorkshire. The rapist took shoes from his victims, presumably as trophies, which earned him the nickname 'the shoe rapist'. Twenty years later, Operation Nightingale, a cold case review, obtained DNA profiles from samples taken at the time of the rapes, but there was no match in the National DNA database. However, a search for near matches from people living in the area identified 43 possible close relatives. The third person approached was a woman who had given a sample after she had been arrested for a drink driving offence. She phoned her brother, James Lloyd, to tell him that the police would be visiting to take a DNA sample. Realising the net was closing in he unsuccessfully tried to hang himself. He pleaded guilty to six of the assaults and was sentenced to an indefinite prison sentence.

Case 3 'Wearside Jack' Ripper hoaxer caught after 25 years

One of the most notorious mass murderers in British history was Peter Sutcliffe, known as the Yorkshire Ripper. Between 1975 and 1980 he murdered 13 women and seriously assaulted several others. The murders continued even while a large manhunt was in progress. Police efforts were seriously hampered by two hoax letters and a tape claiming to come from the murderer. The opening words of the tape were 'I am Jack', spoken in a distinctive accent that suggested that the man came from the Wearside area of north east England, hence his nickname, Wearside Jack. Altogether during the manhunt Peter Sutcliffe was interviewed on nine occasions; in retrospect, there were forensic leads that pointed in his direction and one of his friends even reported suspicions to the police. However, because he did not sound like the voice on the tape, Sutcliffe was not seriously suspected. At least three women were murdered because of the misdirection of police efforts by the hoax. In 2005 a fragment of an envelope in which a hoax letter was sent was subject to LCN DNA analysis and a match was found in the National DNA database to John Humble from Wearside who provided a sample in 2000 for an unrelated incident. He confessed and was sentenced to 8 years in prison for four counts of perverting the course of justice.

to infer ethnicity from crime scene samples. We saw in Chapter 6 how certain SNP alleles are associated with red hair and freckling. Similar associations have been identified with respect to eye colour. In Chapter 9 we saw that there is an association between surnames and Y chromosome haplotypes. None of these phenotypes could be reliably inferred from DNA at the moment but, given the speed of progress in associating complex traits with SNP haplotypes, it is not difficult to imagine that it will be possible in the future. Storage of the original samples is permitted in the UK but these further tests are not permitted at present.

11.4 Difficulties in the legal interpretation of DNA profiles

11.4.1 Statistical issues

In 1999 a man was convicted in the UK of a burglary on the basis that the profile of DNA recovered from the crime scene was matched to his profile, which he gave after police investigated an incident where he hit his daughter in a family row – although he was never charged or convicted of this offence. At that time six STRs were used and the probability of a match by chance was stated to be one in 37 million. There was no other evidence linking him with the burglary. The crime was committed 200 miles from where the man convicted lived. He suffered from advanced Parkinson's disease and was so disabled that he could not drive; indeed, he had difficulty dressing himself. Moreover, he had a corroborated alibi for the time the burglary was committed. When the SGM+ panel was introduced using 10 STR markers, the defendant's lawyers requested further testing that conclusively proved his innocence.

The important point about this example is that the original tests were indeed correct; although a false match seemed unlikely, it had actually occurred on this occasion. The odds of one in 36 million do not mean that there will be only one match in 36 million, and when a match occurs the crime scene sample and the database sample must have come from the same person. It is perfectly possible that there could be several matches among a population of 36 million. The population of the UK is about 60 million. Given that the probability of a match by chance was one in 37 million, then if every single person living in the UK had been tested on this occasion, the chances of a false match by chance were high. Of course, the whole population was not tested, only the 660 000 people who at that time had samples in the DNA database. Nevertheless, it is perfectly possible for there to be a match by chance in that smaller sample, especially considering that one in 36 million is only a crude estimate and could easily be out by an order of magnitude. The use of the SGM+ panel has increased the odds of a random match to approximately 10^{-13}; however, a match by chance is still possible. Does this give a blanket defence against DNA profile evidence? The fallacy in this argument is that not every person in the UK is equally likely to have committed the crime. What is necessary is a way of incorporating additional evidence to that provided by DNA profiling. This can be done using Baysian probabilities, which incorporate all the relevant data and calculates the relative likelihood of the different defence and prosecution scenarios.

DNA profiling provides intelligence as to who is likely to have committed the crime. These arguments show the need for additional corroborating evidence to obtain a conviction – something that was conspicuously missing in the burglary case cited above. Often, when confronted with the DNA profiling evidence, the suspect will confess. In the example of the woman who was raped in her own home described in Box 11.1, the DNA profile identified a man living in a nearby street. Since there was evidence that the perpetrator lived locally from the other attempted attacks, this information clearly adds weight to the evidence from the match in the database. In fact, the match of his fingerprints in the house where the rape took place provided a further level of certainty.

Some situations may increase the chance of a random match. One issue that received considerable attention when DNA profiling was first introduced was that particular alleles may be more common in particular ethnic groups. To safeguard against this, the allele frequencies of all the STRs are carefully documented in different ethnic groups and match probabilities are adjusted in a conservative fashion, i.e. in the defendant's favour. The chance of a false match may also be high if the suspect and perpetrator are related.

11.4.2 Sample contamination

PCR is a powerful tool that will amplify tiny amounts of DNA. It is therefore important that the samples are appropriately handled to prevent cross-contamination. This issue is particularly important in a technique called **low copy number** (LCN) amplification. This technique attempts to amplify DNA that is present from only a small number of cells that might be left by simply touching an object. The extra power is obtained by increasing the number of PCR amplification cycles. It is a very difficult and unreliable technique, which is prone to two particular problems: cross-contamination and allele drop out, which is where some of the STRs fail to amplify. Allele drop out increases the odds of a spurious match because fewer STRs can be used. Because such a tiny amount of DNA is detected, cross-contamination may occur by DNA being transferred from a suspect to a piece of evidence via a third party such as the investigating police officer. So samples used for LCN analysis have to be treated with scrupulous care. How the sample was recovered from the crime scene, who recovered it, who subsequently handled it and what precautions were taken to prevent cross-contamination all have to be fully documented. Because of the technical problems, LCN is not used as evidence in court in the USA but is used only for intelligence purposes to mark a suspect for further investigation. In the UK, LCN is admissible in court, but it was strongly criticised by a judge in a high-profile trial of Sean Hoey, accused of making bombs used in a series of terrorist bombings in Northern Ireland, including the Omagh car bombing which killed 29 people and injured hundreds of others. A central piece of evidence against Sean Hoey was that in some of the bombings, LCN had identified his DNA profile on pieces of bombs recovered after the explosions. However, the defence successfully demonstrated that the samples used for LCN analysis had not been handled with sufficient care and there was no documentation as to who had handled the samples before the LCN analysis was carried out. Moreover, the written judgment from the trial stated that the technique had not been sufficiently validated to meet the high evidential standards necessary for a conviction in a court of law. According to the judgment, validation means that reliability and accuracy of the procedure had been assessed, that the laboratory procedures necessary to obtain reliable results had been determined and that the limitations of the procedure had been defined. However, a subsequent Home Office enquiry found that LCN was acceptable as legal evidence.

11.5 Oversight of forensic DNA profiling

Clearly, procedures for generating DNA profiles and the way that they are assessed has to be standardised so that the scientific procedures are carried out in such a way that they meet the high standard of evidence required in criminal courts and are not open to dispute in every criminal case. In the UK, DNA profiles are generated in the laboratories of the Forensic Science Service (FSS), a government-owned company, and a commercial company called Orchid Cellmark. The FSS runs the DNA database on a day-to-day basis, overseen by a custodian in the Home Office, who also supervises the standards and laboratory procedures used during the generation of profiles. Strategic oversight of the DNA database is provided by a board consisting of interested parties such as the Home Office, the police and the Human Genetics Commission representing lay interests. The Strategic Board produces a report each year. At the time of writing (August 2008), the latest report available was 2005–06, and statistics and some case studies quoted in this chapter are taken from this report.

In the USA, the FBI CODIS unit (combined DNA index system) runs the DNA database, provides quality assurance for the collection of DNA profiles in local crime laboratories and advises the American government on legislation relating to DNA profiling. International collaboration is fostered by CODIS and bodies such as the European Network of Forensic Science (ENFSI). Such collaboration exchanges information on STR allele frequencies and seeks to develop a common set of core STRs to allow the international exchange of information and profile matching.

11.6 Summary

- DNA profiling is based on PCR amplification of short tandem repeats (STRs) and separation of the products on DNA sequencing gels. The result is a series of peaks, the position of each being dependent on the number of repeats at each STR allele. This information can be readily digitised.

- The chances that two people will share the same profile is very low, so if DNA is left by the perpetrator at the scene of a crime, its profile potentially identifies the perpetrator.

- National DNA databases have been set up storing the DNA profiles of criminals (subject profiles) as well as DNA profiles recovered from crime scenes. In the UK, the DNA profile of any person arrested, regardless of subsequent conviction, is also stored in the database.

- Matches between crime scene DNA and subject DNAs have now solved thousands of crimes, including old crimes where profiles have been recovered from archived evidence.

- Although the probability of a match by chance appears low, it can and does happen. Thus it is important to consider carefully whether this is likely to have happened and, if possible, DNA profiling evidence should be supported by non-DNA evidence.

- Crime scene evidence has to be handled carefully to avoid cross-contamination. This is especially important in low copy number amplification where the amount of DNA amplified is so small that it can be transferred by touch.

Further reading

JOBLING, M.A. and GILL, P. (2004) Encoded DNA evidence: DNA in forensic analysis. *Nature Reviews Genetics*, **5**, 739–751.

A detailed review of DNA profiling, its application to forensic science and problems which must be addressed when it is used as evidence in criminal proceedings. Also contains examples of other uses of DNA profiling.

National DNA Database report 2005–06.

http://www.homeoffice.gov.uk

Contains statistics of the UK National Database as well as many case studies, illustrating its use in solving current crimes and cold cases.

JEFFREYS, A. (2005) Genetic fingerprinting. *Nature Medicine*, **11**, 1035–1039.

A personal account of how he discovered the technique and its early application to immigration and paternity disputes, and forensics.

MCCARTNEY, C. (2008) LCN DNA: proof beyond reasonable doubt? *Nature Reviews Genetics*, **9**, 325.

Describes the Sean Hoey case and the criticisms of LCN by the presiding judge.

Online links

STRBase
http://www.cstl.nist.gov/biotech/strbase/index.htm

Contains information about the STR panels used in different jurisdictions.

http://www.forensic-evidence.com/site/EVID/EL_DNAerror.html

Describes the case of the false conviction based on a chance match to the database and discusses the associated statistical issues.

FBI CODIS

http://www.fbi.gov/hq/lab/html/codis1.htm

Information and statistics of the US National DNA Index.

CHAPTER TWELVE

Human genetics and society

Key topics

- Potential dangers of genetics to human society
- Gene testing
  - ○ Prenatal testing
  - ○ Neonatal screening
  - ○ Diagnosis of genetic disease in children after birth
  - ○ Presymptomatic testing for late-onset diseases
  - ○ Presymptomatic testing for predisposition to complex diseases
- Human rights
  - ○ Insurance and employment
  - ○ Privacy and ownership of genetic information
- Patents
- Gene therapy
- Genetic determinism and individual responsibility

12.1 Introduction

This book began with a consideration of the influence that our genetic constitution has over our health. We discovered that single-gene disorders were responsible for much serious childhood ill health, resulting in reduced life expectancy and quality of life for those affected and immense suffering, distress and sorrow both to sufferers and to their families and carers. Furthermore, there are few diseases of later life not influenced in some part by genetic factors. The main part of this book has been concerned with an epic period of scientific advance where, in a few years, we have sequenced the human genome and equipped ourselves with the technology to understand the molecular defects that cause these diseases, to devise efficient means for their diagnosis and to realistically contemplate novel ways to prevent or treat them. Over the next few decades this may well revolutionise medicine and fundamentally alter the way we think of ourselves.

At the same time, these advances in human genetics will have a profound impact on society; therefore, as a society, we must consider carefully how we will respond, ensuring that we use our new-found knowledge to relieve suffering and promote good health whilst avoiding the very real possibility that it could have negative consequences. This chapter explores the nature of these ethical, legal and social issues. It is important to realise that many of these issues are not in any way abstract or far off. They affect us either now or in the very near future.

This chapter is about the ethical issues that arise from human genetics. There has been considerable controversy about the ethical and social implications of human cloning. Although an issue that relates to human genetics, it has developed from separate fields, namely reproductive physiology and developmental biology. It does not use any of the techniques or knowledge described in this book. Although it could be combined with somatic gene therapy at some point in the future, this is not an immediate possibility. Therefore, the possibility of human cloning will not be considered here.

12.1.1 Research and reports into ethical, legal and social issues

There have been a number of formal investigations into these problems by groups comprising scientists, medical practitioners and moral philosophers. One of the goals of the Human Genome Project is to carry out research on the ethical, legal and social issues (ELSI). This has resulted in a number of reports on aspects such as gene testing, employment rights and privacy. These reports are available online from the HUGO website (see Further reading). In the United Kingdom, the House of Commons Select Committee has produced an excellent general report on human genetics and a second specific report on genetic testing and life insurance. Both of these are also available online (see Further reading).

12.1.2 Regulatory and advisory frameworks

In the United Kingdom an extensive framework has been established to provide independent advice to the government on developments in human genetics, to regulate the use to which these developments are put and to provide a mechanism to channel input from different sections of society and interest groups. It is hoped that this will avoid the distrust and problems that were associated with the introduction of genetically modified foods. Details are given in Box 12.1. In the United States, regulation of activities connected with human genetics is less formal, falling largely on local ethics committees. However, the Food and Drug Administration will act to oversee the safety of new treatments as they are introduced.

BOX 12.1: THE REGULATORY AND ADVISORY FRAMEWORK IN THE UK

The Human Genetics Commission (HGC). Provides advice to government on over-all issues relating to human genetics. It has a broad remit, incorporating the impact of human genetics on healthcare and the social, ethical, legal and economic implications of developments in human genetics. It also acts to coordinate the activities of other relevant advisory and statutory bodies, described below. Each of the bodies below includes a member from the HGC. The HGC superseded an earlier body called the Human Genetics Advisory Commission (HGAC) and incorporated a previously independent body that advised separately on gene testing, the Advisory Committee on Gene Testing (ACGT).

The Gene Therapy Advisory Committee (GTAC). Provides advice to government and regulates gene therapy. All proposals for gene therapy must be approved by GTAC.

The Genetics and Insurance Committee (GAIC). Develops criteria for the use of genetic tests in life insurance and examines proposals from the insurance industry to use specific tests. So far (August 2001), it has approved only one test, which was for the diagnosis of Huntington's disease. Further tests for hereditary breast cancer and early-onset Alzheimer's disease are expected in the near future.

Although not directly concerned with human genetics, the following bodies are also relevant and coordinate with the HGC:

The Human Embryology and Fertilisation Authority (HFEA). Regulates *in vitro* fertilisation and related issues such as human cloning.

The Committee on Safety of Medicines. Oversees the safety of medicines, which will be relevant as new treatments based on human genetics are developed.

The United Kingdom Xenotransplantation Interim Regulatory Authority (UKXIRA). Regulates the transplantation of animal organs into humans.

Further details of the terms of reference of these bodies may be obtained from the HGC web site: http://www.hgc.gov.uk/about_regulatory.htm

12.1.3 Eugenics

Deliberations in this area are coloured by the historically bad record that the discipline of genetics has had in the way it has advised and influenced society. In the early part of the twentieth century, there was widespread concern that the genetic fitness of western societies was being eroded by the differential reproduction of what were labelled as 'feeble minded' or 'morally defective' classes. The answer to this was seen to be the improvement of the genetic stock by selective breeding, an approach called **eugenics**. We now appreciate that such philosophies are without any scientific justification and are universally regarded with abhorrence. Changes in the human gene pool can take place only over many generations, in response to selective pressures such as plagues, endemic infections and climatic conditions.

It is difficult to appreciate now the extent to which eugenics held sway among both the scientific and the political classes. Several academic journals of genetics included the word 'eugenics' in the title. Quotations from many renowned statesmen and scientists show that belief in eugenics was considered normal or even responsible for leading members of society. It was not only in Nazi Germany that eugenics was practised. In the 1920s and 1930s a number of states in the USA passed laws for the forcible sterilisation of the 'feeble minded' and tens of thousands of such operations were actually performed. Recently it has been revealed that Sweden carried out over 60 000 sterilisations for eugenic reasons, ceasing only in the 1970s. The eugenics episode serves as an illustration that the ill-advised application of genetics can be exceedingly harmful and emphasises the importance of careful deliberation about the possible consequences of recent advances.

12.2 Genetic testing

Genetic testing refers to any procedure designed to ascertain the status of any aspect of someone's genotype for purposes of diagnosing or predicting disease wholly or partly caused by hereditary factors. This covers tests based on DNA, RNA or chromosomes but may be extended to include protein-based tests or even more broadly based medical examinations. Genetic tests may be carried out in a variety of contexts:

- prenatal testing;
- neonatal screening;
- diagnosis of genetic disease in children after birth;
- screening prospective parents for the carrier status of genetic disorders common in a particular population;
- presymptomatic testing for a serious late-onset disorder;
- presymptomatic testing for genetic predisposition to a complex disease.

Before we consider these different situations in detail, there are some general points that can be made.

1. It is essential that the tests are reliable and the diagnosis clear. This is not as straightforward as it may at first appear. Some genetic tests use linked markers rather than following the mutation itself. Such tests may be falsified by recombination between the marker and the mutation. In addition, many tests look for common mutations and may fail to detect rarer ones. For example, it is straightforward to detect the *CFTR* ΔF508 mutation (see Chapter 8), but there are hundreds of rarer mutations that might affect the gene. Thus a person who may appear to be a ΔF508 heterozygote may in fact be homozygous for inactivating

mutations in the *CFTR* gene. As new tests are developed, there must be a mechanism to ensure that they are subject to independent review so that their effectiveness is known to doctors and genetic counsellors who will order their use. There is also a need for quality assurance in laboratories that carry out the tests. They will be marketed and carried out by private companies for commercial gain, and we may expect a large number to appear over the coming years. Ensuring quality and reliability and preventing exaggerated claims are major priorities.

2. Experimentation on human subjects must be carried out only with the informed consent of the person being tested. This is an obligation enshrined in international law, known as the Nuremberg Code. It was established after evidence in the Nuremberg trials after World War II in response to revelations of medical experimentation in the concentration camps of Nazi Germany. There is a distinction between research and an established medical procedure. Once the validity of a genetic test has been established by research and it is used to diagnose or counsel individual patients, it is no longer experimental and is not covered by the Nuremberg Code. Nevertheless, the consensus is that the requirement for informed consent should still apply when genetic tests are used in a clinical context. Informed consent requires that the subject fully understands the nature of the test and the consequences that may arise from it. This leads to a general need for counselling, both before the test is undertaken and to explain the results when they are available. In some cases, such as a young child or where there is mental impairment, the person most affected may not be in a position to understand the reasons for testing. There is thus the need for the parents or guardian or some other representative to be able to give consent on their behalf. This consent should only be given for immediate medical reasons. Determining the carrier status of a young girl for a sex-linked disease would be an example where this condition is not met. Children's views can be taken into account, and as they get older and their understanding grows their wishes should be given increasing importance.

   Clear counselling is also essential for informed consent in research procedures. This is particularly important in phase I trials of new procedures. These are designed to test the safety of the procedure and to establish a safe dose. They are not designed to test the efficacy, so the subject is unlikely to derive any personal benefit. This may not be clear to the sufferer of the disease, who may undertake the trial in the hope of being the first to receive a new treatment. Because they are designed to test safety, there can be no guarantee beforehand that the trial will be safe. This was tragically emphasised by the death of a volunteer in a phase I trial of a gene therapy procedure (see Section 9.4).

3. Tests should be carried out only when there is some positive action that can be taken as a result. We shall see that these tests can sometimes result in significant detriment to the person tested. Therefore tests can be justified only if they allow the prevention or better treatment of a disease or provide some other genuine benefit.

4. Everybody has the right to know about information that affects their
   health and has the right to expect that this information will be kept
   confidential.

4. Everybody has the right to know about information that affects their
   health and has the right to expect that this information will be kept
   confidential. Thus access to genetic information held on any sort of file
   must be secure and at the same time made available to the individual
   concerned. Alongside the right to know is the right not to know. This
   not only underlines the need for informed consent to any test but also
   emphasises the need for vigilance that unwelcome information is not
   inadvertently obtained or passed on to the individual concerned.
5. Genetic tests have implications for the whole family and may extend
   beyond the immediate nuclear family. Some instances, discussed below,
   may lead to situations where one person's right to confidentiality
   conflicts with another's right to know.

12.2.1 Prenatal testing

Prenatal testing is carried out where there is a known risk of a serious
monogenic disorder or in older women where there is an increased risk of
Down's syndrome and other chromosome abnormalities. This would normally be done in order to provide the opportunity for termination in the
event of an adverse result. Such testing empowers parents, who may not
otherwise risk having a seriously affected child, to reproduce. For example,
where they have already had one affected child, the woman can become
pregnant in the knowledge that the pregnancy can be terminated if the
embryo is affected. The positive value of such tests is clear. However, there
are a number of issues connected with embryo testing.

Prenatal tests can be invasive and carry a certain risk of miscarriage
(1% for amniocentesis). The potential benefit gained from the test must be
commensurate with this risk. From this it follows that tests with a
significant risk should be carried out only where the condition investigated
is serious and the possibility that the embryo is affected is significant.

If prenatal tests are made with a view to termination of pregnancy, it is
important that the parents fully understand this. There is little point and
potentially much harm in testing an embryo if the parents would not wish
to terminate the pregnancy whatever the result. This means that the parents
should undergo counselling before the tests are carried out. Of course, parents retain the right to proceed with the pregnancy in the event of an
unfavourable outcome to the test. Because prenatal testing implies the possibility of a termination, there may be an assumption that members of
certain ethnic or religious groups would not wish such testing because of an
objection in principle to such procedures. It is for the individual concerned
to make such a decision and it is important that the option is always presented.

A number of tests are made routinely in the antenatal clinic. It is possible, even likely, that in such circumstances the prospective mother will not
fully understand the nature of the tests and the possible consequences. She
may then be unprepared for the unwelcome knowledge from an adverse
outcome. This would be particularly serious if she objected in principle to
termination and thus would be forced to complete the pregnancy in the

knowledge that the child would be seriously disabled. At the very least, she may have wanted the opportunity to decide whether she wanted the test before it was carried out. This is not to say that prenatal tests should not be carried out, but that it is important that time is taken to explain their nature to the mother.

Prenatal testing with a view to termination makes a value judgement about the worth of the unborn child. This has two important consequences. First, if it is agreed that certain conditions are sufficiently serious to warrant termination, then there is a danger that this could be construed as making a judgement about the worth of people who have been born who suffer from the disease. Second, since there is a range of disablement that results from genetic disorders, how do we decide what constitutes a sufficiently serious disablement to warrant termination? Most people, but by no means all, agree that a serious condition that will result in a short lifespan with much suffering are sufficient grounds for termination. Many parents would not wish to have a child affected by Down's syndrome. However, parents with such children often enjoy a rich and loving relationship with them. The Down's Syndrome Association in the UK does not consider it as grounds for termination. This seems to be a decision that should properly be left to the parents. How, though, do we react to a diagnosis of a late-acting autosomal dominant disorder where the unborn child can expect to enjoy 40–50 years of normal life? The consensus here is that this is not grounds for termination. If this is the case, it becomes improper to apply the test in the first place, because self-evidently an embryo cannot give informed consent but will have to live many years with the consequences.

The decision to terminate a pregnancy as a result of a genetic test must always be taken by the parents. It is important that the parents do not come under any form of pressure. One such pressure may come from the physician treating them. Another pressure may be the financial consequences of the healthcare needed by the disabled child when it is born. Healthcare is financed in different ways in different countries. In some, such as the UK, it is financed by the National Health Service; in others, such as the USA, it is financed by an insurance-based system. Whatever the system, will it consider the ill health of a child born in such circumstances to be caused by the actions of the parents and thus restrict healthcare?

12.2.2 Neonatal screening

Neonatal screening is carried out for the occurrence of a common genetic disorder. Screening for phenylketonuria is already widespread, as is screening for haemoglobin disorders in communities where they are endemic. Screening for CF is starting. The benefit of screening is that early detection leads to early treatment, which would prevent the onset of the disease in the case of phenylketonuria, or can greatly improve the prognosis in the case of CF and the haemoglobin disorders. Like all testing, informed consent is ideal; however, such tests are routine and in practice it is unlikely that sufficient time can be devoted to any form of counselling. While this is regrettable, the benefits of such programmes are generally thought to outweigh this problem.

Carrier screening can be very effective in reducing the population incidence of a disorder. This may come about by avoiding marriage between carriers, a decision of two carriers not to have children or using prenatal testing and termination of affected embryos. Screening for β thalassaemia in Cyprus and Sardinia has led to a marked reduction in the number of cases. Carrier screening is also carried out for Tay–Sachs disease in a number of Ashkenazi Jewish communities. In all these cases, the programme is actively supported by the relevant religious leaders. In some cases, the local religious leader may act as a clearing house by receiving the results of the tests and acting only where both partners are carriers.

The major problem with carrier screening is that it may lead to discrimination against those diagnosed as carriers. This may be in the form of social stigmatisation in finding marriage partners or lead to difficulties in obtaining insurance or employment. A compulsory screening programme for sickle cell anaemia in several American states in the 1970s is widely held to have been disastrous because it led to discrimination against carriers. A common problem in population screening for recessive disorders is a failure to realise that carrier status does not affect the health of the person being tested but will affect the health of any children resulting from union with a partner who is also a carrier. Thus the discrimination against sickle cell carriers was not only unjust but also without any basis in reality.

The need for a clear understanding of the consequences of carrier status means that counselling must be provided to those who are carriers. Since population screening will be undertaken only for common disorders, it is likely that the frequency of carriers will be high, up to 1 in 20 for CF in Europeans. This causes a problem of resources. One solution that has been suggested is to inform people of their carrier status only when their partner is a carrier as well, because only then is there actually a problem. However, in this scenario most carriers will continue to be unaware of their status and, because they have been tested without any adverse result, may assume that they are not carriers. This could cause difficulties if they subsequently had children with another partner who was a carrier.

12.2.3 Diagnosis of genetic disease in children after birth

Genetic tests may be carried out in a child, less commonly in an adult, to confirm a diagnosis of a genetic disease or in some cases to exclude it. Such use does not differ in principle from any other test used to aid medical diagnosis and does not present any ethical problems.

12.2.4 Presymptomatic testing for late-onset diseases

Presymptomatic testing may be carried out for serious late-onset diseases, such as HD, familial Alzheimer's disease, familial colon cancers and familial breast cancer. It is important to distinguish these sorts of test from novel tests that will become available for predisposition to common diseases, which are considered separately below. For serious late-onset diseases, the presence of the mutation is highly likely to lead to the onset of the disease. However, this is not necessarily inevitable. The penetrance of BRCA1 is only 50% by age 50 and 85% throughout life. Thus some women may have the mutation but not suffer the disorder. Equally, even if a woman does not carry the BRCA1 mutation there is still a significant lifetime risk that she will develop breast cancer from other causes. The situation is further compounded by current uncertainty about the penetrance of the BRCA2 mutation (discussed in Chapter 5). Even where the eventual onset of the disease is virtually certain, as is the case with HD, there is considerable uncertainty about the age of onset and the severity of symptoms.

These tests are unusual because they predict future disease in people who are currently healthy. There can be a number of real and serious consequences from an adverse result. First, such a result can inflict severe distress and psychological damage on a person arising from the knowledge that at some point in the future he/she is likely to suffer from a devastating disorder. Even members of an affected family who test negative can suffer psychologically from feelings of guilt that they have escaped. Second, affected people may find it more difficult to obtain insurance and employment. The use of genetic tests in employment and insurance is discussed below.

Because of the possibility of harmful consequences, there have to be clear positive reasons for undertaking such tests. Such reasons do exist. First, there may be some clear action that can be taken to prevent or ameliorate the disorder. In the case of the colon cancers, removal of the colon has been shown to be an effective prophylactic treatment. Here the reason for undertaking the test is obvious. The situation in familial breast cancer is less clear. BRCA1 and, to a lesser extent, BRCA2 predispose to ovarian cancer at a high frequency and other cancers at a lower frequency. At the very least, prophylactic surgery would require oophorectomy (removal of the ovaries) as well as radical mastectomy. Even then we still do not know whether this will be effective: it is difficult to completely remove all breast tissue and there is still the risk of other cancers. Furthermore, it is the nature of this disease that we are unlikely to know for some years whether prophylactic surgery is effective. In other diseases, such as HD and familial Alzheimer's disease, there is no treatment and it is difficult to see any treatment becoming available in the foreseeable future.

A second advantage from foreknowledge of disease is that it may allow appropriate life choices to be made. This may include decisions about marriage and having children, financial decisions and career choices. People who know they are at risk may make very different decisions if they know for certain whether or not they are affected. One choice that is special to breast cancer is that a woman may choose to have her children at an early age, followed by prophylactic surgery.

In this context of life choices, another ethical dilemma arises. These choices are only open to someone who is relatively young and who may therefore decide to be tested. They may have a middle-aged parent who is at risk but who has not yet developed any symptoms. Since there is no positive value to the parent of knowing their status, they may well decide not to be tested. The problem is that if the child is diagnosed as having the disease, this diagnoses the parent as well. How is a balance to be struck between the right of the child to know and the right of the parent not to know?

12.2.5 Presymptomatic testing for predisposition to complex diseases

Presymptomatic testing for predisposition to common complex diseases is not yet widely available but is likely to become available in the next few years. We have seen in Chapter 6 that type 1 diabetes has a strong genetic component and that these components are being successfully identified. However, type 1 diabetes also has a strong environmental component, since the concordance between identical twins is only 30%. It is not difficult to envisage a situation soon where neonatal screening could identify children at genetic risk, whose parents could then be advised on how to avoid environmental risk factors. Alternatively, children could be intensively monitored for early markers of the disease, such as autoantibodies to β cells. This could then lead to early and aggressive treatment to prevent the full-blown disease from developing.

Research is also well developed into genetic factors that predispose to heart disease. One dominantly acting mutation, affecting about 1 in 500 of the population, has been shown to cause familial hypercholesterolaemia. Untreated, this is likely to lead to heart disease by the age of 40–50 years. However, this can be prevented by changes in diet and lifestyle or by lipid-lowering drugs. A large number of other alleles have also been identified that alter the risk of heart disease but that can be modified by drugs or by diet and other lifestyle changes.

An important difference between this type of test and tests for late-acting autosomal dominants is that the latter carry either a very high risk or virtual certainty that a disease will occur. In contrast, predisposition indicates an elevated risk but no objective certainty about the outcome. This distinction is very important. Apart from this, many of the issues that were considered above for late-acting dominant disorders apply to predisposition as well. Diagnosis of risk factors comes with potential negatives, such as psychological harm and discrimination in insurance and employment. Normally there should be a clear positive reason for undertaking the test. However, most of these tests are likely to become widespread only if they are coupled to an effective course of action for the prevention of the disease or intensive medical surveillance leading to early and more successful treatment.

The complexity of the balance sheet between the benefits and potential harm arising out of both types of presymptomatic testing underscores the need for effective counselling before the tests are carried out and to explain the implications of the results once they are available. Several research studies have shown that the proportion of at-risk individuals who wish to undertake such tests is reduced following counselling.

The difficulty of providing effective counselling is one reason why there is considerable concern at the prospect of private mail-order testing. The incentive to use such a service is that the result may be kept more confidential than would be possible if it was arranged through a doctor. The desire for confidentiality comes from the penalties that may accrue in respect of insurance and employment.

12.3 Human rights

12.3.1 Insurance

Life insurance pays a benefit to the dependants of the insured upon his or her death. In health insurance the benefit pays the costs of medical treatment in the event of the insured suffering ill health. The size of the premium that the insured pays depends on the risk of the insured event happening. Over the years, insurance companies have built up great expertise in calculating the size of these risks, based on information such as the insured person's age and occupation and lifestyle risks such as smoking. The system cannot work if the insured person knows their personal risk is higher than the insurance company realises. A simple example would be someone who tests positive for HD insuring their life for an abnormally large sum without declaring the result of the test. For the insured, this is betting on a certainty; for the insurance company, this is a certain loss which, if it happens on a large scale, will inevitably lead to higher premiums for all.

To prevent such fraud, insurance companies would like to know the result of any genetic tests that have been carried out. For the individual this is a disincentive to take the test in the first place. The undesirability of this outcome may be seen from consideration of a test for hypercholesterolaemia. A person who tests positive for this condition may save their life by appropriate treatment and lifestyle change, but they will not be insurable. In other words they may take the test and live or not take the test and be insured; but they are unable both to live and be insured at the same time.

There is thus a conflict between the needs of society in general and the interests of insurance companies. Genetic tests have yet to be used on a large scale. Indeed, only one test, for HD, has so far been approved in the UK by GAIC (Box 12.1). Nevertheless, many robust tests for single-gene disorders now exist (see Chapter 8) and we may expect tests for susceptibility alleles to come on-stream in the near future. So this issue will grow in importance. In May 2001 the House of Commons Select Committee on Science and Technology issued a report criticising the attitude of the insurance industry to the use of genetic tests and expressed concerns over the constitution and funding of GAIC. It recommended a moratorium on the use of genetic tests. As a consequence, the HGC advised the UK government that there should be a moratorium on the use of genetic tests while a regulatory framework is developed and more data become available on the predictive value of the tests. There were two exceptions to this moratorium. First, those cases where a test could be used to exclude the possibility that someone suffered from a disease for which they were at risk because of family history. Second, where cover of over £500 000 was being sought. The UK insurance industry accepted this moratorium voluntarily.

In the United States, 46 states have legislation prohibiting the use of genetic information. Repeated attempts to extend this to a federal level have been blocked in Congress. However, a federal committee called the Secretary's Advisory Committee on Genetic Testing (SACGT) advises the Department of Health and Human Services (DHHS) on the medical, scientific, ethical, legal and social issues raised by human genetics. It recommended in 2000 that there should be legislation to prohibit a requirement for genetic testing by life insurance companies. In 2001 the DHHS issued advice to the insurance and health industries that genetic information should not be used to assess medical care and life insurance premiums.

12.3.2 Employment

An employer may have an interest in genetic tests of prospective or current employees for a number of reasons. First, employers may not wish to invest resources and time in training someone who in the future will be unable to perform their job because of ill health. Ability to perform a job at the present time and not at some point in the future is an implicit assumption of employment laws in most countries. For example, it is not permissible to discriminate against a woman on the grounds that she may become pregnant at some point in the future. It thus seems fairly straightforward to prohibit genetic discrimination on these grounds.

A second reason an employer may wish to apply a genetic test is for safety or other reasons of public interest. An often-quoted example is that of an airline pilot predisposed to sudden heart disease. It is not difficult to imagine other similar examples. When public safety is an issue it may be necessary to override the rights of the individual. However, it is important that if other means are available then they should be used. For example, an employee doing a job that required high-quality eyesight can be checked by simple eyesight tests rather than a genetic test.

Third, an employer may wish to test for sensitivity to some environmental hazard such as a carcinogen, since there is clear evidence that sensitivity to xenobiotics (chemical compounds not found normally in the body) is influenced by genetic factors. We would expect that every practical method would be used to reduce exposure in the first place. Nevertheless, there may be some industries where this is not possible. It would surely be sensible to screen out those who may be particularly sensitive. If this is so, we would then face the problem of how to react to employees who refused such tests. Clearly it would be wrong to force someone to take a test against their will. However, an employer may also wish to be protected against the consequences if the employee who refused the test was to consequently suffer the ill health that could have been avoided.

12.3.3 Confidentiality

Everyone has the right to keep information about their genetic constitution private. There will be many occasions when this right to privacy may come under attack. For example, we have already seen in recent years sensational treatment in the media concerning the HIV status of well-known people. There is far more scope for this sort of problem in genetic testing, e.g. the future health prospects of a politician. Clearly, genetic information will need some form of legal protection.

A different problem arises out of the fact that genetic testing is concerned with families as much as individuals. When one member of a family is diagnosed as being a carrier of a genetic disease, it becomes possible that other members of a family are as well. Consider, for example, a woman who has a boy affected by a sex-linked disease and it is confirmed that she is indeed a carrier and that this was not a mutation that arose during gametogenesis in her mother. There will be a 50% possibility that her sister will be a carrier as well. Suppose the sister is about to become pregnant. She clearly needs to be aware of this information so she can ascertain her carrier status. With this knowledge she can make her own decisions about her reproductive choices. She may decide not to have children, or she may go ahead and become pregnant but use prenatal testing to ensure the embryo is healthy. What happens if the first sister refuses permission for the information to be passed on? Here we have a clear conflict between the right of the second sister to know and the right of the first sister to confidentiality. We may hope that counselling will persuade the first sister to divulge the knowledge, but this does not always occur in practice.

12.3.4 Individual responsibility

Research into personality disorders may well lead to the conclusion that certain antisocial actions, such as substance abuse or violent behaviour, are influenced by genetic factors. There has already been a demonstration that a history of violence and criminal activity in an extended Dutch family is due to a single mutation in the monoamine oxidase gene. This leads to two contrasting ethical problems. On the one hand, individuals predisposed to such antisocial behaviour may seek to absolve themselves of responsibility for their actions. Genetic predisposition to criminal behaviour has already been used as a defence in a murder trial in the USA. An opposite problem is that society may seek to stigmatise those found to carry such traits. Since such behaviour may be more common in certain sections of the community and in certain ethnic groups, it is but a short step to label those groups as genetically inferior. The scientific fallacies inherent in this form of genetic determinism are discussed below. It is sufficient to note here that such an approach ignores the overwhelming influence of the environment and poverty on violence and drug addiction.

Another area in which individual responsibility will become important is how we react to warnings that we may be genetically predisposed to a complex disease. For example, it is not difficult to imagine in the future that neonatal screening may be carried out for predisposition to type 1 diabetes. Research may also have identified environmental risk factors that trigger the disease or have led to medical treatments that prevent its onset. Will we hold parents responsible for ensuring that a child avoids the environmental risks by keeping to a diet or undergoes suitable medical treatment? Will a disease be judged self-inflicted if people ignore warnings that they are susceptible? There is a growing trend to restrict treatment where disease is self-inflicted as a result of cigarette smoking or substance abuse. We may note here that the justice of such practices may become even more questionable if it is shown that some individuals are more susceptible to addiction than others.

12.4 Patents

Much of the current impetus for the advances in human molecular genetics comes from the private investment of the biotechnological and pharmaceutical industries. This investment is possible only if any invention of commercial value can be protected by patent, which prevents others from using the invention for commercial purposes without the permission of the patent holder. This permission is given in the form of a licence that usually involves the payment of a royalty. In return for the issue of a patent the inventor must disclose full details of the invention. In this way society gains the benefit of the knowledge generated by the research, while the rights of the inventor to exploit the invention commercially are protected.

Most patents relating to human genetics are filed in one of three patent offices: the United States Patent Office (USPTO), the European Patent Office (EPO) or the Japanese Patent Office. International respect for patents comes through international agreements such as the World Trade Organization Agreement on Trade-Related Aspects of Intellectual Property Rights. Although each patent office has its own policy for granting patents, generally to be successful a patent application must meet the following criteria:

1. The invention must have 'utility', i.e. it must have a use that is clearly specified in the application.
2. The invention must be novel. In the United States, an invention may be discussed publicly for a year before filing but in Europe an invention may not be revealed publicly before the patent application. As a result there is a difference between the methods used for deciding priority: in the United States priority goes to whoever was the first to invent; in Europe it is the first to file. The importance of demonstrating priority in the United States has led to the practice of notarised laboratory notebooks in commercially funded research.
3. The invention must not be obvious. This means that it is not an improvement that would be routinely made by an experienced and well trained person in the field. The phrase 'skilled in the art' is often used to describe such a person.

4. The patent must be sufficiently detailed for the invention to be reproduced by a person skilled in the art.
5. There are often further restrictions connected with morality and public good. For example, the European Patent Convention prohibits the issue of patents to inventions that offend morality or public order; for example, a letter bomb or an instrument of torture. In the United States, Congress prohibited patents on nuclear weapons.

A patent application may be provisional. In such cases the applicants have up to a year to make a full application, in which they must substantiate claims made in the provisional application. The patent officer examines the application to see if it meets the above criteria and, if satisfied, the patent is granted. The USPTO pronounces in 3 years, the EPO within 18 months. The decision of the patent officer may be challenged in court. Once granted, a patent lasts for 20 years from the date of application, which in practice means 17 years if the time taken for the USPTO to pronounce on the patent is taken into account. Drug companies point out that the time taken for development and safety trials is often as much as 10 years, so that the time they have to commercially exploit their invention is limited. However, they often find ingenious ways to extend their period of protection by applying for additional patents relating to a successful drug.

In 1980 in the United States a precedent was set for patenting living organisms in the case of Diamond vs Chakrabry, which was decided by the Supreme Court. The case concerned a bacterium genetically modified to dissolve oil. The court ruled that it was patentable as it did not occur in nature. Since the Diamond vs Chakrabry case, over 3 million genome-related patents have been filed in the United States, but only a few patents have actually been issued by the USPTO. Some of the applications relate to genes that have been cloned, fully sequenced, their biological role established and a clear utility described, such as a diagnostic test or a screen for a drug that may interact with the encoded protein to modify its function. On the whole these have not been controversial. However, most of the applications have related to ESTs that are only gene fragments – they encompass as little as 20% of the gene from which they were derived – and for which the function could only be guessed. As genome sequencing developed there was widespread concern that many parts of the human genome would be patented before it came into the public domain. This was one of the main driving forces for the accelerated completion of the publicly funded genome project and the central ethos that sequence should be placed on the internet within 24 hours of it being obtained. By placing the sequence in the public domain it was hoped to limit the scope for patents based solely on sequence without further work to establish function and the development of a useful application. Further applications related to SNPs for which patents were sought on the basis that they could be used in diagnostic tests (see Chapter 6). The SNP consortium patented all SNPs identified but only on a defensive basis, to prevent others from patenting. The declared policy of the consortium is not to enforce these patents.

There is general agreement that it is right that commercial companies should be able to patent in the field of human genetics. However, the attempt to patent gene sequences without further work to establish their function and utility has provoked a fierce controversy. A number of issues have been raised:

1. The rush to patent genomic sequences has been likened to a land grab, where commercial organisations rush to stake claims on parts of the human genome without having fully characterised them or establishing utility.
2. A patent on a gene sequence that forms the basis of a diagnostic test gives the patent holder a monopoly position on all future tests based on that gene. If it turns out that the gene sequence has a totally different application than that envisaged in the initial patent, the patent holder still controls the use of the sequence for the new application, even if this application was quite unimaginable at the time of patenting.
3. Cloning a gene and determining its nucleotide sequence is a discovery, not an invention. Patent lawyers argue that cloning a gene or cDNA is like purifying a substance from nature that has value, which is recognised in patent law. Opponents of this view argue that the value lies in the gene sequence, not the physical DNA whose isolation is incidental to the process.
4. The essence of human genetic research is that it is cooperative. The process of cloning a disease gene may start with individual family members collecting information, which may be assembled into family trees by local research laboratories and physicians. Mapping, and finally cloning the gene, requires the use of dense maps and probably nucleotide sequence data obtained from publicly accessible databases. Both the maps and the databases were assembled with great effort by many different laboratories. It is often remarked that genes seem to be independently cloned by several groups within a few weeks. There is a reason for this: once the information that allows the gene to be located has been collected, there are many competent laboratories that can make the last step. Granting a patent to any one of them provides a reward to one party for the collective labours of many others.
5. The practice of seeking patents for gene fragments means there may be a conflict when the gene is fully characterised by others. If a patent is granted for an EST sequence, does that extend to sequences in the gene that were not present in the original EST? Moreover, different patent applications may unwittingly overlap. For example, a patent for an SNP may involve a sequence that is part of an EST. This may mean that there are a number of different patents controlling the use of a single gene. This causes what is known as patent stacking, which multiplies the royalties that must be paid by a third party wishing to use the gene. This might ultimately discourage product development because of the royalty costs.
6. Patent applications remain secret until the patent is granted. This may mean that companies may invest in a product only to find later that the patent is held by another party.

Despite the philosophical objections to patenting gene sequences, it is clear that patent law allows it in principle and that this is unlikely to change in the foreseeable future. However, in December 1999 the USPTO altered its policy to tighten the requirement to identify utility. Patents relating to gene sequences must have 'substantial real-world utility'. It was envisaged that this would disqualify many EST applications.

12.5 Gene therapy

Gene therapy promises for the first time to cure a genetic illness rather than ameliorate the symptoms. It may be carried out using two fundamentally different strategies. In somatic gene therapy, the genetic constitution of a person is modified or supplemented in such a way that the change is limited to that person's own body and is not passed on to the next generation. In germline therapy, a person's germ cells, or cells that will become germ cells, are changed so that the change is passed on to the next generation.

Somatic gene therapy is theoretically no different from any other form of medical treatment that seeks to rectify a physiological dysfunction. The view that it is similar to an organ transplant is not entirely accurate because gene therapy requires no donor other than the cells from which the original DNA was cloned. Provided that the gene therapy is directed towards treating a disease, rather than enhancing natural characteristics, somatic gene therapy requires no further control than would normally be applied to any experimentation involving human subjects, the ethical principles of which are well established. If the research is successful, then marketing of gene therapy products will be subject to the same controls as any new drug, i.e. clinical trials will be needed to prove that it is both safe and effective.

Germline gene therapy in principle offers many advantages over somatic gene therapy. Once a reliable method is found of manipulating the germline, then that method would be generic for all genes, whereas somatic therapy relies on different *ad hoc* methods of introducing genes to different tissues according to which is affected by a particular mutation. For example, somatic gene therapy for CF requires genes to be delivered to cells lining the airways of the lung. Even if this is successful, other affected cell types such as those of the vas deferens or the pancreatic duct are not treated. Somatic gene therapy for DMD requires genes to be delivered to muscle cells and for HD would require genes to be delivered to brain cells. Each disease presents its own special problems and will require its own research programme to solve them.

Germline therapy would make it much simpler for a couple at risk of having a child affected by a genetic disease to have normal children. At present, each pregnancy must be tested and affected embryos aborted. Sometimes the laws of chance operate unkindly and it is not unknown for a series of pregnancies to have an unfavourable outcome. It has been argued that it is better to fix the problem before conception than to terminate the life of an embryo. In this context it should be noted that preimplantation testing (see Chapter 8) may provide a more realistic way of solving this problem than germline gene therapy.

There are considerable practical difficulties to be overcome before germline gene therapy can be contemplated. Somatic gene therapy aims to supplement the defective gene by introducing the functioning gene in such a way that it does not integrate into the genome. Germline therapy will require the replacement of the defective gene with the functioning copy. If the new gene is not precisely targeted to the location of the defective gene, two deleterious consequences may follow. First, the original mutation may reappear in subsequent generations due to reassortment of the genetic material at meiosis. Second, insertion of new DNA at another site could be mutagenic: it may either inactivate another gene with unforeseeable consequences or change the pattern of expression of another gene perhaps with oncogenic consequences. Techniques are not currently available for this type of targeting. A second technical difficulty is the current lack of any method for delivering genes to the germline. These difficulties are likely to prevent germline gene therapy for some years. Nevertheless, we may reasonably expect these problems to be solved at some time in the future.

There are multiple ethical objections to germline gene therapy. First, it is perceived by many to be 'playing God' by interfering with the basic genetic constitution of the human species. Second, it opens up the road to manipulation for eugenic reasons, i.e. manipulation designed to enhance characteristics such as intelligence, musical or sporting ability, etc. Third, it is irreversible; if for some reason a change turns out to be in error it cannot be removed from the human gene pool. Fourth, its development will require a large amount of experimentation on human embryos and, at some point, will lead to experimental humans who will be at considerable risk of suffering from a genetic dysfunction. By definition, such humans will be unable to give their prior informed consent.

The combined practical and ethical objections have meant that germline therapy is prohibited in all countries that have the research capability to attempt it. Nevertheless, there are many scientists, genetic interest groups and even religious leaders who are of the opinion that the benefits outweigh the drawbacks. Should the technical problems be solved, the controversy is likely to resurface.

12.6 Genetic determinism

One of the themes that runs throughout this book is that genetic factors influence the risk of common diseases and psychiatric disorders. It is also highly probable that our personalities and abilities are also influenced by genetic factors. We have seen that the means have been developed to identify these genetic factors and that we may expect that over the next few years these techniques will be applied to the analysis of the causes of many common diseases and psychiatric disorders. Inevitably they will also be applied to an attempt to improve our understanding of the origin of fundamental human characteristics.

In doing this, there is a danger that we will come to regard our health, personalities and abilities to be mechanically determined by our genetic constitution. Such a view is scientifically wrong. It is important to be fastidious about the relationship between genotype and phenotype because it is easily caricatured by both sides of the political spectrum. Susceptibility to common diseases and the development of phenotypic traits are likely in each particular case to be controlled by several genes that interact with each other and with the environment. These interactions will be complex and are unlikely ever to yield simple predictions about the outcome. Moreover, because the contribution of the environment is nearly always significant, the outcome will usually be modifiable by environmental change.

It is worth considering what we know about the genetic predisposition to IDDM because it is instructive in these matters. Recall that the concordance in identical twins is 30%. Immediately, it is clear that the environment plays the larger part in the onset of the disease. It is not easy to be sure what this environmental contribution could be, since we would imagine that identical twins would be exposed to the same obvious environmental factors, such as diet, etc. It may be very subtle; for example, an important factor in the development of multiple sclerosis, another auto-immune disease with a clear genetic component, is thought to be a virus infection. The locus labelled *IDDM*1 in the HLA region is thought to contribute about 40% of the genetic risk. In European populations, the *IDDM*1 genotype *DR3/DR4* predisposes towards diabetes, but this can vary in other ethnic groups. The concordance of *DR3/DR4* is very high in affected sib pairs. However, whereas 2% of European populations have the *DR3/DR4* haplotype, only 0.1% succumb to diabetes. Moreover, there are some who have IDDM who do not have the *DR3/DR4* genotype. Thus the great majority of people who have the risk genotype do not succumb to the disease and some who have the disease do not have the predisposing genotype. It would clearly be wrong to label *IDDM*1 as the 'diabetes gene'. Nevertheless, many independent lines of research confirm that *IDDM*1 contributes towards the risk of diabetes and we can even start to formulate plausible models of its action.

12.7 Summary

- It is important that advances in human genetics lead to the relief of suffering and the promotion of good health. There are clearly foreseeable ways in which they may have a negative impact on society through genetic discrimination and loss of privacy.

- Genetic tests allow diagnosis of genetic illness. They can be used in a variety of situations to prevent, or allow the early treatment of, diseases that have a genetic component.

- Genetic testing can have negative consequences. It can cause psychological harm or it may lead to actual discrimination in insurance or employment.

- Genetic tests should be used only when a clear benefit can be derived. They must be used only after the person tested has given their informed consent. Informed consent requires counselling.

- Everyone tested has the right to expect confidentiality. This can lead to some conflicts of interest when there are good reasons for other family members to know the result of a test.

- With the right to know comes the right not to know. Care must be taken to ensure that unwelcome information is not given inadvertently.

- Insurance companies wish to know about the results of a genetic test to avoid fraud. This results in a disincentive to genetic tests that may be important for someone's health.

- Employers may wish to discriminate on the basis of genetic tests to avoid hiring or promoting those with poor health prospects. Genetic tests may sometimes be justified on the grounds of public safety or to protect those who are especially sensitive to environmental hazards.

- Confidentiality may be compromised by media interest.

- Predisposition to antisocial behaviour raises issues of individual responsibility for criminal actions or substance abuse. Equally, it may lead to particular groups being labelled as genetically inferior. In such arguments, we must never lose sight of the large contribution made by the environment.

- Gene therapy offers the possibility of a cure for genetic diseases. Somatic gene therapy is not thought to raise any ethical problems. Germline gene therapy may have considerable advantages over somatic gene therapy. However, practical difficulties prevent it being used at present and there are powerful ethical arguments against it ever being used; at present germline gene therapy is prohibited.

- Patents in human genetics are required to allow private investment to fund research. A debate over whether genes and DNA sequences can be patented has been settled in favour of allowing such patenting.

- Genes influence our health, personalities and aptitudes, but genetic determinism is misguided. These characteristics come about through the interaction of many genes with each other and with the environment. These interactions are complex and unlikely ever to yield simple predictions about their outcome.

Further reading

General

House of Commons Science and Technology Committee (1995) *Third Report – Human Genetics: the Science and its Consequences. Volume 1. Report and Minutes of Proceedings.* HMSO, London.

An excellent and well balanced introduction to the issues. Includes first-hand evidence from all interested parties including patient groups and physicians.

The HUGO web site
http://www.ornl.gov/hgmis/research/research.html

Follow the link to ethical, legal and social issues. Presents good introductions to issues such as patenting (see below), and contains many links to other articles and reports.

The Human Genetics Commission
http://www.hgc.gov.uk/

Describes the regulatory network in the UK and links to reports from government advisory committees.

Gene testing

Promoting Safe and Effective Genetic Testing in the United States. Principles and Recommendations. Task Force on Genetic Testing of the NIH–DOE Working Group on Ethical, Legal, and Social Implications of Human Genome Research
http://www.med.jhu.edu/tfgtelsi/promoting/

Adequacy of Oversight of Genetic Tests. Preliminary Conclusions and Recommendations of the Secretary's Advisory Committee on Genetic Testing.
http://www4.od.nih.gov/oba/sacgt/reports/adequacy_of_oversight_of_geneti.htm

This report was requested by the US Surgeon General and published in April 2000.

Genetics and insurance
The House of Commons Science and Technology Committee. *Fifth Report – Genetics and Insurance.*
http://www.parliament.the-stationery-office.co.uk/pa/cm200001/cmselect/cmsctech/174/17402.htm

HGC statement on the use of genetic information in insurance.
http://www.hgc.gov.uk/business_publications_statement_01may.htm

KEEFER, C.M. (1999) Bridging the gap between life insurer and consumer in the genetic testing era: the RF proposal. *Indiana Law Journal*, **74**, 1375–1395.
http://www.ornl.gov/hgmis/resource/keefer.html

Analyses of the legal position of genetics in insurance in the United States
REILLY, P.R. (1997) Fear of genetic discrimination drives legislative interest. Ownership, predisposition major issues. *Human Genome News*, **8** (3&4).
http://www.ornl.gov/hgmis/publicat/hgn/v8n3/01fear.html

Patents
BOBROW, M. and THOMAS, S. (2001) Patents in a genetic age. *Nature*, **409**, 763–764.

ENSERINK, M. (2000) Patent office may raise the bar on gene claims. *Science*, **287**, 1196–1199.

HUGO project information – Genetics and Patenting
http://www.ornl.gov/hgmis/elsi/patents.html

Glossary

λ 1. Relative risk, see Section 5.5.

      2. Bacteriophage λ. A commonly used vector for gene cloning.

ab initio Analysis solely from theoretical considerations without the use of experimental data.

acrocentric A chromosome where the centromere is nearer one end.

additive interaction Where the effect on phenotype of multiple alleles is the sum of their individual effects.

admixture Gene flow between populations as a result of migration.

aetiology The causes of a disease.

affected pedigree member Two individuals affected by a disease in a pedigree who are genetically related.

affected sib pair Two sibs affected by a disease.

age-related macular degeneration (AMD) A condition where damage to the centre of the retina (the macula) leads to loss of vision, mostly in older patients.

allele One of several alternative forms of a gene or DNA sequence.

allele frequency The frequency in a population of each allele at a polymorphic locus.

allele-specific oligonucleotide An oligonucleotide used as a PCR primer in a genetic test that will anneal only to DNA carrying either the wild type allele or a specific mutant allele but not to both.

Alu The most common repeated sequence in the SINE category.

amniocentesis Sampling of cells for genetic testing from the amniotic fluid surrounding the developing fetus. The procedure involves inserting a large tube through the mother's abdomen. These cells are cultured *in vitro* for about 3 weeks to increase numbers. The procedure cannot be attempted before the sixteenth week of pregnancy.

anaemia Blood disorder characterised by deficiency of red blood cells.

angiogenesis-promoting Any agent that causes the formation of new blood vessels from existing ones.

annotate Identify the features of a genome, including the location of genes and information concerning their likely function.

anticipation The effect by which the severity or penetrance of a disease apparently increases with each succeeding generation.

antigen The protein recognised by an antibody.

anti-oncogene See tumour suppressor.

anti-sense mRNA An RNA molecule or oligonucleotide that is complementary to an mRNA molecule. It can thus form a duplex with the mRNA molecule, interfere with its translation and provoke its destruction by ribonucleases specific for double-stranded RNA molecules.

APC (adenomatous polyposis coli) The gene that is dysfunctional in familial adenomatous polyposis.

ApoE Lipoprotein found in blood plasma. Different alleles at the *ApoE* gene have been demonstrated to be risk factors for both cardiovascular disease and Alzheimer's disease.

apoptosis Programmed cell death provoked by irreparable genetic damage.

APP (amyloid precursor protein) Precursor of the amyloid protein found in senile plaques in the brains of Alzheimer's disease patients.

aptamer Short RNA, DNA or peptide molecules that recognise and bind to specific targets.

archaic humans Members of the species *Homo sapiens* that predated anatomically modern humans.

argonaute family A family of RNA-binding proteins that performs the catalytic step in both RNAi and miRNA action.

ARMS (amplification refractory amplification system) A commercial multiplex PCR system for genetic testing.

arrhythmia A group of life-threatening conditions where the electrical output from the heart is abnormal. It may be too fast, too slow or irregular.

Ashkenazi Jews A particular population group that originated in eastern France in the early Middle Ages, who subsequently migrated to Eastern Europe and later to the USA and Israel. By virtue of culture and religion they have been largely reproductively isolated from surrounding populations.

ASO (allele-specific oligonucleotide) An oligonucleotide that will only anneal to one allele, either mutant or wild type.

association Simultaneous occurrence of a disease phenotype and an allele at a frequency that is statistically significant.

autoantibody An antibody produced by an individual's own immune system that recognises antigens in their own body.

autoantigen The protein or part of a protein recognised by the body's immune system in an autoimmune disease.

autoimmune Attack by an individual's own immune system on some part of the individual's body.

autosomal dominant A trait showing a characteristic pattern of segregation indicating that the gene in question is located on an autosome and that the mutant allele is dominant to the wild type allele. Thus the disease is manifested when one copy of the mutant allele is inherited.

autosomal recessive A trait showing a characteristic pattern of segregation indicating that the gene in question is located on an autosome and that the mutant allele is recessive to the wild type allele. Thus the disease is manifested only when two copies of the allele are inherited.

autosome A chromosome that is not an X or Y chromosome. There are 22 autosomes in the human chromosome complement.

BAC (bacterial artificial chromosome) A cloning vector that uses the origin of replication of the F plasmid. Can take inserts of 100–300 kb and is thought to be less prone to cloning artefacts, such as chimeric inserts or rearrangements, upon propagation within the host.

balanced polymorphism A polymorphism that is maintained in the population because the heterozygous state has a greater Darwinian fitness than either homozygous state. Normally, one homozygote is at a severe selective disadvantage, but the allele is maintained in the population at a significant frequency because of the selective advantage of heterozygotes.

balanced translocation When translocated chromosomes are inherited together so there is no change in genomic information content, apart from at the breakpoint.

banding Any process that produces a discrete pattern that can be used to identify individual chromosomes and chromosomal regions. See G-bands and Q-bands.

biallelic A locus with two alleles.

bivalent A chromosome with two chromatids.

biomarkers Molecules that are used to detect some biological state. For example, the presence or absence of a biomarker may suggest a disease.

biotin A small molecule that binds very tightly to the substance streptavidin. By attaching a biotin molecule to a protein or a nucleic acid, that molecule can be isolated or attached to another compound by binding it to streptavidin.

BLAST (basic local alignment of sequences tool) A computer program that searches for similarity of an amino acid or nucleotide query sequence to sequences stored in a database.

buoyant density ultracentrifugation A process of centrifugation that separates molecules according to density rather than mass. Analysis of

DNA molecules involves mixing the DNA with a solution of CsCl whose concentration is chosen to closely match the average density of DNA. The solution is centrifuged at very high speed (40 000–50 000 g) for 48 hours. The centrifugal force generates a gradient in CsCl concentration. The DNA molecules move to an equilibrium position where their density is matched by the density of the CsCl at a particular point in the gradient. The density of DNA is determined by its base composition. Any fraction of the genome whose DNA content differs from the average will form a separate or satellite peak.

Burkitt's lymphomas A type of cancer involving a malignancy of B cells (a type of white blood cell).

CAAT box The sequence GGCCAATCT that can be found in the promoter region of genes and that binds the transcription factors CTF and NF1 to stimulate transcription.

cap The covalent modification of the 5′ end of an mRNA molecule during transcription.

capsid The protein coat of a virus particle.

case control study Study of the association of a disease with particular alleles in which each case of the disease has an unaffected control matched for age, sex and ethnicity.

cDNA library Gene library composed of cDNA inserts synthesised from mRNA using reverse transcriptase.

Celera Genomics Private company founded by Craig Ventnor to carry out genomic research. Undertook the sequencing of the human genome.

cell cycle engine Molecular mechanism that controls the passage of cells through the cell cycle in eukaryotic cells. It consists of one or a family of protein kinase subunits whose activity is regulated by association with different regulatory subunits called cyclins.

centimorgan Unit of genetic map distance corresponding to a recombination fraction of 0.01. Named after Thomas Hunt Morgan.

centromere A constriction visible in metaphase chromosomes where the chromosome is attached to the mitotic or meiotic spindles.

CEPH families Set of reference families for genetic mapping maintained by the Centre d'Étude Polymorphism Humain in Paris.

checkpoint A mechanism that prevents a cycle event from occurring if a previous event is not completed.

chorionic villous sampling The chorion is derived from the zygote but is not part of the developing embryo. It is a membrane with projections or villi that surround the embryo. Chorionic villous sampling is carried out by introducing a catheter through the vagina until it touches the chorion. It can be safely carried out in the eighth or ninth week of pregnancy. It thus has an advantage over amniocentesis because it allows an earlier termination of pregnancy should this be indicated by the results of the genetic test.

chromatid A chromosomal strand consisting of a complex of a single DNA molecule and its associated protein and RNA components.

chromatin The complex of DNA, RNA and protein that makes up a chromosome.

chromosome jumping A method used in the course of a chromosome walk that allows tens or hundreds of kilobases to be covered in a single operation.

chromosome walking A method of moving from a linked marker to a gene.

chronic lymphocytic leukaemia (CLL) A type of cancer of B cells (a type of white blood cells), usually affecting adults, that progresses very slowly.

cis See phase.

cladistic Method of phylogenetic classification that traces organisms back to a common ancestor using qualitative characters to determine the order of descent.

clone-by-clone Strategy adopted by IHGSC for sequencing the human genome in which the genome was cloned into BAC vectors with an average insert size of 100–200 kb. The clones were ordered by *Hin*DIII fingerprinting. Each BAC clone was then shotgun-sequenced. The final sequence was then assembled by merging the contigs from adjacent BAC clones.

clone library A collection of clones that ideally contains all possible sequences from a donor genome.

clone map A physical map based on determining overlap between clones.

coalescence time The average number of generations to the common ancestor of any two random members of a population.

coding sequence A length of DNA or RNA whose sequence determines the sequence of amino acids in a protein.

Common Disease Common Variant (CDCV) A model of complex disease that considers the risk is caused by the combined action of many alleles, each of which is common in a population.

comparative genome hybridisation (CGH) An experimental technique that uses the intensity with which genomic DNA hybridises to a clone or oligonucleotide array to scan for segments that are present with an altered copy number.

complement fixation The binding of a group of blood globulin proteins to cause lysis of foreign cells after they have been coated with antibody.

complex disease Disease whose aetiology consists of a mixture of environmental and genetic factors.

compound heterozygote An individual that lacks a gene function because each copy of the gene is inactivated by different mutations.

concordance The percentage of twins where both suffer from a disease or display a phenotypic trait when that disease or trait occurs in one of the pair.

congenital A disorder that is present at birth.

congenital bilateral absence of the vas deferens (CBAVD) Often one of the symptoms of cystic fibrosis; results in male sterility.

consanguineous The individuals concerned are related and therefore have a proportion of their genes in common.

consensus sequence An idealised sequence that represents the nucleotides most often found at each position when a number of different sequences are compared.

contig A DNA region represented by a group of clones whose relationship one to another is defined by overlap between the sequences of the inserts.

copy number variants (CNVs) Segments of DNA that are present in an individual with a copy number that is different from the reference genome.

cosmid Cloning vector that takes the form of a plasmid that contains the *cos* site from phage λ. This allows recombinant DNA molecules of ~40 kb to be packaged into λ phage particles *in vitro*. Upon infection into a host cell the recombinant molecule replicates as a plasmid.

coverage The average number of times a nucleotide is sequenced.

CpG island A region of approximately 1 kb in which the CpG dinucleotides are not methylated and occur at higher frequency than that found in the rest of the genome, being equal to that expected from the percentage (G+C) of human DNA. CpG islands often span the promoter region of transcriptionally active genes and sometimes the start of the coding sequence.

cystic fibrosis transmembrane conductance regulator (CFTR) The protein affected by mutation in cystic fibrosis patients. Also used to describe the encoding gene.

cytogenetic A characteristic of the karyotype that is revealed by examination of metaphase chromosomes.

cytokines Hormone-like factors that regulate the activity of the immune system.

Darwinian fitness The number of an organism's offspring reaching reproductive maturity.

deletion mutations The loss of one or more nucleotides.

derived The allele in a polymorphism which is different from the ancestral state.

diabetes mellitus Syndrome caused by failure of cells to take up glucose.

differentially expressed RNA transcripts present in one sample with a different abundance compared with a second sample.

dinucleotide Two successive nucleotides on the same DNA strand, written with the 5′ nucleotide first.

distal Away from the centromere compared to the chromosomal point of reference.

dizygotic twins Twins derived from different fertilised eggs. Otherwise known as non-identical twins.

DNA chip A high-density miniaturised array of oligonucleotides attached to a silica or glass substratum. They are being developed to scan genes for the presence of mutations. It may be possible to use them for automatic sequencing.

DNA methylation The addition of a methyl group to cytosine to form 5-methylcytosine in CpG dinucleotides.

DNA profile Features of a DNA sample that can be used to identify the individual from which it originated. Differs from genetic fingerprinting in that it does not purport to uniquely identify the individual from all other humans.

domain Part of a protein where a continuous length of the polypeptide chain folds to form a discrete globular structure that may have a defined function.

dominant negative mutation A mutation that prevents another wild type protein in the same cell from functioning. Commonly acts by producing an altered polypeptide that prevents the assembly of a multimeric protein.

downstream A region of the DNA molecule that lies 3′ to the point of reference.

draft sequence Preliminary form of sequence from the Human Genome Project where there are gaps and the sequence is not edited to ensure the highest possible accuracy.

driver mutation A mutation in a cancer cell line that contributes to the cancer phenotype.

dynamic mutation A trinucleotide repeat expansion mutation that changes in size during meiosis or, in some cases, mitosis.

dystrophia myotonica See myotonic dystrophy.

dystrophia myotonica protein kinase (DMPK) Implicated in myotonic dystrophy, as the trinucleotide repeat expansion responsible for the disease occurs in the 3′ untranslated region of this gene.

dystrophin The protein affected by mutation in Duchenne muscular dystrophy. Also used to describe the encoding gene.

endocytosis A process by which eukaryotic cells take in material from the outside. The cell membrane invaginates to form intracellular vesicles, which then fuse to form the endosome.

endonuclease An enzyme that cleaves RNA or DNA molecules at an internal position rather than progressively from either the 5′ or 3′ end.

endosome See endocytosis.

enhancer A *cis*-acting DNA sequence that increases the expression of a gene upon binding a transcription factor. The action of an enhancer is not critically dependent on its position or orientation and in many cases exerts its influence at a considerable distance.

Ensembl Genome browser and database of sequenced genomes jointly maintained by the European Bioinformatics Institute and the Wellcome Trust Sanger Institute. Accessed on the web by the URL www.ensembl.org.

ePCR Using a computer to carry out a conceptual PCR reaction using a known nucleotide sequence.

episome A DNA molecule that can stably replicate. The significant point is that it does not have to be integrated into a chromosomal location to be propagated in daughter cells.

epistasis Originally defined as the interaction between two genes so that the expression of one is controlled by the other. Now more loosely applied to situations where the effect on phenotype of alleles at two genes is differ-ent from that which would be expected by combining the individual effect of each allele.

erythropoiesis Production of new red blood cells.

euchromatin The main fraction of chromosomal DNA that is not hetero-chromatin. It is uncoiled during interphase, and contains transcriptionally active regions.

eugenics 'Improvement' of the human genetic stock by selective breeding.

exon The part of a gene that is transcribed and remains in an mRNA mole-cule after splicing. This mainly consists of protein-coding regions, but it also includes 5′ and 3′ untranslated regions of the mRNA molecule.

exon trapping Special technique used to search for exons in genomic clones.

expressed sequence map A map that plots the position of DNA sequences expressed in mRNA. Based on EST markers.

expressed sequence tag A special form of STS that is generated from cDNA clones and thus identifies sequences expressed in mRNA.

expression signature A set of gene expression levels that are characteristic of a certain cell type or disease state.

expressivity The difference in the severity of a disorder in individuals who have inherited the same disease alleles.

***ex vivo* gene therapy** Where cells are removed from the body, manipulated *in vitro* and then reintroduced to the patient.

familial Where a disease is transmitted in families as opposed to its sporadic occurrence.

familial adenomatous polyposis Cancer of the colon and rectal areas. Inherited as an autosomal dominant trait due to mutations in the *APC* gene.

fingerprint clone contig Contig formed by BAC clones using the *Hind*III fingerprint of each clone to detect overlap.

fingerprinting Any method that identifies unique features of a clone that can be used to determine overlap between other clones in a library.

finished sequence Final form of sequence from the Human Genome Project containing less than 1 error in 10 000 bp.

flow cytometer Apparatus that allows cells to be sorted on the basis of some physical or optical characteristic, often the level of fluorescence from a fluorescent chemical with which the cells have been labelled.

fluorescence *in situ* hybridisation *In situ* hybridisation in which the probe is labelled with a fluorophore. Upon examination with a fluorescence microscope, the chromosomal region to which the probe is binding can be visualised.

fluorophore A chemical moiety that fluoresces when stimulated by light of a particular wavelength, usually in the ultraviolet range.

forward genetics A form of analysis where organisms are mutated at random. Individuals showing phenotypes of interest are isolated and the mutated genes identified.

founder effect The effect that results in the genetic structure of the population that is derived from the founder group being different from the ancestral population.

founder group A small group of individuals that splits off from the main population and settles in a previously uninhabited territory.

founder mutation A disease-causing mutation on chromosomes in different individuals that has descended from an ancestral chromosome on which the original mutation occurred.

fragile site Non-staining gap in a chromosome visible in metaphase spreads of cells cultured under conditions such as folate deprivation or chemical inhibition of DNA synthesis.

fragmentation A term used in genome annotation where two separate genes were predicted but were eventually judged to be the same gene.

frameshift mutation Deletion or insertion of bases that alter the reading frame.

framework See scaffold.

functional genomics The study of the function of genes in a genome.

G_1 (Gap 1) The period between nuclear division and DNA synthesis (S phase) in the cell cycle.

G_2 (Gap 2) The period between DNA synthesis (S phase) and nuclear division.

gametogenesis The process by which gametes are produced.

ganglioside A group of complex glycolipids found chiefly in nerve membranes, containing sphingosine, fatty acids and an oligosaccharide chain containing at least one acid sugar such as N-acetylglucosamine, N-acetyl neuramic acid or N-acetylgalactosamine.

gap junction A small area where the plasma membranes of two adjacent cells are separated by a narrow gap.

G-bands Bands produced in metaphase chromosomes using Giemsa.

GC box The sequence GGGCGG found in the promoter region genes, which binds the transcription factor SP1 to stimulate transcription.

gene A DNA sequence that contributes to the phenotype of an organism in a way that depends on its sequence. Normally this refers to a protein-coding sequence, together with any sequences that are required for expression. Some genes encode RNA that is directly functional and is not translated.

gene families Similar but not identical genes that have arisen by a process of gene duplication and divergence. Gene families can be dispersed through the genome or exist as gene clusters at one location.

gene library See clone library.

gene structure The exons in the genome that together constitute a gene.

gene therapy Treatment of a disease by genetic modification of the patient's cells.

general transcription factor A protein required for the formation of the transcription complex. Normally designated in the form TFIIX, where II indicates the RNA polymerase (in this case RNA polymerase II) and X denotes the particular protein.

Genethon Laboratory in Paris funded by the French Muscular Dystrophy Association. Constructed the standard genetic map of the human genome.

genetic drift The change in allele frequencies from one generation to another due to sampling effects.

genetic fingerprinting A process that uniquely identifies an individual according to some feature of their genome.

genetic heterogeneity Where apparently clinically similar disorders are caused by mutations in different genes (non-allelic), or where mutations in the same gene result in clinically diverse conditions (allelic).

genetic map A genomic map based on the order of genetic mapping markers and the genetic distance between them measured by the recombination frequency.

genome The sum total of the genetic material.

genome scan A systematic survey to discover if a phenotypic trait or genetic disease is linked to a genetic mapping marker.

genome-wide association study (GWAS) An epidemiological strategy to identify alleles contributing to the risk of a complex disease by scanning whole genomes in many individuals for alleles that are associated with the complex disease.

genomics Study of a genome by considering all the genes together rather than the properties of individual genes.

genomic imprinting A mechanism whereby cells preferentially express an autosomal gene inherited from one parent.

genotype relative risk The risk of a complex disease in individuals with a particular genotype compared with the general population.

germline The cells that will give rise to gametes. Mutations in germline cells will be inherited by the next generation.

germline gene therapy Where the genetic modification may affect the germline and be passed on to succeeding generations.

Giemsa A dye that binds DNA.

glycolipid A lipid containing a carbohydrate chain.

Golgi apparatus A membranous cellular compartment that forms part of the secretory pathway.

growth factor A specific factor that must be present in the environment before a cell can proliferate. Usually a protein or peptide that interacts with a growth factor receptor at the cell surface.

haematopoietic stem cells The progenitor cells for red and white blood cells and platelets.

haemoglobinopathies A group of genetic diseases that affect the production or function of haemoglobin.

hairpin A structure that is formed by RNA when inverted repeats fold back and base pair to each other forming a structure known as the stem. Any sequence connecting the two halves of the stem is known as the loop.

haplotype A set of closely linked alleles that tend to be inherited together, i.e. not separated by recombination at meiosis.

Hardy–Weinberg distribution The mathematical description of the distribution of locus genotypes in a population.

hemizygous The presence of only one copy of a gene. Usually applied to genes on the X chromosome in males.

hereditary breast cancer (HBC) A subgroup of breast cancer cases where the cancer is caused by the inheritance of an autosomal dominant allele of genes such as *BRCA*1 and *BRCA*2.

hereditary non-polyposis colorectal cancer (HNPCC) Cancer of the colon and rectal areas. Inherited as an autosomal dominant trait due to mutations in genes mediating mismatch repair.

heritability The proportion of total phenotypic variation in a population that is due to genetic variation.

heterochromatin Chromosomal regions that are late replicating, transcriptionally inactive and stain more densely with Feulgen, a dye that binds DNA.

heterogeneous nuclear RNA (hnRNA) Primary transcripts before processing containing introns and exons. Initially named because of its variable size and location.

heteroplasmic Where both mutant and wild type mitochondrial genomes are found in the same individual.

heterozygote An individual who has two different alleles at a specified locus.

heterozygous advantage When an allele that is deleterious in the homozygous state is advantageous in the heterozygous state.

highly repetitive DNA Simple sequences repeated up to 10^6 times per genome.

histones Highly conserved basic proteins that bind DNA and organise the formation of nucleosomes.

hnRNA (heterogeneous nuclear RNA) Populations of nuclear RNA molecules of heterogeneous size that are not present in spliced mRNA.

Hodgkin's lymphoma A type of cancer of B cells characterised by the presence of giant cells known as Reed–Sternberg cells.

homologs Genes that are similar by virtue of descent from a common ancestor.

homoplasmic Where the mitochondrial genomes within an individual are all identical.

homozygote An individual with identical alleles at a locus.

housekeeping gene A gene that encodes a protein performing a basic function common to most cells.

human leucocyte antigens (HLA) Antigens found on the surface of cells that signal whether the cell is self or non-self.

huntingtin The protein affected by mutation in Huntington's disease. Also used to describe the encoding gene.

hybridisation The formation of a hybrid DNA molecule (or RNA/DNA duplex) by the base pairing of complementary strands that originated from different sources.

hydrops fetalis A condition lethal at or before birth caused by a complete absence of α-globin.

hypertrophic cardiomyopathy A disease where the muscle of the heart is thickened without any obvious cause and is a leading cause of sudden unexpected cardiac death.

hypertrophy The increase in size of an organ or an area of tissue because of an increase in cell size.

IDDM See insulin-dependent diabetes mellitus.

identical by descent (IBD) A situation where two individuals in a kindred have inherited the same alleles at a locus from a common ancestor.

identical by state (IBS) A situation where two individuals in a kindred have the same alleles at a locus but it cannot be demonstrated that they are identical by descent.

immortal The property of a cancer cell line that enables it to be propagated indefinitely in culture.

informative meiosis A meiosis where it is possible to distinguish between parental and recombinant chromosomes in the progeny.

informative missingness A systematic error in genome-wide association studies. Occurs when a particular genotype is more likely to be rejected by quality control procedures in microarray analysis so that the genotype is underrepresented in a particular group of affected or control individuals.

initial gene index Name given by the IHGSC to the collection of sequences identified or suspected of forming genes.

in silico Analysis carried out by computer.

in situ hybridisation Hybridising a nucleic acid probe to chromosomes spread on a slide. Usually a metaphase spread is used but recently also carried out with interphase spreads to increase resolution.

insulin-dependent diabetes mellitus Also known as type 1 diabetes. Form of diabetes mellitus that responds to insulin. Usually shows juvenile onset and is caused by the autoimmune destruction of the islets of Langerhans in the pancreas.

interferon response An innate antiviral response involving the release of interferon proteins which trigger a transcriptional shutdown and eventually cell death.

intermediate repeated DNA DNA sequences that are repeated between 100 and 10^5 times per genome.

interphase The period of the cell cycle between nuclear divisions.

intron The portion of a gene that is transcribed but removed during splicing.

intronic miRNA MicroRNAs that are encoded by sequences found in the intron of other genes.

in vivo gene therapy Where the transgene is introduced directly into the target cells of the patient's body.

karyotype The number, size and shape of chromosomes in a somatic cell.

kb Kilobase. A thousand base pairs.

ketone bodies Group of chemical compounds, such as β-hydroxybutyrate, acetoacetone and acetone, that are metabolites of acetyl coA and which accumulate in diabetes.

ketonaemia Accumulation of ketone bodies in the bloodstream in diabetes.

ketonuria Accumulation of ketone bodies in the urine in diabetes.

ketosis Accumulation of ketone bodies in diabetes.

kindred A group of people that are related genetically or by marriage.

kinetochore A protein structure that assembles at the centromere to which the mitotic or meiotic spindle attaches.

knockout mice Mice where specific genes have been deleted.

Kozak rule Used to identify the position in an mRNA molecule where translation starts. Derived from a consensus sequence that spans the AUG start codon and facilitates the initiation of translation. The consensus sequence is gccRccAUGG where R = a purine and upper-case letters are the most critical for efficient translation initiation.

LINE1 (L1) A repeated sequence in the LINE category.

lariat A splicing intermediate where the intron resembles a stem–loop structure or lasso.

liability The combined total of genetic and environmental effects that predispose an individual to a multifactorial disease.

linkage The propensity of two genetic markers to be co-inherited through meiosis. It indicates that the markers are located close together on the same chromosome.

linkage disequilibrium The tendency of particular alleles at one locus to occur with particular alleles at a second closely linked locus.

lipofection The process of introducing DNA or RNA into a cell using liposomes.

lipoplex Complex formed between cationic lipids and DNA used for gene therapy.

liposome Spheres consisting of lipid molecules surrounding an aqueous interior. Used to encapsulate DNA for gene therapy.

locus A unique chromosomal region that corresponds to a gene or some other DNA sequence.

locus control region A region located 50–60 kb upstream of both the α- and β-globin gene clusters and that controls their expression.

LOD score The logarithm of the ratio of odds that two loci are linked with a specified recombination fraction θ, to the odds that they are unlinked. A LOD score of 3 or more is required for linkage to be significant.

long interspersed elements (LINEs) A category of intermediate repeated sequence.

long-range restriction map See rare cutter.

long terminal repeat (LTR) Repeated sequences at each end of a retrovirus.

loss of heterozygosity (LOH) When a region of a chromosome becomes homozygous for alleles that were heterozygous.

low copy number amplification A technique used to obtain a DNA profile from a very small sample of DNA – perhaps just a small number of cells that might be left on an object after it has been touched. It works by increasing the number of PCR cycles in the amplification of the DNA sample.

luciferase An enzyme which catalyses the oxidation of a pigment known as luciferin, leading to the production of light. The most commonly used form of luciferase was isolated from fireflies.

lymphomas Cancers of lymphocytes, a type of white blood cell, including T cells, B cells and natural killer cells.

Lyonisation See X chromosome inactivation.

lysosome A membrane-bound organelle that contains digestive enzymes that break down material taken in by phagocytosis or recycle cellular components.

major disorders Disorders that are both relatively common and severe in their consequences.

mature miRNA A short, single-stranded RNA molecule, generally 21–23 nucleotides long, that are partially complementary to target mRNAs.

maximum likelihood score (MLS) The most likely of a series of alternatives judged by which gives the highest LOD score. For example, the MLS for different values of recombination fraction, θ, indicates the most likely map distance between two loci.

Mb Megabase. One million base pairs.

meconium ileus Obstruction of the bowel in the newborn. Often one of the symptoms of cystic fibrosis.

meiosis A process consisting of two successive nuclear divisions which results in reduction of chromosome number from the diploid chromosome complement to the haploid complement. During the first meiotic prophase recombination takes place between homologous chromosomes.

Mendelian inheritance The pattern of segregation of a disease in a family or kindred that conforms to Mendel's laws of inheritance. Normally taken to indicate that the disease is monogenic.

Mesolithic A period of the Stone Age between the Paleolithic and Neolithic periods.

metacentric A chromosome that has a centrally located centromere, such that both arms are of equal size.

metastatic Describes a tumour that has spread from the tissue of origin to other organs of the body.

methylguanosine cap An altered guanosine nucleotide added to the 5' end of mRNA via an unusual 5' to 5' bond (rather than the usual 5' to 3' bond). This structure stabilises the structure and allows recognition by translation machinery.

microRNA families microRNAs that share the same seqeuence from bases 2 to 7 (the seed sequence) are grouped together into families. microRNAs in the same family are expected to target many of the same mRNAs.

microsatellites Tandem repeats of a short sequence 2–4 nucleotides in length found at many different locations in the genome.

minisatellites A class of highly repetitive sequences that consist of sequences between 10 bp and 100 bp long repeated in tandem arrays that vary in size from 0.5 kb to 40 kb. They tend to occur near telomeres although they have been found elsewhere. Also known as variable number tandem repeats (VNTRs).

minor allele frequency The frequency in a population of the less common allele, usually of a biallelic polymorphism.

(mi)RISC – (micro)RNA-induced silencing complex The protein complex that catalyses the silencing of genes in the miRNA and RNAi pathways.

missense mutation Nucleotide substitution that results in an altered amino acid sequence in the encoded protein.

mitogen A substance that stimulates cell division.

mitosis The process of nuclear division in which the daughter cells contain the same number of chromosomes as the mother cell.

mitotic crossing over Recombination that occurs in mitosis of somatic cells. Results in homozygosity in daughter cells of the region of the chromosome distal to the crossover point.

mobile genetic element A sequence of DNA that can move from one chromosomal location to another by transposition. Characteristically they are flanked by short direct repeats.

model organism Any organism used because features of its biology, life cycle or existing knowledge make experimentation easier than a more complex organism. Use of model organisms is based on the supposition that fundamental biological processes are conserved in evolution.

MODY (maturity onset diabetes of the young) A form of insulin-dependent diabetes that affects teenagers and young adults.

monogenic See single-gene defect.

monosomic A somatic cell with one copy of a chromosome.

monozygotic twins Derived from the same fertilised egg – otherwise known as identical twins.

mortal The propensity of human cells to die after about 50 generations in tissue culture.

mosaicism When not all cells in the body are genetically identical. This may come about through a mutation in early development and may result in either the germline or somatic cells being affected.

multi-drug resistance (MDR) protein A protein which pumps a range of molecules out of the cell, including cancer drugs.

multifactorial disease A disease caused by the interaction of the environment and polygenes or oligogenes.

multiplex PCR Where several PCR reactions are carried out together using more than one pair of primers.

multiplicative interaction Where the effect on phenotype of multiple alleles is the product of their individual effects. Multiplicative interaction is a special case of epistasis.

multipoint mapping Testing for linkage between genetic markers and a disease by using more than one marker at a time.

multiregional evolution hypothesis Theory of human evolution that states that *Homo sapiens* evolved independently from *Homo erectus* on multiple occasions in different parts of the world.

myoblasts Mononucleate cells that are normally quiescent but can divide and fuse with muscle fibres to repair damage.

myotonic dystrophy (MD) Neuromuscular disorder caused by a trinucleotide repeat expansion.

nanoparticles Very small, ordered structures of up to 100 nm. Nanoparticles made of positively charged polymers and peptides can be used for the transfer of DNA and RNA into cells.

Neolithic Sometimes called the New Stone Age. A period of human history which started about 10 000 years ago, marked by the appearance of agriculture, settled communities and polished stone tools.

neovascularisation The formation of new networks of blood vessels not branched from old vessels.

neural crests A ridge of ectoderm that forms above the neural tube during early embryogenesis. Subsequently the cells migrate and develop into the dorsal root ganglia of the sensory nervous system, the adrenal medulla and some skeletal elements of the face.

neutral allele One of the alleles of a neutral polymorphism.

neutral molecular polymorphism A polymorphism for a DNA sequence that has no effect on the genotype.

neutral polymorphism A polymorphism in which the different alleles do not have a discernible effect on the fitness of the organism.

neutral theory of evolution Theory that postulates that the main mechanism that generates nucleotide substitution is the fixation of neutral alleles through the effect of genetic drift.

non-disjunction The failure of homologous chromosomes to segregate at meiosis, resulting in one daughter cell with two copies and one daughter cell with no copies of the chromosome in question.

non-insulin-dependent diabetes mellitus Form of diabetes that does not respond to insulin (type 2 diabetes). Typically shows middle-aged onset.

non-obese diabetic mouse (NOD) A mouse that spontaneously develops a disease that closely resembles IDDM in humans.

non-parametric A method of mapping disease genes that does not require the formulation of models with specified parameters.

nonsense codon A codon that does not encode an amino acid; causes translation to terminate.

nonsense mutation Nucleotide substitution that results in a nonsense codon that results in a truncated protein. Also results from frameshift mutations, as the incorrect reading frame will contain nonsense codons.

non-sequence-specific off-target effects Any effect of RNAi on a transcript other than the intended target, which is not dependent on the sequence of the RNAi inducing agent.

non-synonymous Nucleotide change in a coding sequence which changes the amino acid residue inserted into the encoded protein.

northern blotting A procedure similar to Southern blotting but RNA not DNA is fractionated and hybridised to the probe. Allows the size and expression of mRNA molecules to be studied.

nucleocapsid The genome and protein coat of a virus.

nucleosome Basic unit of DNA packing in which DNA winds around a tetramer of the histones H1, H2A, H2B, H3 and H4 to form a beads-on-a-string appearance in the electron microscope.

Oct box The sequence ATTTGCAT that can be found in the promoter region genes, which binds the transcription factors Oct-1 or Oct-2 to stimulate transcription.

odds ratio (OR) Used in epidemiological studies to compare the risks of a complex disease in two population groups with different genotypes.

off-target effect Any direct effect of RNAi on levels of transcripts other than those that are the intended target.

oligogene A gene which makes a large contribution to the aetiology of a disorder.

oligogenic disorder A disorder caused by the action of a few genes.

oligonucleotide A polymer of nucleotides of small size (5–100 nucleotides).

oligonucleotide probe A short piece of RNA or DNA that is used to detect the presence of a complementary sequence.

oncogene A gene the mutation of which is implicated in the aetiology of cancer.

oncomiRs An miRNA that acts as an oncogene.

oogonia The diploid precursors of female gametes.

open reading frame (ORF) A significantly long sequence of codons in one of the six possible reading frames in a DNA sequence that does not contain a nonsense codon.

ordered clone library A clone library with overlapping inserts where overlap between all the clones has been determined.

ortholog Genes in different organisms that are similar by virtue of having descended from a common ancestor. Usually assumed to carry out similar functions.

OTTO System developed by Celera Genomics to identify sequences in the genome that are genes.

out-of-Africa hypothesis Theory of human evolution that states that modern humans are descended from a small group that emerged from Africa 100 kya.

oxidative phosphorylation The production of ATP at the inner membrane of the mitochondria. Electrons are passed along the respiratory chain in the mitochondria, eventually interacting with oxygen and protons to form water. The energy released in the process is used to pump protons across the inner mitochondrial membrane. As the protons re-enter the mitochondrial matrix, the energy of the proton gradient drives the synthesis of ATP.

p arm The shorter arm of a chromosome in a metaphase spread.

P **value or** *P* **trend** In genome-wide association studies a parameter that reports the probability of obtaining a deviation from random association at least as great as that observed if the null hypothesis is true that there is no association.

P1-derived artificial chromosome A cloning vector based on generalised transduction using phage P1. Can accept inserts of approximately 100 kb and is liable to fewer cloning artifacts than YAC vectors.

Paleolithic Sometimes called the Old Stone Age. Phase of human history prior to the advent of agriculture and settled communities. Marked by a hunter–gatherer lifestyle and chipped stone tools.

paralog Gene copies formed by duplication within the genome of the same organism.

parametric A model of inheritance that can be tested by LOD score analysis that requires the specification of parameters.

parasitic DNA See selfish DNA.

passenger mutations Mutations documented in a cancer cell line that do not contribute to the cancer phenotype.

pathophysiology The biochemical or cellular changes that cause the symptoms of a disease.

pedigree A representation of the ancestral relationship between individuals related genetically or by marriage.

penetrance The frequency with which a particular genotype manifests itself in the phenotype.

pericentric Around the centromere of a chromosome.

phase Specifies whether particular alleles at adjacent loci are on the same (*cis*) or different (*trans*) chromosomes.

phenocopy The occurrence of a disorder casued by an environmental factor with the same symptoms as an inherited disorder.

phycoerythrin A fluorescent protein found in cyanobacteria and red algae. It is often attached to other molecules to make them fluorescent.

phylogenetic tree An attempt to describe the history of populations since they diverged from a common ancestral population. The order in which the branches split describes the order in which populations diverged, and the length of the branches describes the time since the divergence.

physical map A genomic map based on the actual location of DNA sequences.

polyadenine tail (poly-A tail) A string of adenine residues added to the 3′ of mRNAs after transcription.

polycistronic miRNAs miRNAs that are co-expressed on the same primary transcript.

polygene One of several genes each of which makes a small contribution to the aetiology of a disorder.

polygenic disorder A disorder caused by the action of several genes.

polymerase chain reaction (PCR) A technique for amplifying DNA *in vitro* using a thermostable DNA polymerase and primers that anneal at sites which flank the region to be amplified.

polymorphic A locus that is heterozygous at a significant frequency in a population.

polymorphism information content (PIC) Measure of the informativeness of a genetic marker.

population-attributable risk (PAR) The fraction of the overall population incidence of a disease that would be prevented if the allele was not present in the population.

population bottleneck When a previously large population contracts to a small number and then expands to its previous numbers. As a result the gene pool may be less diverse, and gene frequencies may be altered because of the effects of sampling and genetic drift.

population stratification The presence in a population of different ethnic subgroups that may have different susceptibilities to complex disease.

positional candidate cloning Cloning a disease gene by identifying a biologically relevant gene located near the map position of the disease gene.

positional cloning The cloning of a gene using no other information apart from its chromosomal location.

power The probability that a study will identify the involvement of a risk allele, with a given locus-specific λ or genotype relative risk, to a defined statistical threshold.

precursor miRNA (pre-miRNA) A hairpin structure, excised from the primary miRNA transcript. The mature miRNA is contained on one side of the stem.

premutation A mutation which has no phenotypic effects but which increases the odds of a trinucleotide repeat expansion in a subsequent generation.

primary miRNA (pri-miRNA) The initial transcript that enocdes a miRNA.

primary RNA transcript The RNA molecule before splicing that contains both introns and exons.

primers Oligonucleotides that anneal to template DNA to prime synthesis mediated by DNA polymerase.

principal component analysis Mathematical procedure to represent multi-variate data by producing a synthetic map.

private mutation A mutation that is unique to a particular family or individual.

proband An individual suffering from a genetic disease who is the first member of a pedigree to come to medical attention.

processed pseudogene A pseudogene that lacks introns and is therefore thought to have arisen by retrotransposition.

prodrug A form of a drug which requires some kind of modification before it is active.

promoter Part of the gene upstream of the transcribed region that contains the elements necessary for a basal level of transcription.

protein kinase An enzyme that phosphorylates another protein and so regulates its activity.

proteome The complete set of proteins that can be encoded by a genome.

proximal Towards the centromere compared with the chromosomal point of reference.

proxy In genome-wide association studies an allele that is in linkage disequilibrium with a second allele. The genotype of the proxy allele allows that genotype of the second allele to be inferred without actually directly determining its genotype.

pseudogene A DNA sequence that closely resembles a functional gene but has been rendered non-functional by an inactivating mutation.

pseudouridine An isomer of uridine, it is found in all classes of RNAs and is formed by the specific modification of uridine residues, post-transcriptionally.

pulsed field gel electrophoresis (PFGE) A special adaptation of agarose gel electrophoresis, where the direction of the electric field is periodically changed. Used to construct long-range restriction maps. See also rare cutter.

q arm The longer arm of a chromosome in a metaphase spread.

Q-bands Bands produced in metaphase chromosomes using the fluorophore quinacrine.

quantitative trait locus (QTL) A locus contributing towards a quantitative phenotype.

quinacrine A dye that binds DNA in chromosomes and fluoresces in UV light.

radiation hybrid mapping Method of mapping that is based on the frequency with which markers are co-retained on the same genomic fragment after fragmentation by X-rays.

rare cutter A restriction enzyme that has target sites in human DNA that are more widely spaced than normal. Used in conjunction with PFGE to construct long-range restriction maps; that is, maps that extend over hundreds of kilobases.

ras A 21-kilodalton protein with GTPase activity that acts as a molecular switch in signal transduction pathways involving external mitogenic stimulation. The *ras* gene is a proto-oncogene because mutations that lock it into its active state cause cellular transformation. Mutations to this gene have been implicated in a wide variety of cancers.

read In DNA sequencing the nucleotide sequence obtained from one sequencing item, i.e. a single lane in capillary sequencing using the classic Sanger dideoxy method or from a single cluster in next generation sequencing technologies.

reassociation The reformation of a duplex DNA molecule after the original molecule has been denatured by heat or other physical treatment.

recombinant protein A protein produced by the expression of a gene in a heterologous host cell.

recombination fraction The proportion of chromosomes that are recombinant, i.e. of a different constitution from parent chromosomes.

recurrence risk See relative risk.

relative risk (λ) The ratio of the frequency of a multifactorial disease in the relatives of an affected person compared with its rate in the general population.

replacement substitution A nucleotide substitution that does result in an amino acid substitution in the encoded protein.

respiratory syncytial virus (RSV) A common single-stranded RNA virus which causes respiratory tract infections in patients of all ages, but particularly infants and children.

response element A region upstream of a gene that activates its transcription in response to a general environmental influence such as heat, serum or cAMP.

restriction fragment length polymorphism (RFLP) A genetic marker based on the presence or absence of a target for a restriction enzyme due to a polymorphism at a single base pair.

retinoblast Retinal epithelial cells that give rise to neuroblasts and neuroglial precursor cells.

retinoblastoma A tumour of the eye originating from retinal cells. May be sporadic or inherited due to mutation of the *RB* gene.

retrotransposition A process of transposition in which RNA serves as a template for reverse transcriptase. The resulting cDNA is then inserted into the genome at a new location.

retrotransposon A mobile genomic element that transposes when an RNA transcript is converted to DNA by reverse transcriptase and the resultant cDNA reinserted into the genome at a new location.

retrovirus A virus that has an RNA genome. The mature virion contains reverse transcriptase encoded by the RNA genome. This copies the RNA into cDNA which is then inserted into the host genome.

reverse transcriptase (RNA-dependent DNA polymerase). An enzyme that copies RNA into DNA; the product is known as cDNA.

RH mapping See radiation hybrid mapping.

risk allele Allele that increases the risk of a complex disease.

ribozyme An RNA molecule with a catalytic activity, such as endonuclease activity.

RNA interference (RNAi) A phenomenon whereby the introduction of double-stranded RNA into a cell causes the cleavage of cellular RNA which is complementary.

RNAi screening Performing a large number of RNAi experiments to find which genes affect a process when knocked down.

RT-PCR A form of PCR that amplifies RNA by first converting it to cDNA using reverse transcriptase.

run In DNA sequencing the combined sequence data obtained in the whole sequencing operation, i.e. the combined data from all the capillaries in Sanger dideoxy sequencing or the combined data from all the clusters in a next generation sequencing technology.

S phase The period in the cell cycle during which nuclear DNA is replicated.

sarcolemma The flexible sheath that surrounds muscle fibres.

satellite DNA See highly repetitive DNA.

satellite peak See buoyant density ultracentrifugation.

scaffold (in sequencing) Use of long-range maps or other features to link together adjacent contigs and locate them along the length of a chromosome.

segregation The transmission of alleles through a pedigree in a way that follows Mendel's laws of inheritance.

selection The process which leads to an increase in the frequency of a favourable allele and a decrease in the frequency of a deleterious allele.

selfish DNA DNA that does not contribute to the phenotype of the organism, but evolves to increase its copy number in the genome by transposition.

sequence-specific off-target effects Off-target effects of RNAi where the identity of the transcripts affected is dependent on the sequence of the RNAi inducing agent.

sequence tagged site (STS) A site that is identified from its sequence using a PCR reaction. Can be used as a marker in physical, genetic and genomic maps.

serotype The subdivision of bacteria or viruses based on their antigenic properties.

severe acute respiratory syndrome coronavirus (SARS-CoV) A single-stranded RNA virus causing a serious respiratory infection which led to the deaths of 774 people in the only major recorded outbreak between November 2002 and July 2003.

short-hairpin RNAs (shRNAs) Hairpin RNA structures similar to precursor miRNAs transcribed from artificially introduced DNA vectors. shRNA are processed within the cell to release siRNAs.

short interfering RNAs (siRNAs) Double-stranded RNAs of 19–21 base pairs which trigger the cleavage of complementary mRNAs via the RNA interference mechanism.

short tandem repeat (STR) An array of short sequences each normally 2–4 nucleotides in length. See also microsatellite.

shotgun sequencing Method of sequencing DNA by first generating random fragments. These are sequenced separately before assembly of the whole sequence by overlapping the sequences in the individual fragments.

sibs Individuals who have the same parents, e.g. brother and sister.

silencer A region upstream of a gene that turns off transcription in response to a particular signal.

silent substitution A nucleotide substitution that does not result in an amino acid substitution in the encoded protein because of the redundancy in the genetic code.

SINES Short interspersed elements, a category of intermediate repeated sequence.

single-gene defect A disorder or disease that is caused by a mutation in a single gene.

single nucleotide polymorphism (SNP) Single nucleotide that shows significant variation (above 1%) in a population. They occur approximately every 1.3 kb in the human genome.

single sequence DNA A sequence of DNA that is not repeated in the genome.

somatic Cells in the body that are not part of the germline.

somatic gene therapy Where the attempt to repair the genetic defect is restricted to the somatic cells of a person affected by the disease.

somatic mutation A mutation that does not affect the germline and so will not be passed on to the next generation. In cancer studies a mutation that has been acquired after fertilisation of the zygote and so was not inherited from either parent.

Southern blotting A form of hybridisation where DNA from the target is digested with restriction enzymes and fractionated by agarose gel electrophoresis. The DNA is transferred by capillary action to a nitrocellulose or nylon membrane. The membrane is then hybridised to a radioactively labelled probe and the molecular weight of the hybridising fragments revealed by autoradiography.

spacer region Tracts of DNA that apparently have no function and which separate one gene from the next.

splice acceptor site The junction between the dinucleotide AG at the end of an intron and the start of the next exon.

splice donor site The junction between the end of an exon and the dinucleotide GT at the start of the next intron.

splicing The process by which introns from the primary transcript are removed in the nucleus. The product of splicing of an RNA polymerase II transcript is the mRNA molecule, which only contains introns.

streptavidin A tetrametric protein purified from a bacterium. Its primary feature is the strength with which it binds the small molecule biotin.

stringency During Southern hybridisation duplex formation can still occur even if there is a mismatch between the sequence of the probe and the target sequence. The amount of mismatch that can be tolerated is dependent on the physicochemical conditions during the hybridisation process. For probes longer than about 200 bp low salt (0.1 × SSC) and high temperature

(65°C) are high stringency conditions which allow duplex formation only if the match between the probe and target is perfect or near perfect. Higher salt and lower temperature allow a greater degree of mismatch to be tolerated and are said to be low stringency conditions. If the probe is an oligonucleotide the exact conditions for high stringency hybridisation are calculated from its length and base content.

structural genomics The study of the structure of the proteins encoded by a genome.

STS content map A map based on the order of STSs in an ordered clone library.

submetacentric A chromosome with the centromere nearer one end compared to the other. Often appears J-shaped in metaphase spreads.

susceptibility allele See risk allele.

synonymous codon A change in the DNA sequence that results in a new codon that codes for the same amino acid as in the original codon. Thus there is no change in the sequence of amino acids in the encoded protein.

synonymous mutation A mutation in a coding sequence which does not lead to a change in the amino acid sequence of the encoded protein.

synteny Conservation of gene location on chromosomes of different organisms.

synthetic map See principal component analysis.

tag In genome-wide association studies a genotyped allele that identifies the particular haplotype carried by the individual concerned.

tandem repeat array An identical or near identical sequence repeated so that each copy lies immediately adjacent to the next and arranged so that the end of one unit abuts the start of the next.

TATA-binding protein (TBP) The protein that recognises the TATA box to initiate formation of the transcription complex.

telocentric A chromosome where the centromere is at one end.

telomerase The enzyme that adds the telomere. It contains an RNA molecule that serves as the template for DNA synthesis.

telomere A special structure at the ends of chromosomes that in humans consists of a tandem repeat of the sequence TTAGGG and ends in a 3' extension.

thalassaemias Anaemias caused by an imbalance of globin synthesis.

tiling path or array The set of clones that represents the sequence of a region or entire chromosome with the minimum overlap.

tissue microarray An array of tissue samples, usually prepared from patients, inserted at regular intervals into a block of paraffin.

toll-like receptors A class of receptor molecules that recognise broad types of molecules that are shared by pathogens, but not found in hosts, such as lipopolysaccharides (polysaccharide sugars joined to lipid molecules).

topology The geometrical arrangement of sequences in clones.

trans See phase.

transcription factor A protein that binds upstream of a gene to facilitate or stimulate its transcription.

transcriptome The sum total of all the DNA sequences in a genome that are transcribed.

transgene A DNA molecule introduced into a cell to alter its genetic constitution.

transgenic An organism that has a genome modified by genetic engineering.

translocation An exchange of segments between non-homologous chromosomes.

transmission disequilibrium test (TDT) Ascertains whether a parent who is heterozygous for an associated and a non-associated allele transmits the associated allele more often to affected offspring.

transplacement An organism where genes have been replaced with genes carrying specific mutations.

transposition The process by which a mobile genetic element copies itself and inserts the copy in a new location.

transposon A sequence of DNA that has the capacity to move to a new position in the genome. Usually the process involves duplication, so that a copy of the sequence remains in the original location. Thus transposons are normally repetitive elements.

trinucleotide repeat expansion (TRE) A mutation caused by the increase in the number of copies of a repeated trinucleotide.

trisomic Describes a somatic cell with three copies of a chromosome.

trisomy 21 A karyotype with an extra copy of chromosome 21 leading to Down's syndrome.

tropism The propensity of certain viruses to infect particular cell types.

tumour suppressor Genes which negatively regulate cell division. Mutations to the genes are recessive at a cellular level but show a dominant pattern of inheritance.

type 1 diabetes See insulin-dependent diabetes.

type 2 diabetes See non-insulin-dependent diabetes.

type I shRNAs The original type of shRNA, type I shRNAs consist of two short repeats connected by a few bases.

type II shRNAs A more advanced type of shRNA. Type II shRNAs consist of the precursor sequence of a miRNA, with the active sequence removed and replaced with a sequence complementary to the target.

3′ untranslated region (3′ UTR) The region of an mRNA transcript that is after the end (i.e. 3′) of the open reading frame. It is not translated into protein.

unbalanced translocation The non-reciprocal duplication of a chromosome segment.

univalent Describes a chromosome with one chromatid.

untranslated region The region of the mRNA molecule that is not translated. The 5′ UTR is the part of the mRNA molecule upstream of the AUG start codon. The 3′ UTR is the part of the mRNA molecule downstream of the stop codon.

upstream A region of the DNA molecule that lies 5′ to the point of reference.

variable number tandem repeats An alternative term for minisatellite loci.

variance A parameter that measures the total phenotypic variation in a population.

vehicle The physical means by which a transgene is introduced into a target cell in gene therapy.

virion A complete virus particle.

X chromosome One of the pair of sex chromosomes. XX individuals are female, XY are male.

X chromosome inactivation The clonal inactivation of one of the X chromosomes in females during development.

X-linked A trait showing a characteristic pattern of segregation indicating that the gene in question is located on the X chromosome. See Section 1.3.

yeast artificial chromosome (YAC) A cloning vector using a yeast host cell that can accept very large inserts of DNA (~1 Mb).

zoo blotting Southern blotting where human DNA is used to hybridise DNA from a range of other mammals.

Index